Modeling, Analysis and Control Processes of New Energy Power Systems

Modeling, Analysis and Control Processes of New Energy Power Systems

Editors

Haoming Liu
Jingrui Zhang
Jian Wang

Basel • Beijing • Wuhan • Barcelona • Belgrade • Novi Sad • Cluj • Manchester

Editors

Haoming Liu
College of Energy and
Electrical Engineering
Hohai University
Nanjing
China

Jingrui Zhang
Department of Instrumental
and Electrical Engineering
Xiamen University
Xiamen
China

Jian Wang
College of Energy and
Electrical Engineering
Hohai University
Nanjing
China

Editorial Office
MDPI AG
Grosspeteranlage 5
4052 Basel, Switzerland

This is a reprint of articles from the Special Issue published online in the open access journal *Processes* (ISSN 2227-9717) (available at: https://www.mdpi.com/journal/processes/special_issues/modeling_power_systems).

For citation purposes, cite each article independently as indicated on the article page online and as indicated below:

Lastname, A.A.; Lastname, B.B. Article Title. *Journal Name* **Year**, *Volume Number*, Page Range.

ISBN 978-3-7258-1975-1 (Hbk)
ISBN 978-3-7258-1976-8 (PDF)
doi.org/10.3390/books978-3-7258-1976-8

Contents

About the Editors

Haoming Liu

Haoming Liu (Professor) received his B.S., M.S., and Ph.D. from Nanjing University of Science and Technology, Nanjing, China, in 1998, 2001, and 2003, respectively. He was a Postdoctoral Research Fellow at Southeast University, Nanjing, China, from 2004 to 2005. He was also a Visiting Scientist with the University of Hong Kong, Hong Kong, China, and Tsinghua University, Beijing, China. He is currently a Full Professor with the College of Energy and Electrical Engineering, Hohai University, Nanjing, China. His research interests include new energy generation, distribution system analysis and control, and the electricity market.

Jingrui Zhang

Jingrui Zhang (Associate Professor) received his B.S. and Ph.D. in Automation and Control Science and Engineering from Huazhong University of Science and Technology, Wuhan, China, in 2007 and 2012, respectively. He is currently an Associate Professor at Xiamen University, Xiamen, China. He was a Visiting Scholar with the University of Macau from 2021 to 2022. His research interests include optimization of power system operations, energy management for active distribution networks and microgrids, and swarm evolutionary algorithms.

Jian Wang

Jian Wang (Assistant Professor) received his B.S. and Ph.D. in electrical engineering from Chongqing University, Chongqing, China, in 2015 and 2020, respectively. He is currently an Assistant Professor at Hohai University, Nanjing, China. He served as a Visiting Research Student with the University of Saskatchewan, Saskatoon, SK, Canada, and a Postdoctoral Fellow with the University of Macau, Macau, China. His research interests include active distribution networks and convex optimization.

Preface

With the increase in global energy demand, global climate change, environmental pollution, and energy shortages have become increasingly serious. The development of new energy represented by wind power and photovoltaic power is the key way to achieve low-carbon and green development. The scale of global new energy power generation continues to grow, and the integration of high-penetration new energy will become the basic feature and development trend of power systems worldwide. The power generation principle, control strategy, and grid connection mode of new energy units such as wind power and photovoltaic power are significantly different from those of traditional power units. The fluctuation in new energy and the high proportion of power electronic devices bring new challenges to new energy power systems, including the spatio-temporal mismatch of power supply and load, as well as the stability and security of power systems.

In order to overcome these challenges, some new technologies such as demand response, energy storage, and power electronics devices have been introduced into power systems to promote the integration of new energy. They require that the modeling, analysis, and control methods of power systems can adapt to the transformation of power systems.

This reprint aims to promote state-of-the-art research in this promising area. Seventeen original articles were recommended for acceptance and publication. These published articles mainly cover original research on the economic planning and operation of new energy power systems, the stability analysis and control of new energy power systems, and the modeling of power equipment. We hope that the research achievements presented in the following reprint can contribute to a transition from current power systems to low-carbon and green ones in the future.

We would like to express our thanks to all authors who have contributed to this reprint and, in particular, the editorial staff of Processes for their support and assistance.

Haoming Liu, Jingrui Zhang, and Jian Wang
Editors

Editorial

Special Issue on "Modeling, Analysis and Control Processes of New Energy Power Systems"

Haoming Liu [1,*], **Jingrui Zhang** [2] and **Jian Wang** [1]

1   College of Energy and Electrical Engineering, Hohai University, Nanjing 211100, China
2   Department of Instrumental and Electrical Engineering, Xiamen University, Xiamen 361005, China
*   Correspondence: liuhaom@hhu.edu.cn

Citation: Liu, H.; Zhang, J.; Wang, J. Special Issue on "Modeling, Analysis and Control Processes of New Energy Power Systems". *Processes* **2023**, *11*, 235. https://doi.org/10.3390/pr11010235

Received: 28 November 2022
Accepted: 6 January 2023
Published: 11 January 2023

1. Introduction

In recent years, global climate change, environmental pollution, and energy shortage have become increasingly serious. Countries all over the world regard the development of new energy, represented by wind power and photovoltaics, as the key to achieving low-carbon and green development. The scale of global new-energy power generation continues to grow, and the high penetration of new energy will inevitably become one of the basic features and development trends in future power systems. New energy units, such as converter-interfaced wind power and photovoltaics, significantly differ from traditional power units in the perspectives of the power generation principle, control strategy, and grid connection mode. The variability of new energy and the high proportion of associated power electronic devices have brought profound challenges to the new energy power system, including the spatial–temporal mismatch between variable power supply and load, and the stability and security of electronic-enabled power systems.

In order to overcome these challenges, some new technologies such as demand response, energy storage, and FACTs (flexible AC transmission systems) devices have been introduced into the power systems to promote the integration of new energy. Facilitating these new technologies requires adapting the modeling, analysis, and control methods to the transformation of new energy power systems.

This Special Issue on 'Modeling, Analysis and Control Processes of New Energy Power System' aims to promote state-of-the-art research in this promising area. Seventeen original articles were recommended for acceptance and publication. These published articles mainly cover original research on the economic planning and operation of new energy power systems, the stability analysis and control of new energy power systems, and the modeling of power equipment.

2. Brief Synopsis of Papers in the Special Issue

Lei et al. [1] established a two-stage majorization configuration model to identify and understand how variable energy affects a hybrid energy storage system in active distribution networks. Chen et al. [2] proposed a two-stage layout method for the met mast based on discrete particle swarm optimization zoning and micro-sitting. This study provided a quantitative planning method for met mast layout in practical projects with improved wind-monitoring accuracy. Yang et al. [3] constructed a framework that is suitable for city regional integrated energy systems to participate in the energy market, and proposed an evaluation index system for low-carbon capabilities in the energy market. Li et al. [4] analyzed the life-cycle cost of synchronous condensers and introduced the blind number theory into the cost calculation model to quantify the impacts of various uncertain pieces of information on the cost of the synchronous condenser projects. Li et al. [5] proposed a multi-energy transaction decision-making strategy for a community-level integrated energy system considering user interaction, and the proposed strategy improved both the profit of the community operator and the value-added benefit of energy users.

Yang et al. [6] presented an optimal day-ahead scheduling model for a multi-renewable energy power system with distributed generations while satisfying flexibility constraints. Yuan et al. [7] proposed a time-of-use pricing strategy for integrated energy suppliers and integrated energy users in the integrated energy systems based on game theory.

Hu et al. [8] studied the transient behavior and stability issues of a direct-drive wind turbine during fault recovery in a DC-link voltage control timescale. Zhu et al. [9] defined the static voltage stability assessment problem as a regression problem and constructed an artificial neural network for online assessment. Fu et al. [10] proposed a double-layer fault diagnosis model for the main bearing of a wind turbine that combines the auxiliary classifier generation adversarial network and the deep residual shrinkage network. Zhang et al. [11] used the virtual vector-based model predictive current control to select the optimal virtual vector and apply it to five-phase induction motors. Liu et al. [12] proposed a predictive commutation failure suppression strategy considering multiple harmonics of commutation voltage considering the distortion characteristics of AC voltage of HVDC systems. Yang et al. [13] considered the transmission loss reduction of the HVDC system and established a multi-order fitting function of transmission loss under joint impacts of line-commutated converter stations, voltage source converter stations, and DC lines.

Chen et al. [14,15] proposed an improved magnetic-noise prediction model of a five-phase induction motor through large-slot opening and pole-slot schemes. Luo et al. [16] used an extended Kalman filter algorithm for the parameter identification of the five-phase squirrel cage induction motor. Finally, Xue et al. [17] established an analytical model of an unequal-pitch linear phase-shifting transformer by combining the distributed magnetic circuit method and the Schwartz–Christopher transformation.

Author Contributions: Writing—Original draft preparation, J.W.; Writing—Revision, H.L. and J.Z. All authors have read and agreed to the published version of the manuscript.

Funding: This research received no external funding.

Conflicts of Interest: The authors declare no conflict of interest.

References

1. Lei, G.; Huang, Y.; Dai, N.; Cai, L.; Deng, L.; Li, S.; He, C. Optimization Strategy of Hybrid Configuration for Volatility Energy Storage System in ADN. *Processes* **2022**, *10*, 1844. [CrossRef]
2. Chen, W.; Qian, G.; Qi, W.; Luo, G.; Zhao, L.; Yuan, X. Layout Method of Met Mast Based on Macro Zoning and Micro Quantitative Siting in a Wind Farm. *Processes* **2022**, *10*, 1708. [CrossRef]
3. Yang, Z.; Wang, X. Research on Low-Carbon Capability Evaluation Model of City Regional Integrated Energy System under Energy Market Environment. *Processes* **2022**, *10*, 1906. [CrossRef]
4. Li, C.; Liu, M.; Guo, Y.; Ma, H.; Wang, H.; Yuan, X. Cost Analysis of Synchronous Condenser Transformed from Thermal Unit Based on LCC Theory. *Processes* **2022**, *10*, 1887. [CrossRef]
5. Li, Y.; Wang, X. Community Integrated Energy System Multi-Energy Transaction Decision Considering User Interaction. *Processes* **2022**, *10*, 1794. [CrossRef]
6. Yang, L.; Huang, W.; Guo, C.; Zhang, D.; Xiang, C.; Yang, L.; Wang, Q. Multi-Objective Optimal Scheduling for Multi-Renewable Energy Power System Considering Flexibility Constraints. *Processes* **2022**, *10*, 1401. [CrossRef]
7. Yuan, X.; Guo, Y.; Cui, C.; Cao, H. Time-of-Use Pricing Strategy of Integrated Energy System Based on Game Theory. *Processes* **2022**, *10*, 2033. [CrossRef]
8. Hu, Q.; Xiong, Y.; Liu, C.; Wang, G.; Ma, Y. Transient Stability Analysis of Direct Drive Wind Turbine in DC-Link Voltage Control Timescale during Grid Fault. *Processes* **2022**, *10*, 774. [CrossRef]
9. Zhu, Z.; Zhang, P.; Liu, Z.; Wang, J. Static Voltage Stability Assessment Using a Random under Sampling Bagging BP Method. *Processes* **2022**, *10*, 1938. [CrossRef]
10. Fu, Z.; Zhou, Z.; Yuan, Y. Fault Diagnosis of Wind Turbine Main Bearing in the Condition of Noise Based on Generative Adversarial Network. *Processes* **2022**, *10*, 2006. [CrossRef]
11. Zhang, Q.; Zhao, J.; Yan, S.; Xiong, Y.; Ma, Y.; Chen, H. Virtual Voltage Vector-Based Model Predictive Current Control for Five-Phase Induction Motor. *Processes* **2022**, *10*, 1925. [CrossRef]
12. Liu, X.; Cao, Z.; Gao, B.; Zhou, Z.; Wang, X.; Zhang, F. Predictive Commutation Failure Suppression Strategy for High Voltage Direct Current System Considering Harmonic Components of Commutation Voltage. *Processes* **2022**, *10*, 2073. [CrossRef]
13. Yang, Z.; Gao, B.; Cao, Z. Optimal Current Allocation Strategy for Hybrid Hierarchical HVDC System with Parallel Operation of High-Voltage and Low-Voltage DC Lines. *Processes* **2022**, *10*, 579. [CrossRef]

14. Chen, H.; Zhao, J.; Xiong, Y.; Luo, X.; Zhang, Q. An Improved Model for Five-Phase Induction Motor Based on Magnetic Noise Reduction Part I: Slot Opening Width. *Processes* **2022**, *10*, 1496. [CrossRef]
15. Chen, H.; Zhao, J.; Xiong, Y.; Yan, S.; Xu, H. An Improved Model for Five-Phase Induction Motor Based on Magnetic Noise Reduction Part II: Pole-Slot Scheme. *Processes* **2022**, *10*, 1430. [CrossRef]
16. Luo, X.; Zhao, J.; Xiong, Y.; Xu, H.; Chen, H.; Zhang, S. Parameter Identification of Five-Phase Squirrel Cage Induction Motor Based on Extended Kalman Filter. *Processes* **2022**, *10*, 1440. [CrossRef]
17. Xue, J.; Zhao, J.; Yan, S.; Wang, H.; Zhou, C.; Yan, D.; Chen, H. Modeling and Analysis of New Power Devices Based on Linear Phase-Shifting Transformer. *Processes* **2022**, *10*, 1596. [CrossRef]

Article

Optimal Current Allocation Strategy for Hybrid Hierarchical HVDC System with Parallel Operation of High-Voltage and Low-Voltage DC Lines

Zhichao Yang, Bingtuan Gao * and Zeyu Cao

School of Electrical Engineering, Southeast University, Nanjing 210096, China; seuyangzhichao@seu.edu.cn (Z.Y.); tozy_study@126.com (Z.C.)
* Correspondence: gaobingtuan@seu.edu.cn; Tel.: +86-025-8379-2260

Abstract: For long-distance and bulk-power delivery of new energy, high-voltage direct current (HVDC) is a more effective way than high-voltage alternative current (HVAC). In view of the current capacity disparity between line commutated converter (LCC) and voltage source converter (VSC), a hybrid hierarchical HVDC topology with parallel operation of 800 kV and 400 kV DC lines is investigated. The optimal current allocation method for hybrid hierarchical HVDC is proposed distinct from the same rated current command configuration method of high-voltage and low-voltage converters in traditional topology. Considering the transmission loss reduction of the HVDC system, a multi-order fitting function of transmission loss including LCC converter stations, VSC converter stations and DC lines is established. To minimize the transmission loss and the voltage deviation of key DC nodes comprehensively, a multi-objective genetic algorithm and maximum satisfaction method are utilized to obtain the optimal allocation value of rated current command for high-voltage and low-voltage converters. Through the optimization model, an improved constant current controller based on the current allocation strategy is designed. The hybrid hierarchical HVDC system model is built in PSCAD software, and simulation results verify the effectiveness of the proposed topology and optimal current allocation strategy.

Keywords: hybrid hierarchical HVDC; high-voltage and low-voltage converters; transmission loss function; voltage deviation; optimal current allocation

Citation: Yang, Z.; Gao, B.; Cao, Z. Optimal Current Allocation Strategy for Hybrid Hierarchical HVDC System with Parallel Operation of High-Voltage and Low-Voltage DC Lines. *Processes* **2022**, *10*, 579. https://doi.org/10.3390/pr10030579

Academic Editor: Haoming Liu

Received: 24 February 2022
Accepted: 14 March 2022
Published: 16 March 2022

Publisher's Note: MDPI stays neutral with regard to jurisdictional claims in published maps and institutional affiliations.

1. Introduction

HVDC is a critical technology for grid interconnection and massive energy transmission with the increasing conversion of fossil fuel-based grids to renewable energy-based grids [1–3]. In 2020, global renewable energy increased by 260 gigawatts (GW), mainly composed of wind power (127 GW) and photovoltaic power (111 GW) [4]. To deliver new energy from resource-rich areas to load-concentrated areas, HVDC is more cost-effective than HVAC. Throughout the development of HVDC technology over the last decade, it has overcome many technical obstacles faced by HVDC grids [5,6].

The application of HVDC can realize the interconnection of multiple new energy power grids. In Figure 1, the new energy power system based on HVDC is composed of thermal power plants, photovoltaic power stations, wind farms, sending-end converter stations, DC lines, receiving-end converter stations, and receiving-end grids. Conventional LCC-HVDC technology is widely used, but it presents many disadvantages, such as commutation failure, consumption of a large amount of reactive power, and so on. Although VSC-HVDC can avoid the technical bottlenecks of LCC-HVDC, the short-boards for smaller transmission capacity and lower voltage level also limit its development for new energy integration [7]. Therefore, some scholars proposed hybrid HVDC, which combines LCC-HVDC and VSC-HVDC [8]. However, there are some differences between LCC and VSC in the current capacity and control method. Therefore, it is critical to design the topology and control strategy of the hybrid HVDC, which can make full use of their advantages in this paper.

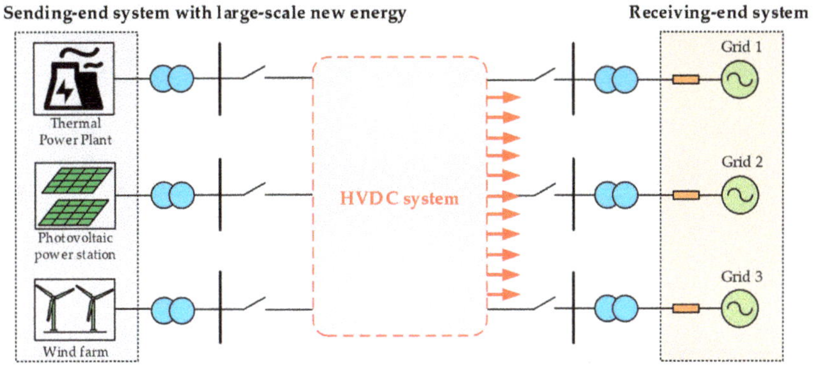

Figure 1. New energy power system via HVDC.

With the development of VSC converter technology, hybrid HVDC transmission technology integrating LCC and VSC converter has attracted extensive attention. The existing literature focuses on the topology of hybrid HVDC transmission systems [9–11], control and protection strategies [12–14], small-signal stability analysis [15–17], experimental platforms [18–20], etc. In November 2020, China approved the Baihetan-Jiangsu HVDC project, which adopts hybrid cascade multi-terminal technology for the first time in the world. The receiving-end station of the transmission project adopts three modular multi-level converters (MMC) in parallel and then connected in series with LCC, which may cause MMC power imbalance and reverse transmission risk after AC or DC faults. The literature [21] proposes an AC fault ride-through strategy for Baihetan HVDC based on a modified DC chopper and enhanced VDCOL (voltage-dependent current order limiter), which can realize stable fault ride-through of HVDC in commutation failure mitigation and overvoltage suppression. In December 2020, the Kunliulong HVDC project was put into operation as the first multi-terminal hybrid HVDC project of 800 kV in the world. Since different converters present various fault characteristics, the literature [22] studied the traveling wave characteristics and verified the adaptability of traveling wave protection in software PSCAD. For HVDC fault detection, the literature [23] designed an improved identification scheme by comparing current polarity with transient current limiting boundary on the basis of full-utilizing capacity of the converter, and intensive simulation is carried out to prove the accuracy and coordination of the proposed scheme.

The above research results enrich the theoretical basis of the hybrid HVDC, but most of the existing topologies cannot make full use of the current capacity of LCC. The cascade technology of the Baihetan HVDC project has high requirements for the current coordinated control of three VSC converters in [24]. To coordinate the current capacity of LCC and VSC, a hybrid hierarchical HVDC is designed in this paper, which can not only make full use of the current capacity of LCC but also improve the ability to connect weak systems by utilizing VSC. The coordination control method is the focus of this paper for its consideration of different converters, different DC lines and loss reduction of HVDC transmission involved in this topology, differing from the same rated current command configuration method of high-voltage and low-voltage converters in traditional topology. The main contributions of this paper are summarized as follows:

(1) A hybrid hierarchical HVDC topology with parallel operation of 800 kV and 400 kV DC lines is designed. By considering the LCC converter station, VSC converter station and DC line, the HVDC transmission loss model is established.

(2) To minimize the transmission loss and the voltage deviation of key DC nodes comprehensively, a multi-objective genetic algorithm and maximum satisfaction method are used to obtain the optimal allocation value of rated current command for high-voltage and low-voltage converters. An improved constant current controller based on optimal current allocation strategy is designed and validated in PSCAD.

The topology of the hybrid hierarchical HVDC system is depicted in Section 2. In Section 3, the HVDC transmission loss model based on the multi-order fitting method is formed in detail. In Section 4, improved HVDC control based on the current optimal allocation method considering the multi-objective optimization method is designed. Simulation results based on PSCAD software and discussion are presented in Section 5. Finally, we conclude this paper in Section 6.

2. Model of Hybrid Hierarchical HVDC System

2.1. Selection of System Topology

To reduce transmission loss and increase transmission capacity, an HVDC with a high-voltage level of 800 kV or higher is utilized [25]. For the 800 kV HVDC system in topology 1 in Figure 2, the rated current of the LCC converter is set as 5 kA, so the DC power is 4000 MW. In consideration of the flexible transformation scheme for integration of new energy, the upper LCC converter is replaced with an MMC converter to form topology 2. Limited by the current capacity (3 kA) of MMC, the transmission power is reduced from the original 4000 MW to 2400 MW. If a 400 kV DC line is added in topology 2, topology 3 is established, and we can calculate that the transmission power of the hybrid HVDC system is improved from 2400 MW to 3200 MW.

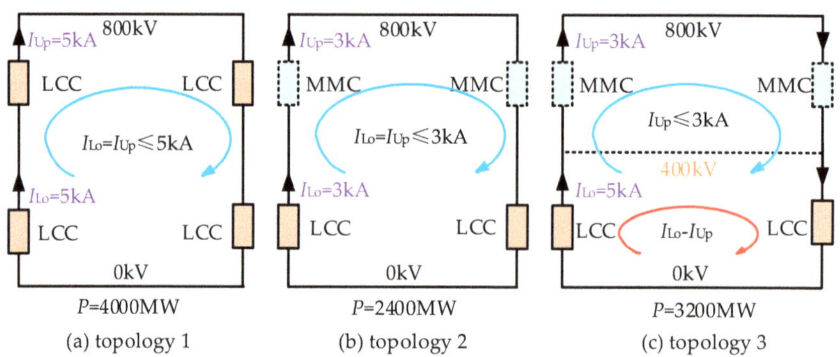

Figure 2. Comparison for different HVDC topologies.

For hybrid hierarchical HVDC, shown in Figure 3, hierarchical structure means that two DC lines are led out from point A at the high-voltage side and point B at the medium-voltage side, respectively. The 800 kV lines and 400 kV lines operate in parallel, and the system contains four DC lines. For ease of description, the converter between point A and point B on the rectifier side is called a high-voltage converter, and the converter between point B and point C is called a low-voltage converter. A 12-pulse LCC converter is adopted for the low-voltage converter, and an MMC converter is adopted for the high-voltage converter. The bipolar HVDC system is symmetrically equipped with the same converter stations.

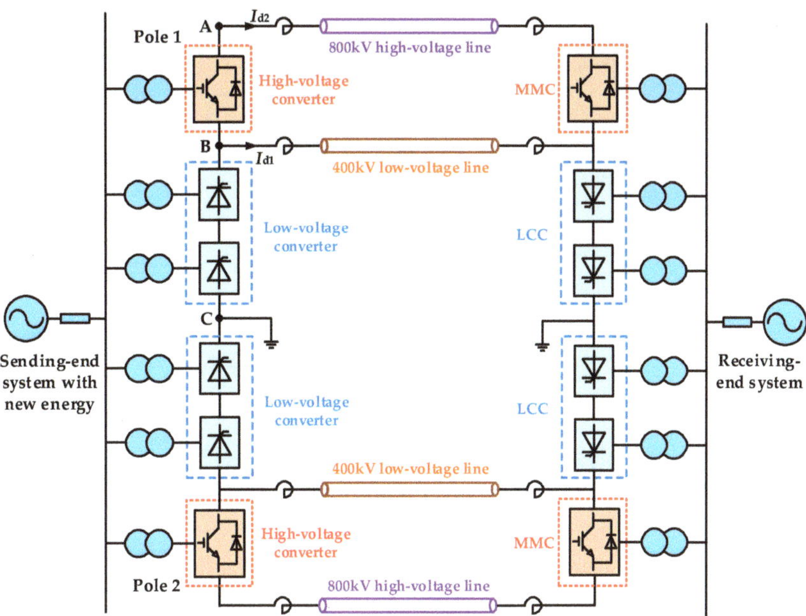

Figure 3. Structure of hybrid hierarchical HVDC system.

2.2. LCC and VSC Station

The low-voltage converter in Figure 3 adopts a 12-pulse converter. The quasi-steady mathematical model of the LCC converter is shown in Figure 4. The formulas of the mathematical model are as follows:

$$U_d = 2.7 E_r \cos \alpha - \frac{6}{\pi} X_d I_d \tag{1}$$

$$I_d = \frac{E_r}{\sqrt{2} X_d}(\cos \alpha - \cos(\alpha + \mu)) \tag{2}$$

$$P_d = U_d I_d \tag{3}$$

$$Q_c = P_d \tan \varphi \tag{4}$$

where E_r is the effective value of line voltage at the rectifier valve side. U_d is the DC voltage, I_d is the DC current, and P_d is the DC power. Q_c is the reactive power absorbed by the converter, and X_d is the commutation reactance in phase. ϕ is the power factor angle. α is the trigger delay angle, and μ is the commutation overlap angle.

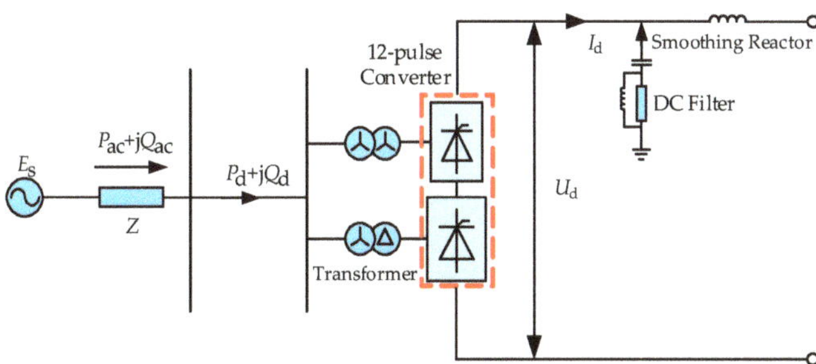

Figure 4. Schematic diagram of 12-pulse LCC converter.

For the high-voltage converter in Figure 3, VSC is adopted. As shown in Figure 5, the VSC converter includes six bridge arms, and each bridge arm is equipped with n sub-modules (SMs). The on, off or locking state of SMs is judged by the on/off signal state of switch T_1/T_2.

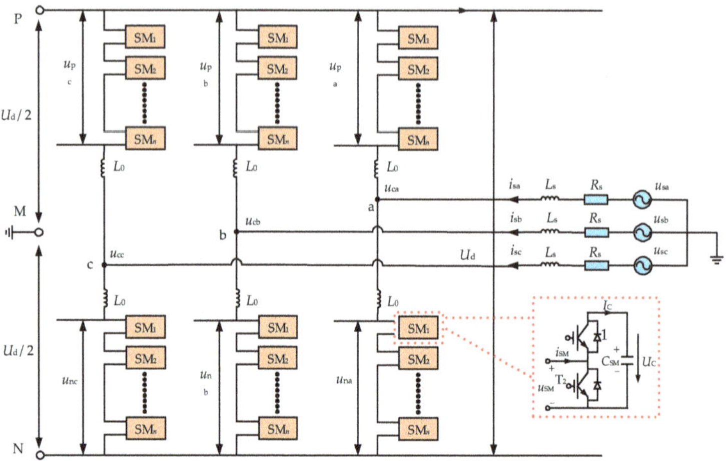

Figure 5. Schematic diagram of VSC.

According to Figure 5, the mathematical model of VSC is described as:

$$\begin{cases} L_s \dfrac{di_{sa}}{dt} = u_{sa} - R_s i_{sa} - u_{ca} \\ L_s \dfrac{di_{sb}}{dt} = u_{sb} - R_s i_{sb} - u_{cb} \\ L_s \dfrac{di_{sc}}{dt} = u_{sc} - R_s i_{sc} - u_{cc} \end{cases} \tag{5}$$

where u_{cm} (m = a, b, c) is the electromotive force in VSC. u_{sm} (m = a, b, c) is the network-side voltage, and i_{sm} (m = a, b, c) is the network-side current. L_s and R_s are the equivalent inductance and equivalent resistance of the VSC station, respectively.

2.3. System Equivalent Circuit

Since a hybrid hierarchical HVDC system contains LCC and VSC converters, it is necessary to establish the system equivalent circuit to study the coordinated control strategy. As pole 1 and pole 2 operate symmetrically, we choose pole 1 as the research object. According to the mathematical models of LCC and VSC, the simplified equivalent circuit of hybrid hierarchical HVDC is established.

In Figure 6, α_1 is the trigger delay angle of LCC at the rectifier side, and β_2 is the trigger advance angle of LCC at the inverter side. X_{d1} and X_{d2} are the commutation reactance of the rectifier and inverter station, respectively. $U_{m1(t)}$ represents the sum of unbalanced voltage and voltage change of bridge arm reactor caused by charging or discharging of the sub-module in the rectifier-side MMC. $U_{m2(t)}$ represents the sum of unbalanced voltage and voltage change of bridge arm reactor caused by charging or discharging of the sub-module in the inverter-side MMC. R_1 and R_2 are the equivalent resistances of 400 kV and 800 kV DC lines, respectively. I_{d1} and I_{d2} are the DC current of 400 kV and 800 kV lines, respectively. I_{Up} and I_{Lo} are the DC current of the high-voltage and low-voltage converters, respectively. U_{dr1} and U_{di1} are the DC voltage of the rectifier-side LCC and inverter-side LCC, respectively. U_{dr2} and U_{di2} are the DC voltage of the rectifier-side MMC and the inverter-side MMC, respectively.

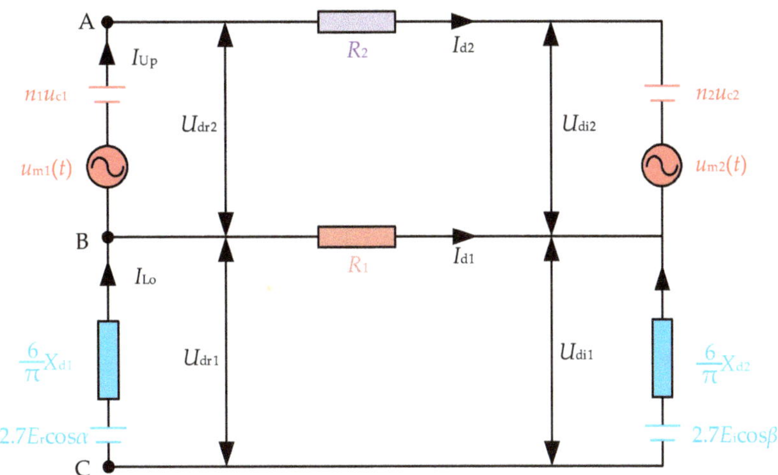

Figure 6. Equivalent circuit of simplified system.

Thus, the steady-state mathematical model of hybrid hierarchical HVDC system is established as:

$$U_{dr} = U_{dr1} + U_{dr2} \approx 2.7E_r \cos \alpha_1 - \frac{6}{\pi}X_{d1}(I_{d1} + I_{d2}) + n_1 u_{c1} \tag{6}$$

$$U_{di} = U_{di1} + U_{di2} \approx 2.7E_i \cos \beta_2 - \frac{6}{\pi}X_{d2}(I_{d1} + I_{d2}) + n_2 u_{c2} \tag{7}$$

$$I_{d1} = (U_{dr1} - U_{di1})/R_1 \tag{8}$$

$$I_{d2} = (U_{dr2} - U_{di2})/R_2 \tag{9}$$

$$P_r = U_{dr1}(I_{d1} + I_{d2}) + U_{dr2}I_{d2} \tag{10}$$

$$P_i = U_{di1}(I_{d1} + I_{d2}) + U_{di2}I_{d2} \tag{11}$$

where N_1 and N_2 are the number of sub-modules of the MMC converter at the rectifier side and inverter side, respectively. u_{c1} and u_{c2} are the voltage of a single sub-module of the MMC converter at the rectifier and inverter, respectively. P_r and P_i are the active power of the rectifier and inverter, respectively.

3. HVDC Transmission Loss Model Based on Multi-Order Fitting Method

To lower the cost of the HVDC system, transmission loss is an important assessment factor, mainly including LCC station loss, MMC station loss and DC line loss.

3.1. Loss Model of LCC Station

The loss model of the LCC station involves a converter valve, converter transformer, smoothing reactor and AC filter. Among them, the converter valve and converter account for more than 85% of the total loss. The specific formulas of various equipment losses are described specifically in existing references. To facilitate the objective optimization, the loss model of the LCC converter station can be expressed as:

$$P_{LCCLoss} = a_n I_d^n + a_{n-1} I_d^{n-1} + \cdots + a_1 I_d + a_0 \tag{12}$$

where $P_{LCCLoss}$ is the loss of LCC station. I_d is DC current of the LCC converter. $a_n, \ldots, a_1, a_0$ are the loss coefficients. n is the order of the fitting polynomial.

Based on the CIGRE model, the rated DC voltage U_{dN} of the LCC converter station is set as 400 kV and the rated DC current I_{dN} is 5.0 kA by selecting 22 operation data between

[0, 1.1 p.u.] in the simulation model, and using the multi-order fitting method to fit this group of data. The 1st-order, 2nd-order, 3rd-order and 4th-order fitting functions are shown in Figure 7. According to the error analysis, the fitting effect of the 4th-order fitting function is the best. Therefore, the LCC loss model adopts the 4th-order fitting function.

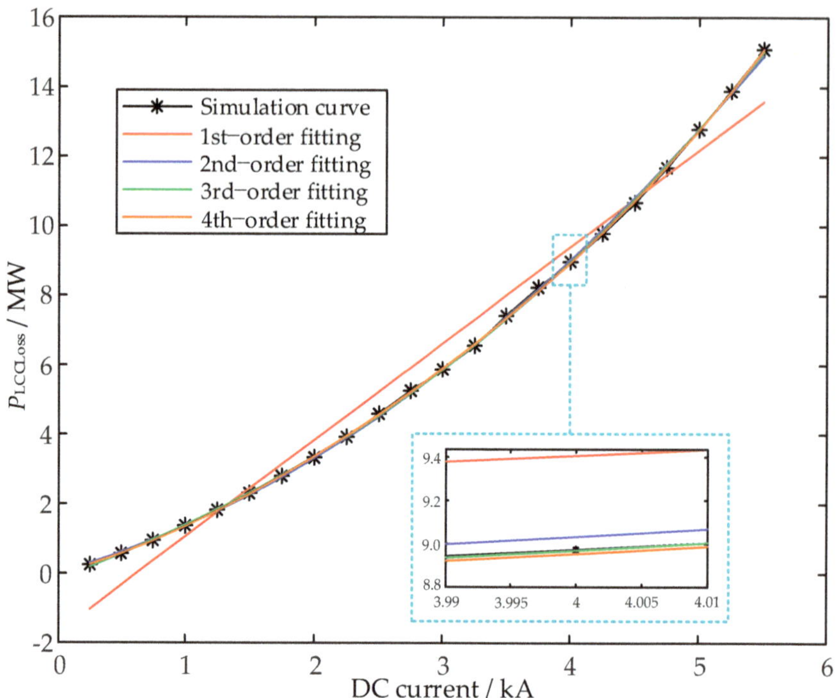

Figure 7. Loss fitting curve of LCC station.

3.2. Loss Model of MMC Station

The loss model of the MMC station involves a converter valve, converter transformer and smoothing reactor. Among them, the converter valve and converter account for more than 80% of the total loss. Reference [19] studies the loss of a 201-level MMC under frequency conditions from 50 to 1000 Hz. The detailed loss model and loss approximate model are adopted, respectively, and the calculation error is less than 6%. The loss approximation model characterizes the quadratic relationship between the MMC converter station and current. In order to improve the fitting effect, the approximate expression of the MMC loss is established as:

$$P_{\mathrm{MMCLoss}} = b_n I_{\mathrm{d}}^n + b_{n-1} I_{\mathrm{d}}^{n-1} + \cdots + b_1 I_{\mathrm{d}} + b_0 \tag{13}$$

where P_{MMCLoss} is the loss of MMC station. I_{d} is the DC current of the MMC converter. $b_n, \ldots, b_1, b_0$ are the loss coefficients. n is the order of the fitting polynomial.

Similarly, referring to the multi-order fitting method of the LCC converter station, the detailed MMC model is traversed and simulated to determine the fitting order of the MMC converter station loss model. As shown in Figure 8, according to the error analysis, the fitting effect of the 4th-order polynomial function is the best, so the MMC loss model adopts the 4th-order fitting function.

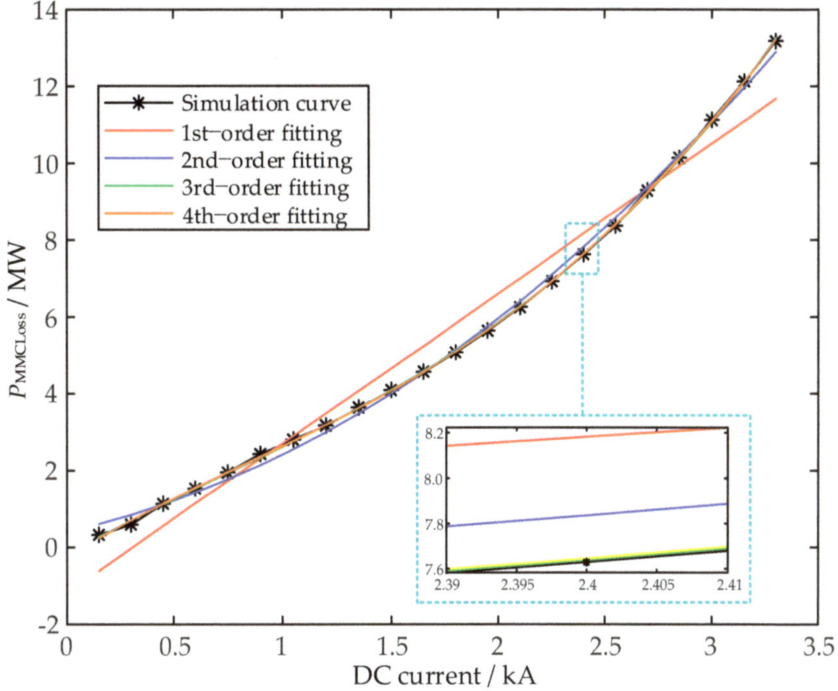

Figure 8. Loss fitting curve of MMC station.

3.3. Loss Model of Transmission Line

Since the corona loss of the DC line is small and negligible, only the resistance loss is considered. The expression of the DC Line is:

$$P_{\mathrm{LineLoss}} = P_{\mathrm{L1Loss}} + P_{\mathrm{L2Loss}} = R_1 I_{\mathrm{d1}}^2 + R_2 I_{\mathrm{d2}}^2 \tag{14}$$

where P_{LineLoss} is the total loss of DC lines. P_{L1Loss} and P_{L2Loss} are the loss of the 400 kV and 800 kV DC lines, respectively.

3.4. Loss Model of Hybrid Hierarchical HVDC

Based on the above analysis, the loss function of the hybrid hierarchical HVDC including the LCC converter station loss, MMC converter station loss and DC line loss is established as:

$$P_{\mathrm{Loss}} = P_{\mathrm{LCCLoss}} + P_{\mathrm{MMCLoss}} + P_{\mathrm{LineLoss}} = \sum_{i=1}^{N_1} P_{\mathrm{LCC}i\mathrm{Loss}} + \sum_{j=1}^{N_2} P_{\mathrm{MMC}j\mathrm{Loss}} + \sum_{p=1}^{N_3} P_{\mathrm{Line}p\mathrm{Loss}} \tag{15}$$

where N_1, N_2, N_3 are the number of LCC converter stations, MMC converter stations and DC lines, respectively. $P_{\mathrm{LCC}i\mathrm{Loss}}$ is the loss of the LCC station i. $P_{\mathrm{MMC}j\mathrm{Loss}}$ is the loss of the MMC station j. $P_{\mathrm{Line}p\mathrm{Loss}}$ is the loss of the DC line p.

4. Current Allocation Strategy considering Multi-Objective Optimization Method

Due to the different resistance values of the high-voltage and low-voltage DC lines, and the loss difference between the LCC station and the MMC station, the same rated current command configuration method for high-voltage and low-voltage converters is unsuitable in the proposed topology. Therefore, considering the perspective of transmission loss reduction and voltage deviation optimization of key nodes, the constant active power controller of the high-voltage converter and the constant current controller of the low-voltage converter are designed based on the optimal current allocation strategy. The strategy introduces the hybrid HVDC loss function and voltage deviation function of the key node, and utilizes a multi-objective genetic algorithm and maximum satisfaction method to determine the optimal compromise value of the DC current for high-voltage and low-voltage converters, so as to ensure the coordinated control and economy of the proposed HVDC system.

4.1. Multi-Objective Optimization Model

A hybrid hierarchical HVDC system includes multiple converter stations and multiple DC lines. Combined with each loss model, the objective function F_1 is determined, and the objective function F_2 is determined by considering the minimum voltage deviation of key nodes. The multi-objective optimization model can be expressed as:

$$\begin{cases} F_1 = P_{\text{Loss}} = \sum_{i=1}^{N_1} P_{\text{LCC}i\text{Loss}} + \sum_{j=1}^{N_2} P_{\text{MMC}j\text{Loss}} + \sum_{p=1}^{N_3} P_{\text{Line}p\text{Loss}} \\ F_2 = \sum_{m=1}^{N} |U_m - U_{m\text{N}}| \end{cases} \tag{16}$$

where U_{m} and $U_{m\text{N}}$ are the actual voltage values and the rated voltage values of the key node j, respectively. N is the number of key nodes.

Active power, DC current, and key node voltage are selected as constraints:

$$\begin{cases} P_{\text{ref}} = U_{\text{dr1}}(I_{\text{d1}} + I_{\text{d2}}) + U_{\text{dr2}}I_{\text{d2}} \\ 0 \leq I_{\text{d1}} \leq I_{1\text{max}}, 0 \leq I_{\text{d2}} \leq I_{2\text{max}} \\ 0 \leq I_{\text{d1}} + I_{\text{d2}} \leq I_{\text{LCCmax}} \\ (1-k)U_{m\text{N}} \leq U_m \leq (1+k)U_{m\text{N}} \end{cases} \tag{17}$$

where P_{Loss} is the total loss of HVDC. $I_{1\text{max}}$ and $I_{2\text{max}}$ are the maximum currents allowed for the 400 kV line and 800 kV lines, respectively. P_{ref} is the rated power. I_{LCCmax} is the maximum current of the LCC. k is the node voltage deviation coefficient.

4.2. Optimal Current Allocation Strategy Based on Multi-Objective Genetic Algorithm

After the Pareto solution set of the optimization model is obtained by a multi-objective genetic algorithm, the optimal compromise values of the DC current (I_{d1opt} and I_{d2opt}) need to be selected according to the operation of the hybrid hierarchical HVDC system. In this paper, the maximum satisfaction method is considered in [26]. The optimization function is Formula (16), and the optimization objective is to minimize the transmission loss and voltage deviation of the key node. Therefore, the partial small satisfaction function is used for the calculation, as shown in Formula (18). For the objective function corresponding to each optimal solution in the Pareto solution set of multi-objective genetic algorithms, the membership function is used to calculate its satisfaction. After standardization, the solution with the largest satisfaction value is the optimal compromise value of the DC current in Formula (19).

$$\mu_n^i = \begin{cases} 1, f_n \leq f_{n\text{min}} \\ \dfrac{f_{n\text{max}} - f_n}{f_{n\text{max}} - f_{n\text{min}}}, f_{n\text{max}} \leq f_n \leq f_{n\text{min}} \\ 0, f_n \geq f_{n\text{max}} \end{cases} \tag{18}$$

where f_n is the value of the nth objective. $f_{n\min}$ and $f_{n\max}$ are the minimum and maximum values in the Pareto solution set, respectively.

$$\mu^i = \frac{\sum\limits_{n=1}^{N} \mu_n^i}{\sum\limits_{i=1}^{M}\sum\limits_{n=1}^{N} \mu_n^i} \tag{19}$$

where μ_i is the standardized satisfaction of the ith optimal solution. M is the number of Pareto optimal solutions, and N is the total number of objective functions.

After determining the optimal values I_{d1opt} and I_{d2opt} of the DC current, we calculate the active power order of the high-voltage MMC converter (P_{Ref_Up}) and the DC current order of the low-voltage LCC converter (I_{LCC_Lo}) according to Formulas (20) and (21):

$$P_{Ref\_Up} = I_{Ref\_Up} \cdot U_{dr2} = I_{d2opt} \cdot U_{dr2} \tag{20}$$

$$I_{Ref\_Lo} = I_{d1opt} + I_{d2opt} \tag{21}$$

Based on the optimal current allocation model for the high-converter and the low-voltage converter, the constant active power controller of the high-voltage converter and the constant current controller of the low-voltage converter are designed, as shown in Figure 9. Firstly, the input parameters are the HVDC power order P_{ref}, rectifier DC voltage U_{dc}, voltage deviation coefficient k, etc. The DC current order of the high-voltage MMC converter (I_{Ref_Up}) and the low-voltage LCC converter (I_{Ref_Lo}) are obtained through the multi-objective optimization model considering the optimization of the hybrid HVDC transmission loss and voltage deviation. The active power order P_{Ref_Up} is obtained according to Formula (15) and then put into the outer loop of the constant active power controller for the high-voltage converter. The measured current value of the LCC converter is compared with the reference value, and then β_{LCC_Up} is obtained by PI modular. Finally, by subtracting β_{LCC_Lo}, the trigger delay angle (α_{LCC_Lo}) of the low-voltage LCC converter can be obtained.

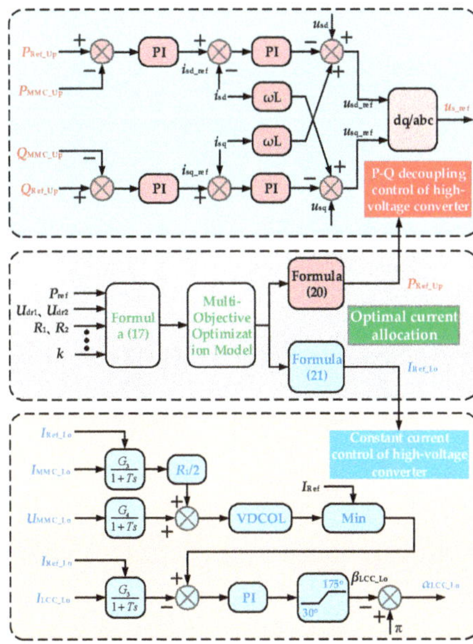

Figure 9. Improved HVDC control based on optimal current allocation.

5. Case Study

5.1. Results of Optimal Current Allocation Strategy

The resistance of the 400 kV line and the 800 kV line in the hierarchical HVDC system are 0.024 Ω/km and 0.018 Ω/km, respectively, and the length of the transmission line is 200 km. Due to the limitation of the current capacity, the maximum current capacity of the MMC converter and LCC converter are 3.125 kA and 5.45 kA, respectively. The DC voltage node, of which the DC voltage is greater than or equal to 400 kV, is regarded as the key node, and the transmission power of the single-pole system is 3000 MW. Based on the multi-objective optimization model, MATLAB is used for optimal calculation. Figure 10 shows the Pareto set result obtained by the multi-objective genetic algorithm.

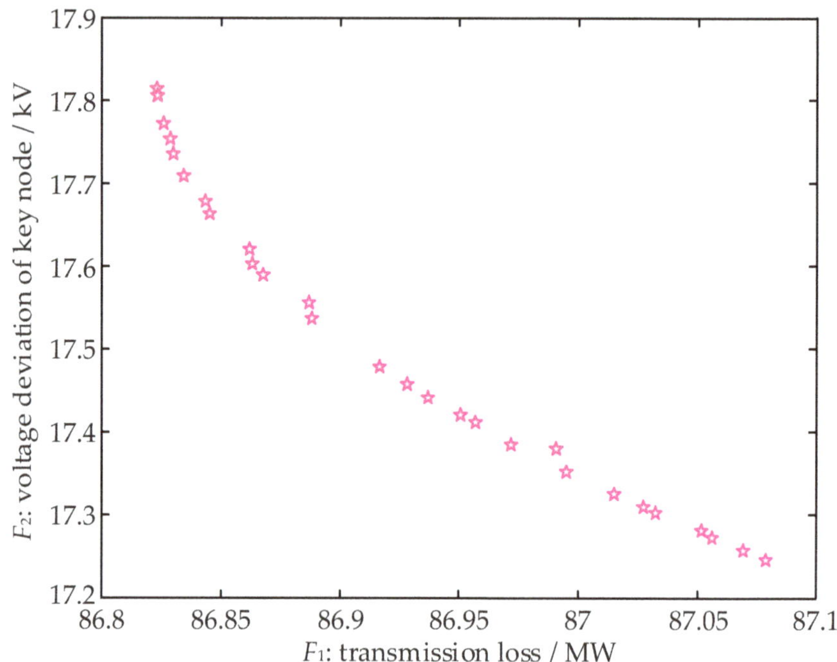

Figure 10. Result of multi-objective genetic algorithm.

According to Formulas (20) and (21), the maximum satisfaction method is used to obtain the optimal compromise value of the DC current. The optimal compromise command value of the DC current flowing through the high-voltage converter and the low-voltage converter is 3.12 kA and 4.38 kA, respectively. A total of 15 groups of typical data were taken for comparison, as shown in Table 1. U_1' and U_2' represent the DC voltage of the low-voltage line and high-voltage line at the inverter side, respectively. With the increase in the DC current of the low-voltage line, the power loss decreases. When I_{d1} increases, the DC voltage of the high-voltage line at the inverter side increases. Compared with the traditional control method (No. 6 in Table 1), the proposed control method (No. 13 in Table 1) can reduce the transmission loss by 7.67% and the voltage deviation drops by 17.71%.

Table 1. Comparison of power loss and voltage variation.

Number	I_{d1} (kA)	I_{d2} (kA)	P_{Loss} (MW)	$U_1{'}$ (kV)	$U_2{'}$ (kV)
1	5.450	2.050	112.38	383.68	776.30
2	5.400	2.100	109.84	384.16	776.60
3	5.300	2.200	105.16	385.12	777.20
4	5.200	2.300	101.01	386.08	777.80
5	5.100	2.400	97.39	387.04	778.40
6	5.000	2.500	94.30	388.00	779.00
7	4.900	2.600	91.75	388.96	779.60
8	4.800	2.700	89.73	389.92	780.20
9	4.700	2.800	88.24	390.88	780.80
10	4.600	2.900	87.28	391.84	781.40
11	4.500	3.000	86.87	392.80	782.00
12	4.400	3.100	86.98	393.76	782.60
13	4.380	3.120	87.07	393.95	782.72
14	4.375	3.125	87.10	394.00	782.75
15	4.300	3.200	87.63	394.72	783.20

5.2. Steady Characteristics

The detailed electromagnetic transient model of the hybrid hierarchical HVDC system, shown in Figure 3, is constructed based on PSCAD software. The rectifier side adopts the 12-pulse converter model of the CIGRE standard system. The specific parameters are shown in the literature [27], and the basic operating parameters of the proposed HVDC system are shown in Table 2. According to the optimization results, the optimal current of the high-voltage MMC converter and the low-voltage LCC converter is 3.12 kA and 4.38 kA, respectively. The optimal current of the high-voltage DC line and the low-voltage DC line is 3.12 kA and 1.26 kA, respectively. The LCC converter on the rectifier side adopts the constant DC current control mode, and the MMC converter on the rectifier side adopts the constant active power and reactive power control method equipped with the optimal current allocation strategy proposed in this paper. The LCC converter on the inverter side adopts the constant DC voltage control mode, and the MMC converter on the inverter side adopts the constant DC voltage and reactive power control method.

Table 2. Basic operating parameters of proposed system.

Parameters	Rectifier	Inverter
AC voltage/kV	380	220
Rated capacity/MW	3000	3000
DC voltage/kV	800	782.72
Converter type	LCC	MMC
Number of sub-modules	/	100
Transformer ratio	380/172	161/220

The simulation time is set as 6 s for depicting the steady-state characteristics of the hybrid hierarchical HVDC with the high-voltage and low-voltage DC lines operating in parallel.

It can be seen from Figure 11 that the start-up process of the HVDC is relatively stable. On the rectifier side, the DC voltage of the high-voltage DC line and low-voltage DC line is 800 kV and 400 kV, respectively. On the inverter side, considering the voltage drop of the converter stations and DC lines, the high-voltage DC line can realize the power transmission of 782.72 kV and 3.12 kA, and the low-voltage DC line can realize the power transmission of 393.95 kV and 1.26 kA, according to Figure 11b,c. According to Figure 10, since the electrical power of the low-voltage converter flows into the high-voltage DC line and low-voltage DC line, the current of the low-voltage converter is 4.38 kA, while the DC current of the high-voltage converter is 3.12 kA. In Figure 11e, the active power of the sending-end system and the receiving-end system is 3000 MW and 2912.93 MW, respectively. The transmission loss is 87.07 MW, which is the result of the optimal allocation of high-voltage and low-voltage converters. During the stable operation, the voltage of SM is 4 kV, according to Figure 11f. The above simulation results verify the effectiveness of the optimal current allocation strategy. The coordination strategy can not only ensure the stable start-up of the system but also achieve the goal of minimizing transmission loss and reducing the withstand voltage requirements of converter devices.

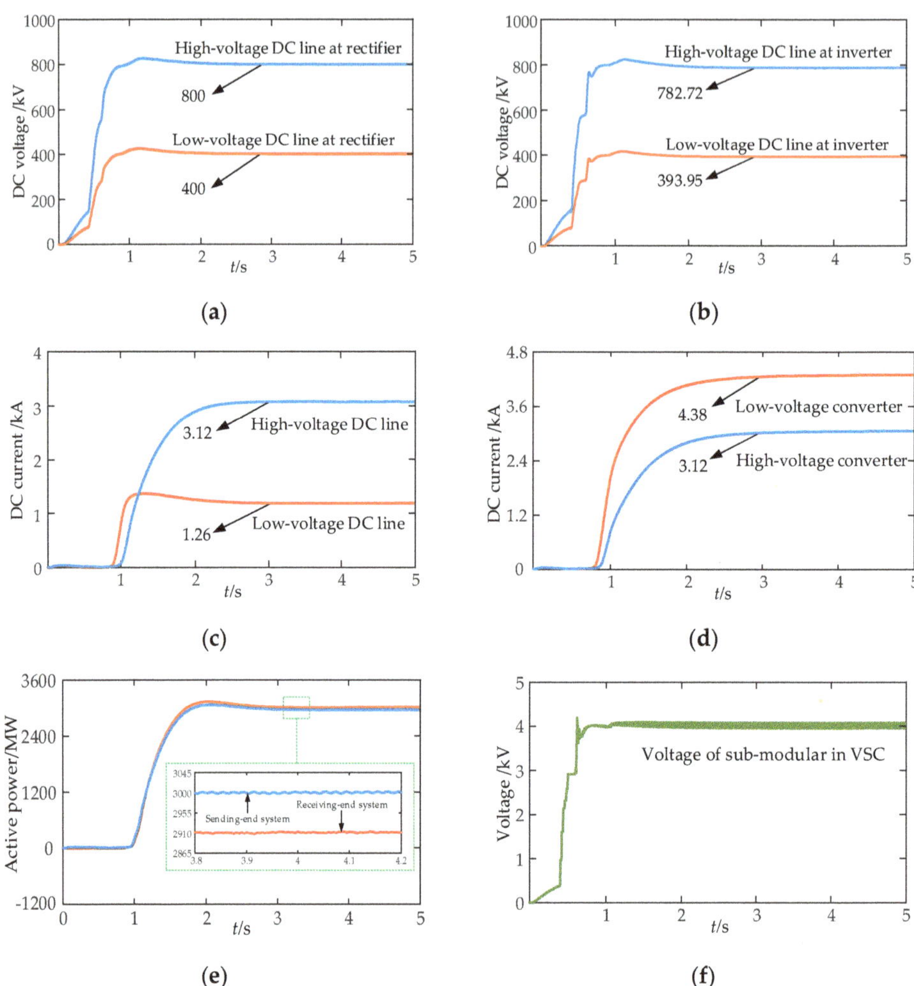

Figure 11. Steady-state waveform of the proposed system. (**a**) DC voltage at rectifier side. (**b**) DC voltage at inverter side. (**c**) DC current of transmission line. (**d**) DC current of converter. (**e**) Active power of HVDC. (**f**) DC voltage of sub-module in VSC.

5.3. Step Characteristics

We set the active power command as 1.0 p.u. (3000 MW), and introduced an active power command module including three stages: (1) the initial power command is 1.0 p.u., and then drops to 0.8 p.u. from 3.0 to 3.5 s; (2) from 3.5 to 4.0 s, the power command remains unchanged; (3) from 4.0 to 4.5 s, the power recovers to 1.0 p.u. Figure 12 is the response result of the hybrid hierarchical HVDC under the variation of power command.

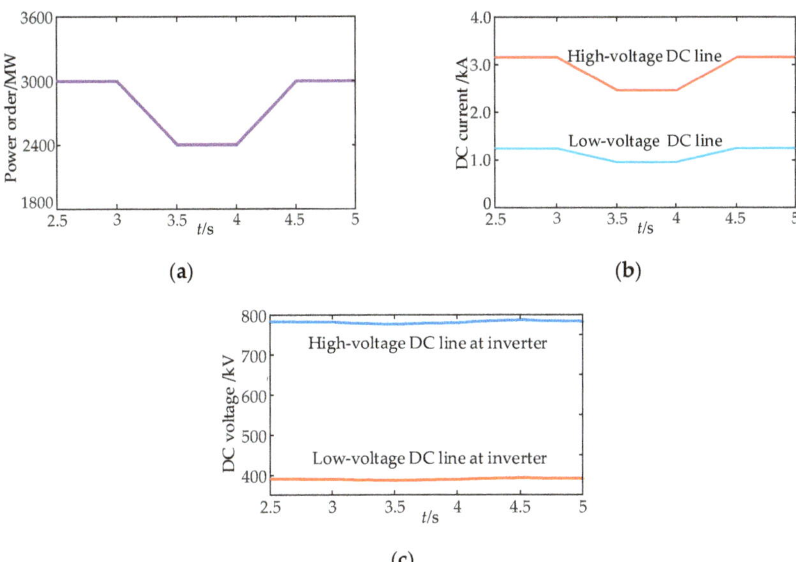

Figure 12. Response under system power command. (**a**) Power command of system. (**b**) DC current of transmission line. (**c**) DC voltage of system at the inverter side.

According to Figure 12, when the active power order issued by the active power command module decreases step by step from 3.0 s, the current commands of the high-voltage line and low-voltage line decrease accordingly. When the active power command increases at 4.0 s, the current command increases. At 4.5 s, the active power order recovers to 1.0 p.u., and the DC current fluctuates with the active power order rapidly. The DC voltage of the high-voltage DC line and low-voltage DC line at the inverter side is 782.72 kV and 393.95 kV, respectively. Therefore, the proposed HVDC system presents good response characteristics.

5.4. AC Fault at Rectifier Side

A three-phase fault occurs at the rectifier side in 3 s with a duration of 0.1 s, and the grounding resistance is 8 Ω. The simulation results are shown in Figure 13.

When the three-phase fault occurs in the sending-end system, as shown in Figure 13a, the AC system at the rectifier side drops to 60% of the normal value, and the DC parameters drop accordingly. In Figure 13b–d, the DC current of the transmission line drops greatly, the high-voltage DC line drops from 3.12 kA to 1.87 kA, the low-voltage DC line drops from 1.26 kA to 0.33 kA. The active power transmitted by the HVDC system decreases from 3000 MW to 1628 MW, which returns to the normal level after 0.33 s. Due to the regulating effect of the DC voltage regulator in the MMC, the drop range of the DC voltage is smaller than that of the DC current, and the rated voltage can be restored in a short time. Therefore, the optimal current allocation method of the hybrid hierarchical HVDC system can respond to the rectifier-side fault in time. Meanwhile, the DC current is always continuous without dropping to zero. After the fault is cleared, the system can quickly return to its normal operation state.

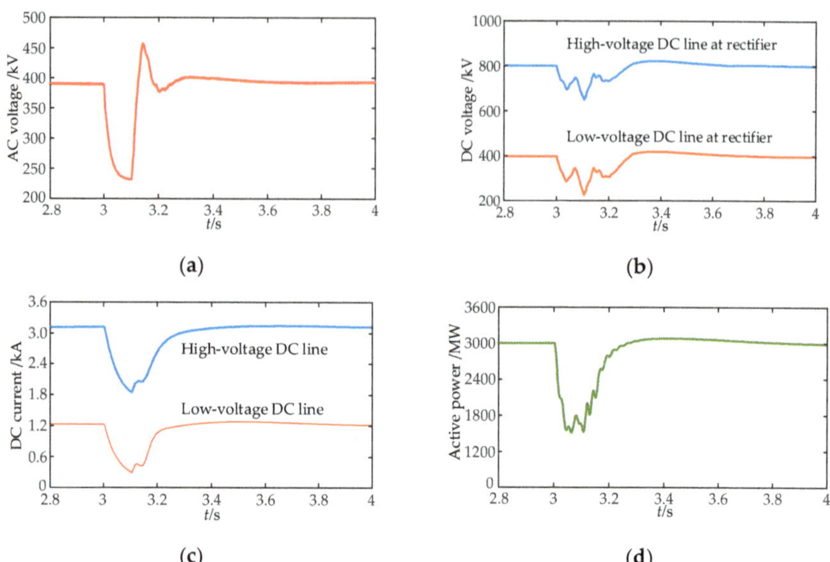

Figure 13. Waveform diagram under fault at rectifier side. (**a**) AC voltage at rectifier side. (**b**) DC voltage at rectifier side. (**c**) DC current of transmission line. (**d**) Active power of system.

5.5. AC Fault at Inverter Side

A three-phase fault occurs at the inverter side in 5 s with a duration of 0.1 s, and the grounding resistance is 8 Ω. The simulation results are shown in Figure 14.

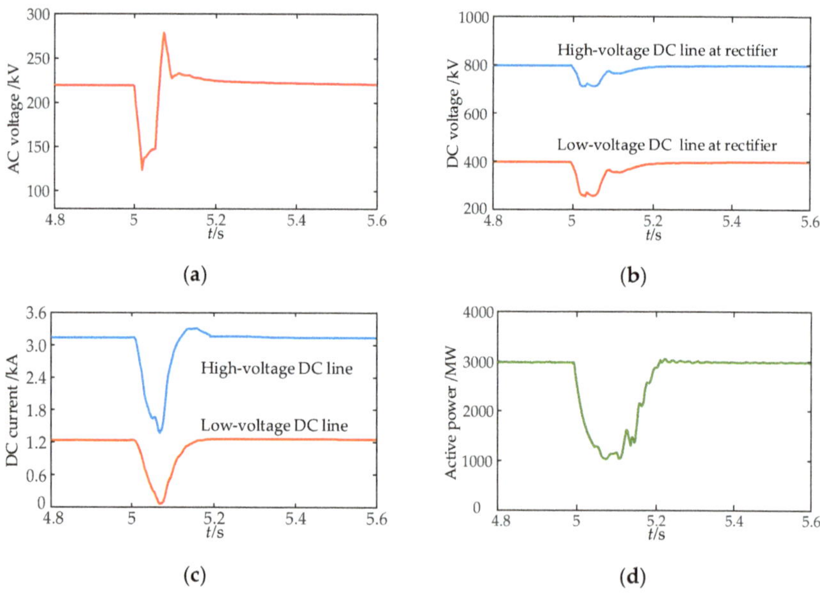

Figure 14. Waveform diagram under fault at inverter side. (**a**) AC voltage at inverter side. (**b**) DC voltage at rectifier side. (**c**) DC current of transmission line. (**d**) Active power of system.

Figure 14a shows the voltage curve of the receiving-end system. When the three-phase fault occurs, the AC voltage at the inverter side drops from the rated value of 220 kV to

132 kV. Figure 14b–d depicts the curves of the DC voltage, DC current and active power during an inverter-side fault, respectively. The AC system fault on the inverter side causes the drop of the AC voltage amplitude and the power transmitted to the receiving-end system decreases. At this time, the MMC converter can activate its DC voltage regulation ability and provide reactive power support. Due to the mismatch of active power between the rectifier side and the inverter side, the unbalanced power enters the capacitance of the sub-module in the MMC, which further reduces the drop degree of the DC voltage. During the whole fault process, the system recovers quickly and returns to normal operation by 0.20 s. In the simulation scenario, there is no commutation failure that occurs in the inverter station. The DC current and voltage of the hybrid hierarchical HVDC system remain well-regulated during the AC fault, which depends on the optimal current allocation strategy.

6. Conclusions

This study proposes an optimal current allocation method for a hybrid hierarchical HVDC with several DC lines with different voltage grades. By considering the current capacity difference between LCC and VSC, the hybrid hierarchical HVDC scheme, with parallel operation of 400 kV and 800 kV DC lines, is designed for massive energy transmission. We establish the multi-order fitting function of transmission loss including LCC converter stations, MMC converter stations and DC lines. The current optimal allocation strategy based on a multi-objective genetic algorithm is proposed to form an improved constant current controller suitable for the proposed HVDC topology to realize the stable operation goal of the high-voltage and low-voltage converters. The transient-state and steady-state characteristics are verified in the PSCAD simulation. According to the simulation results, the current optimal allocation strategy has better performance for reducing HVDC transmission loss by 7.67% and dropping voltage deviation by 17.71% compared to the traditional control method.

The following works will focus on a detailed station loss model considering multiple operating conditions of converter valves and the transient coordinated control strategy of the HVDC system with the integration of massive renewable energy.

Author Contributions: Conceptualization: Z.Y., B.G. and Z.C.; validation: Z.Y. and Z.C.; investigation: Z.Y. and B.G.; methodology: Z.Y.; original draft: Z.Y.; writing—review and editing: Z.Y. and B.G. All authors have read and agreed to the published version of the manuscript.

Funding: This research received no external funding.

Institutional Review Board Statement: Not applicable.

Informed Consent Statement: Not applicable.

Data Availability Statement: The data presented in this study are available in the article.

Conflicts of Interest: The authors declare no conflict of interest.

Nomenclature

General and Abbreviation

HVDC	High-voltage direct current.
HVAC	High-voltage alternative current.
LCC	Line commutated converter.
VSC	Voltage source converter.
DC	Direct current.
kV	Kilovolt.
GW	Gigawatt.
MW	Megawatt.
MMC	Modular multilevel converter.
SM	Sub-module.

E	Effective value of line voltage.
U	DC voltage.
I	DC current.
P	Active power.
Q	Reactive power.
X	Commutation reactance.
α	Trigger delay angle of LCC.
β	Trigger advance angle of LCC.
L	Equivalent inductance.
R	Equivalent resistance.
$PSCAD$	Power systems computer-aided design.
Subscripts	
a, b, c	Quantities of a, b and c-phase.
r, i	Quantities of rectifier and inverter side.
max, min	Quantities of maximum and minimum.
Ref_Up	Quantity of high-voltage converter.
Ref_Lo	Quantity of low-voltage converter.

References

1. Wang, M.; An, T.; Ergun, H.; Lan, Y.; Andersen, B.; Szechtman, M.; Leterme, W.; Beerten, J.; Hertem, D.V. Review and Outlook of HVDC Grids as Backbone of Transmission System. *CSEE J. Power Energy Syst.* **2021**, *7*, 797–810.
2. Luo, X.; Li, F.; Fan, L.; Niu, T.; Li, B.; Tian, L.; Yu, H. Influence of Synchronous Condensers on Operation Characteristics of Double-Infeed LCC-HVDCs. *Processes* **2021**, *9*, 1704. [CrossRef]
3. Ayobe, A.S.; Gupta, S. Comparative Investigation on HVDC and HVAC for Bulk Power Delivery. *Mater. Today Proc.* **2022**, *19*, 958–964. [CrossRef]
4. The International Renewable Energy Agency (IRENA). Renewable Capacity Statistics 2021. Available online: https://www.irena.org/publications/2021/March/Renewable-Capacity-Statistics-2021 (accessed on 1 March 2021).
5. Wang, J.; Huang, M.; Fu, C.; Li, H.; Xu, S.; Li, X. A New Recovery Strategy of HVDC System during AC Faults. *IEEE Trans. Power Del.* **2019**, *34*, 486–495. [CrossRef]
6. Liang, X.; Mehdi, A. HVDC Transmission and Its Potential Application in Remote Communities: Current Practice and Future Trend. *IEEE Trans. Ind. Appl.* **2022**, *58*, 1706–1719. [CrossRef]
7. Sang, Y.; Yang, B.; Shu, H.; An, N.; Yu, T. Fault Ride-Through Capability Enhancement of Type-4 WECS in Offshore Wind Farm via Nonlinear Adaptive Control of VSC-HVDC. *Processes* **2019**, *7*, 540–561. [CrossRef]
8. Zhao, Z.; Iravani, M.R. Application of GTO Voltage Source Inverter in a Hybrid HVDC Link. *IEEE Trans. Power Del.* **1994**, *9*, 369–377. [CrossRef]
9. Chang, Y.; Cai, X. Hybrid Topology of a Diode-Rectifier-Based HVDC System for Offshore Wind Farms. *IEEE Trans. Emerg. Sel. Topics Power Electron.* **2019**, *7*, 2116–2128. [CrossRef]
10. Bakas, P.; Okazaki, Y.; Shukla, A.; Patro, S.K.; Nami, A. Review of Hybrid Multilevel Converter Topologies Utilizing Thyristors for HVDC Applications. *IEEE Trans. Power Electron.* **2021**, *36*, 174–190. [CrossRef]
11. Xiao, H.; Sun, K.; Pan, J.; Li, Y.; Liu, Y. Review of Hybrid HVDC Systems Combining Line Communicated Converter and Voltage Source Converter. *Int. J. Electr. Power Energy Syst.* **2021**, *129*, 1–9. [CrossRef]
12. Li, Z.; He, Y.; Li, Y.; Gu, W.; Tang, Y.; Zhang, X.P. Hybrid Control Strategy for AC Voltage Stabilization in Bipolar VSC-MTDC. *IEEE Trans. Power Syst.* **2019**, *34*, 129–139. [CrossRef]
13. Lee, G.; Hwang, P.I.; Moon, S.I. Reactive Power Control of Hybrid Multi-Terminal HVDC Systems Considering the Interaction Between the AC Network and Multiple LCCs. *IEEE Trans. Power Syst.* **2021**, *36*, 4562–4574. [CrossRef]
14. Chen, X.; Li, H.; Wang, G.; Xu, C.; Liang, Y. A Convolution Power-Based Protection Scheme for Hybrid Multiterminal HVDC Transmission Systems. *IEEE Trans. Emerg. Sel. Topics Power Electron.* **2021**, *9*, 1655–1667. [CrossRef]
15. Ni, X.; Gole, A.M.; Zhao, C.; Guo, C. An Improved Measure of AC System Strength for Performance Analysis of Multi-Infeed HVdc Systems Including VSC and LCC Converters. *IEEE Trans. Power Del.* **2018**, *33*, 169–178. [CrossRef]
16. Zhu, J.; Li, S.; Yu, L.; Bu, S.; Li, Y.; Deng, Z.; Wang, Y.; Jia, H.; Wang, C.; Liu, D. Coherence Analysis of System Characteristics and Control Parameters for Hybrid HVDC Transmission Systems Based on Small-Signal Modeling. *IEEE Trans. Emerg. Sel. Topics Power Electron.* **2021**, *9*, 7436–7446. [CrossRef]
17. He, Y.; Xiang, W.; Ni, B.; Lu, X.; Wen, J. Impact of Strength and Proximity of Receiving AC Systems on Cascaded LCC-MMC Hybrid HVDC System. *IEEE Trans. Power Del.* **2021**, *1*. [CrossRef]
18. Sreedevi, J.; Manohar, P.; Aradhya, R. Dynamic Performance of Hybrid Multiinfeed HVDC System on RTDS. In Proceedings of the Eighteenth National Power Systems Conference (NPSC), Guwahati, India, 18–20 December 2014.
19. Guo, C.; Wang, Y.; Li, J.; Zhao, C.; Fu, C. Design and Development of Experimental Platform for Hybrid HVDC System. *High Volt. Eng.* **2019**, *45*, 3157–3163.

20. Xiao, H.; Duan, X.; Zhang, Y.; Li, Y. Analytically Assessing the Effect of Strength on Temporary Overvoltage in Hybrid Multi-Infeed HVDC Systems. *IEEE Trans. Power Electron.* **2022**, *37*, 2480–2484. [CrossRef]
21. Cheng, F.; Yao, L.; Xu, J.; Chi, Y.; Sun, Y.; Wang, Z.; Li, Y. A new AC Fault Ride-through Strategy for HVDC Link with Serial Connected LCC-VSC Hybrid Inverter. *CSEE J. Power Energy Syst.* **2022**, *8*, 175–187.
22. Liu, Y.; Luo, G.; Yang, Y.; Zhang, X.; Yang, L. Adaptability Analysis of Traveling Wave Protection in Multi-terminal Hybrid DC Transmission Lines. In Proceedings of the 2020 4th International Conference on HVDC (HVDC), Xi'an, China, 6–9 November 2020.
23. Hou, J.; Song, G.; Chang, P.; Xu, R.; Hussain, K. Fault Identification Scheme for Hybrid Multi-terminal HVDC System Based on Control and Protection Coordination Strategy. *Int. J. Electr. Power Energy Syst.* **2022**, *136*, 1–13. [CrossRef]
24. Guo, C.; Wu, Z.; Yang, S.; Hu, J. Overcurrent Suppression Control for Hybrid LCC/VSC Cascaded HVDC System Based on Fuzzy Clustering and Identification Approach. *IEEE Trans. Power Del.* **2021**, *1*. [CrossRef]
25. Qin, Y.; Wen, M.; Yin, X.; Bai, Y.; Fang, Z. Comprehensive Review of Commutation Failure in HVDC Transmission Systems. *Electr. Power Syst. Res.* **2022**, *205*, 1–10.
26. Farina, M.; Amato, P. A Fuzzy Definition of "Optimality" for Many-criteria Optimization Problem. *IEEE Trans. Syst. Man Cybern. A Syst. Hum.* **2004**, *34*, 315–326. [CrossRef]
27. Ouyang, J.; Zhang, Z.; Li, M.; Pang, M.; Diao, Y. A Predictive Method of LCC-HVDC Continuous Commutation Failure Based on Threshold Commutation Voltage under Grid Fault. *IEEE Trans. Power Syst.* **2021**, *36*, 118–126. [CrossRef]

Transient Stability Analysis of Direct Drive Wind Turbine in DC-Link Voltage Control Timescale during Grid Fault

Qi Hu, Yiyong Xiong, Chenruiyang Liu *, Guangyu Wang and Yanhong Ma

National Key Laboratory of Science and Technology on Vessel Integrated Power System, School of Electrical Engineering, Naval University of Engineering, Wuhan 430033, China; qihu0725@163.com (Q.H.); xiongyiyong1989@163.com (Y.X.); hust_wang000@163.com (G.W.); yanhongma0421@163.com (Y.M.)
* Correspondence: lychee0825@outlook.com

Abstract: Transient stability during grid fault is experienced differently in modern power systems, especially in wind-turbine-dominated power systems. In this paper, transient behavior and stability issues of a direct drive wind turbine during fault recovery in DC-link voltage control timescale are studied. First, the motion equation model that depicts the phase and amplitude dynamics of internal voltage driven by unbalanced active and reactive power is developed to physically depict transient characteristics of the direct drive wind turbine itself. Considering transient switch control induced by active power climbing, the two-stage model is employed. Based on the motion equation model, transient behavior during fault recovery in a single machine infinite bus system is studied, and the analysis is also divided into two stages: during and after active power climbing. During active power climbing, a novel approximate analytical expression is proposed to clearly reveal the frequency dynamics of the direct drive wind turbine, which is identified as approximate monotonicity at excitation of active power climbing. After active power climbing, large-signal oscillation behavior is concerned. A novel analysis idea combining time-frequency analysis based on Hilbert transform and high order modes is employed to investigate and reveal the nonlinear oscillation, which is characterized by time-varying oscillation frequency and amplitude attenuation ratio. It is found that the nonlinear oscillation and even stability are related closely to the final point during active power climbing. With a large active power climbing rate, the nonlinear oscillation may lose stability. Simulated results based on MATLAB® are also presented to verify the theoretical analysis.

Keywords: direct drive wind turbine; grid fault; nonlinear oscillation; transient stability; time-frequency analysis; transient switch

Citation: Hu, Q.; Xiong, Y.; Liu, C.; Wang, G.; Ma, Y. Transient Stability Analysis of Direct Drive Wind Turbine in DC-Link Voltage Control Timescale during Grid Fault. *Processes* **2022**, *10*, 774. https://doi.org/10.3390/pr10040774

Academic Editor: Zhiwei Gao

Received: 21 March 2022
Accepted: 12 April 2022
Published: 15 April 2022

Publisher's Note: MDPI stays neutral with regard to jurisdictional claims in published maps and institutional affiliations.

1. Introduction

With an increasing penetration of wind power integrated into modern power systems, the dynamic issue of part grid tends to be dominated by wind turbines instead of traditional synchronous generators. Wind turbines have different dynamic characteristics than synchronous generators, resulting in the system experiencing different dynamic issues. Among different types of wind turbines, direct drives that have superior grid-connected performance are increasingly installed. However, due to the reverse distribution of wind resources and load centers, a large scale of direct drive wind turbines is installed in a weak AC grid, bringing in strong interaction between the wind turbine and AC grid. The strong dynamic interaction significantly challenges the safe and stable operation of the system, necessitating the analysis.

Existing studies have paid much attention to the stability issue resulting from the connection of wind turbines [1]. In [2–8], a small-signal oscillation problem related to wind turbines integrated into a high impedance AC grid is investigated. Due to a wide band control of equipment, oscillation is characterized by a multi-time scale, and oscillation frequency ranges from hundreds of Hz to several Hz. Previous works have carried out detailed analyses about this. However, these works address the small-signal stability issue with disturbance around the equilibrium point, and a linearized system is applicable for analysis. In practice, faults, including wind turbine system faults and grid faults, are common. Fault diagnosis and resilient control for a wind turbine system is a research hotspot that has attracted massive research in recent years [9,10]. Except for this, transient issue analysis during grid fault is also worthy of research. In [11], the rotor angle stability of the synchronous generator affected by the dynamical characteristics of a wind turbine is analyzed. The work is carried out from the viewpoint that synchronous generators are dominant equipment, and the dynamics of wind turbines are only influential factors. This is reasonable at a relatively low penetration of wind power. However, with the increasing penetration of wind power, the dynamic issue of the part grid is dominated by wind turbines instead of synchronous generators. Transient issue faces new challenges and begins to be paid attention to. Transient stability dominated by the control of renewable energy generating units are investigated in [12–17]. Analysis results show that similar transient instability that is common in traditional synchronous generators also exists in PLL-synchronized converters. The transient stability can be explored from the accelerating and decelerating areas method. These analyses are based on a simplified control structure and attempt to reveal transient instability mechanisms. Yet, practical control of wind turbines is complex, even on a single time scale [3,8–11]. In this paper, transient behavior during fault recovery in DC-link voltage control timescale is studied, with complex practical control considered.

A deep understanding of an equipment's characteristics is the precondition of dynamic issue analysis. In order to investigate the dynamic behavior of a system dominated by renewable energy, kinds of equipment models are proposed [18–21]. The impedance model is developed and widely used in small-signal oscillation analysis. External characteristics of equipment are investigated through impedance frequency spectrum with specific control structure packing treatment [22–24]. At the time of bringing convenience, it has some difficulty in mechanism explanation of the relationship between specific control loop and oscillation. Based on this consideration, the motion equation model from the idea of Newtonian mechanics is proposed to deeply study equipment's characteristics [20,21]. By establishing the relationship between unbalanced powers and dynamics of internal voltage, the form of the motion equation model is similar to the rotor motion of a synchronous generator, and equivalent inertia and damping can be obtained. Thus, oscillation with increasing amplitude can be physically explored from the viewpoint of insufficient damping. However, the two models are both applicable for small signal analysis. Under large-signal disturbance, a new model is needed to study equipment's transient characteristics. Based on the advantage of the motion equation model in studying the equipment's characteristics, it is necessary to be popularized for the condition of large-signal disturbance. In this paper, the transient motion equation model in the DC-Link voltage control time scale is developed with transient switch control considered.

Although large signal analysis is difficult due to the non-negligible influence of non-linearity, kinds of meaningful methods are proposed to address the issue [25–32]. The methods based on computational intelligence may be powerful for the analysis and control of a complex, large-scale system [25–27]. However, they may have difficulty explaining the stability mechanism and influence factor concerned by this paper. Time-frequency analysis based on the Hilbert transform is usually employed to analyze low-frequency and sub-synchronous nonlinear oscillation in traditional power systems [28,29]. Based on data from transient simulations, instantaneous attributes of oscillation behavior can be identified. In addition to numerical analysis, the inclusion of higher-order terms is usually used to evaluate accurate modal characteristics that linear analysis can not provide [30,31]. Based on the Normal Form theory, higher-order modal interactions resulting from the influence of nonlinearity can be revealed. By combining the two methods, nonlinear oscillation can be deeply investigated [32]. This paper draws lessons from the two methods and carries out large-signal oscillation analysis during fault recovery.

The rest part of this paper is organized as follows. In Section 2, transient switch control of direct drive wind turbine is investigated. Then motion equation model during fault recovery is developed in Section 3. Based on the developed model, transient behavior analysis in a simple system is carried out in Section 4. Finally, conclusions are drawn in Section 5.

2. Transient Switch Control of Direct Drive Wind Turbine

When grid faults occur, the wind turbine usually undergoes complex transient switch control to support the grid or protect the wind turbine itself. Figure 1 shows the typical auxiliary control and circuit referred to [33,34]. Due to this concerning issue, the control in electromagnetic time scale receives special attention.

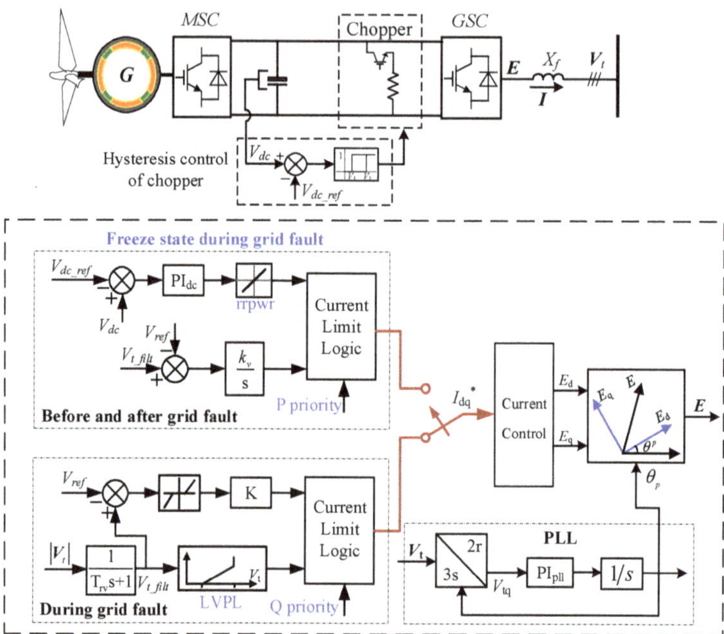

Figure 1. Typical auxiliary control and circuit in response to grid faults.

As shown in Figure 1, the whole process in response to grid faults can be divided into three stages according to that grid faults are detected and then cleared, which is as shown in Figure 2. Each stage employs a different control structure in order to satisfy different requirements.

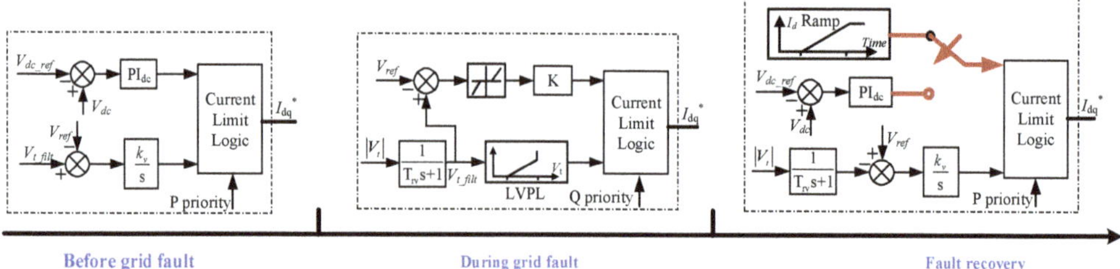

Figure 2. Transient switch control in response to grid fault.

Before grid fault, active and reactive current orders are controlled by DC-Link voltage control and terminal voltage control, respectively. Due to the limited capacity of the grid side converter, active power priority is usually utilized in current limit logic.

In the case that grid faults are detected, states of DC-Link voltage control and terminal voltage control before the grid fault are frozen first. Grid side converter takes the role of supporting grid voltage and injects reactive current as required by grid codes. In order to reduce the system stress during grid faults, the active current order is limited by a cap (upper limit) through Low Voltage Power Logic (LVPL) [34]. In normal operating conditions, there is no cap. When the voltage falls, a cap is calculated and applied. Thus, the dynamics of the active current order are influenced by the amplitude of terminal voltage during a deep grid fault. Referring to possible DC-Link overvoltage resulting from limited active power transfer, a hysteresis controller based on chopper-controlled resistors is employed to stabilize DC-Link voltage in the set narrowband.

When grid faults are cleared, active and reactive current orders are re-controlled by DC-Link voltage control and terminal voltage control, respectively. However, a ramp rate limit is applied to the active current order rate of increase to reduce system stress [34]. Since active current order during the grid fault is usually very small, it increases with time according to the ramp rate limit in a short time during fault recovery. When it reaches about the frozen value before the grid fault, the active current order begins to be adjusted by DC-Link voltage control. As a result, the transient process during fault recovery can be further divided into two stages: during and after active power climbing. A switched system should be employed to portray the transient behavior during fault recovery.

3. Developed Motion Equation Model

Since direct drive wind turbines employ a power electronic converter as a grid-connected interface, their transient characteristic is dominated by complex control. In order to physically study the transient characteristic, a motion equation model based on Newton mechanics is proposed, which establishes the relationship of internal voltage dynamics induced by unbalanced powers. Then the transient characteristic of the direct drive wind turbine can be explored from the equivalent motion driven by unbalanced powers. Concerning the transient switch control during fault recovery, the transient analysis should be divided into two stages: during and after active power climbing. The switched system should be employed to depict transient characteristics. During active power climbing, active current order increases with time according to the ramp rate limit, and DC-Link voltage control does not take effect. After active power climbing, active current order begins to be adjusted by DC-Link voltage control.

3.1. Motion Equation Model in Stage of Active Power Climbing

Based on Figure 1, the dynamics of the wind turbine's internal voltage in the stage of active power climbing are dominated by terminal voltage control and a phase-locked loop, as shown in Figure 3. Since the two control loops are in response to dynamics of terminal voltage and then adjust current orders, the modeling work is mainly composed of two parts. One is that the phase and amplitude dynamics of terminal voltage should be obtained through active and reactive power (P,Q). Based on this, a model can be developed in the form that dynamics are induced by unbalanced powers, and the model has good portability due to no relationship with the information of the network. The other is that internal voltage should be calculated through current orders since the internal voltage is selected to represent the external characteristic of the wind turbine.

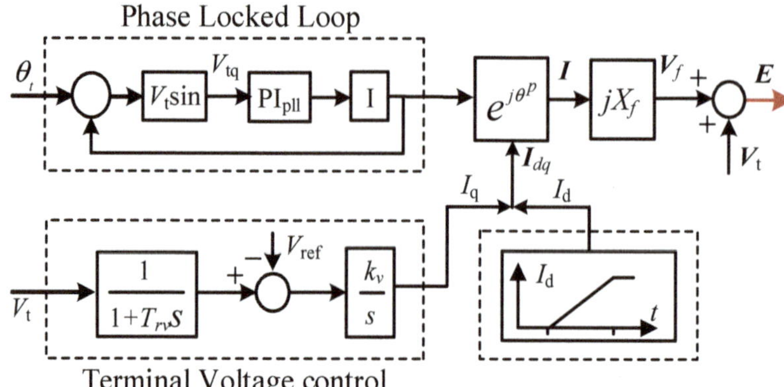

Figure 3. Dynamics of internal voltage in stage I.

First, the relationship of terminal voltage with active and reactive power output is calculated. According to the circuit topology in Figure 1, the output power is represented by

$$P = \frac{EV_t \sin(\theta_e - \theta_t)}{X_f} \tag{1}$$

$$Q = \frac{\left[E^2 - EV_t \cos(\theta_e - \theta_t) \right]}{X_f} \tag{2}$$

Then, combining (1) and (2), phase and amplitude dynamics of terminal voltage can be obtained by

$$\theta_t = \theta_e - \arctan \left[\frac{PX_f}{\left(E^2 - QX_f \right)} \right] \tag{3}$$

$$V_t = \frac{\sqrt{P^2 X_f{}^2 + \left(E^2 - QX_f \right)^2}}{E} \tag{4}$$

Thus, information on terminal voltage can be replaced by internal voltage and active and reactive power output.

Second, the internal voltage should be calculated through current orders. It is known that current orders adjusted by control loops are in the PLL reference frame. Based on the circuit relationship in Figure 1, the *dq* component of internal voltage can be calculated by

$$E_d = V_t \cos \theta_t^p - X_f I_q \tag{5}$$

$$E_q = V_t \sin \theta_t^p + X_f I_d \tag{6}$$

Through polar coordinates transformation, amplitude and phase (that is, relative to d-axis of PLL) of internal voltage can be obtained by

$$E = \sqrt{E_d{}^2 + E_q{}^2} \tag{7}$$

$$\theta_e^p = a \tan \left(\frac{E_q}{E_d} \right) \tag{8}$$

Due to the employed PLL synchronization, the phase of internal voltage is composed of two parts: the synchronous phase provided by PLL and the phase that is relative to PLL, as represented by

$$\theta_e = \theta_e^p + \theta_p \tag{9}$$

Based on the above deduction, the developed motion equation mode during active power climbing is shown in Figure 4. It is clearly seen that the dynamics of internal voltage can be studied from the equivalent motion driven by unbalanced active and reactive power.

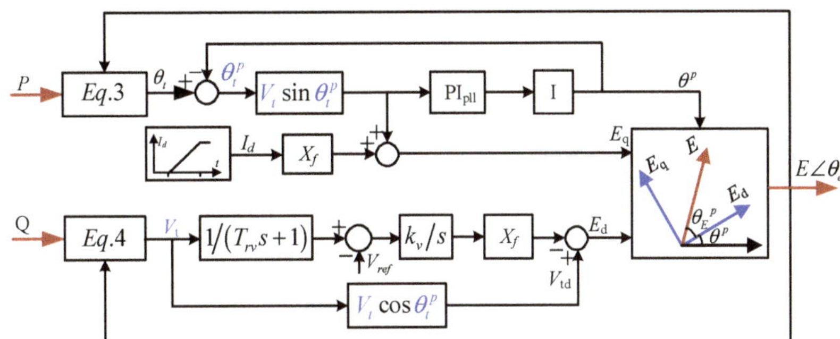

Figure 4. Motion equation mode in stage of active power climbing.

3.2. Motion Equation Model after Active Power Climbing

When the active current order approaches frozen value before grid fault, ramp rate limit will be out of action, and DC-Link voltage control begins to take effect. Thus the influence of DC-Link voltage control on dynamics of internal voltage should be considered after active power climbing.

In this case, when electromagnetic power injected into the power grid is not equal to feed power from the machine side, the DC-link capacitor will go through charging or discharging. Then the active current will be adjusted and thus significantly influence phase dynamics. This indicates that unbalanced active power drives the motion of phase, although the relationship between them is complex. Moreover, when DC-Link voltage exceeds the limit value, the chopper will take effect, and consumed power by the chopper should be taken into account. Based on Figure 4, the motion equation model after active power climbing can be easy to be obtained, as shown in Figure 5.

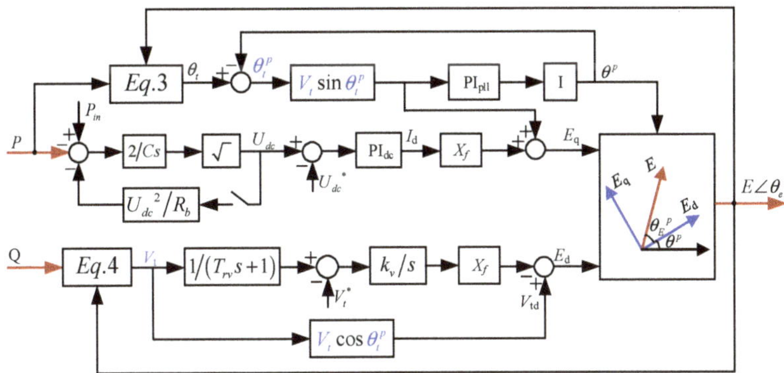

Figure 5. Motion equation model after active power climbing.

3.3. Equipment's Transient Characteristic Analysis

Based on the developed motion equation model in Figures 4 and 5, it is known that wind turbine is very different from traditional synchronous generators and its transient characteristic are much more complex, which can be concluded as

(1) Discontinuity. Unlike the synchronous generators that can employ a unified model for electromechanical transient analysis in different fault stages, the developed

motion equation model of the direct drive wind turbine is discontinuous due to transient switch control.

(2) Nonlinearity. In transient stability analysis of traditional power systems, nonlinearity mainly results from the network, which lies in the power angle curve, and the linear rotor motion model is used to depict the equipment's transient characteristic. However, the wind turbine's motion equation model is characterized by strong nonlinearity, and nonlinearity is mainly embodied in the following three aspects: polar transformation, PLL, and replacing terminal voltage information. The main types of nonlinearity are trigonometric and square functions.

(3) High order. Due to the complex control of wind turbines, the relationship between unbalanced power and internal voltage is characterized by high order. As a result, the inertia that is used to depict the relationship between unbalanced active power and phase dynamics is variable. This is different from a synchronous generator, which has constant inertia.

(4) Strong Coupling. In a wind turbine, phase dynamics are strongly coupled with amplitude dynamics, and the coupling that mainly results from the control of the wind turbine is implemented in an orthogonal coordinate system, while amplitude and phase are obtained from the polar coordinate system. Compared with a synchronous generator that directly controls amplitude and phase, the coupling in a wind turbine is stronger.

4. Transient Analysis in Single-Machine Infinite-Bus System

Based on the developed motion equation model, transient analysis during fault recovery in a typical single-machine infinite-bus (SMIB) system shown in Figure 6 is carried out. A three-phase ground fault is set at one line, and after a certain time, the faulted line is cut off. In this paper, we assume that a stable operating point has been achieved during a grid fault, and transient behavior during fault recovery is mainly concerned.

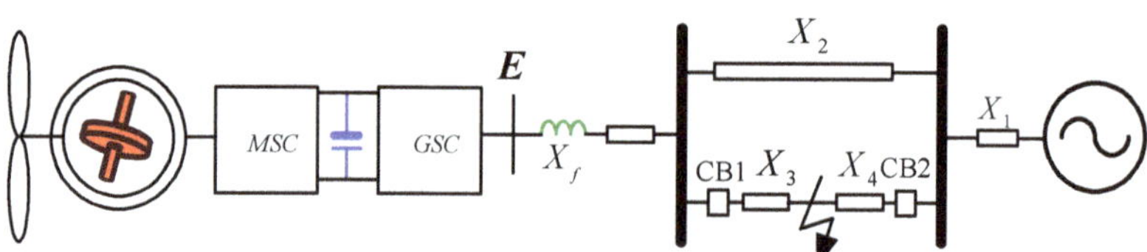

Figure 6. Single type-4 wind turbine infinite-bus (SMIB) system.

The transient analysis idea during fault recovery is shown in Figure 7. It is assumed that the system achieves a stable state at point a during grid fault. When the fault is cleared, the system goes through the transient process from point a to equilibrium point c after grid fault. However, due to the transient switch control introduced by active power climbing, the transient process is divided into two stages: during and after active power climbing. In the two stages, the network equations are the same. However, the motion equation models of direct drive wind turbines are different, resulting in state trajectories that are dominated by different dynamic equations. In the stage of active power climbing, the state trajectory moves from point a driven by the motion equation model and network equation in the stage of active power climbing. When the active current order reaches about the frozen value before grid fault, the stage ends, and the final state is the initial state of the second stage. After active power climbing, the system goes through the transition process from the final state in the stage of active power climbing to a stable equilibrium point after a grid fault. Due to the strong nonlinearity, the dynamic behavior and even stability issue in the second stage is significantly influenced by the final state in the stage of active power climbing. According to the attraction region theory of nonlinear system, there exists an attraction region in state space for the stable equilibrium point. Only if the initial state lies in the attraction region can the system keep transient stable. Otherwise, transient instability will occur. Since the initial state in the second stage during fault recovery is determined by the final state in the first stage, the dynamic behavior in the stage of active power climbing will have much influence on the dynamic behavior and transient stability issue after active power climbing.

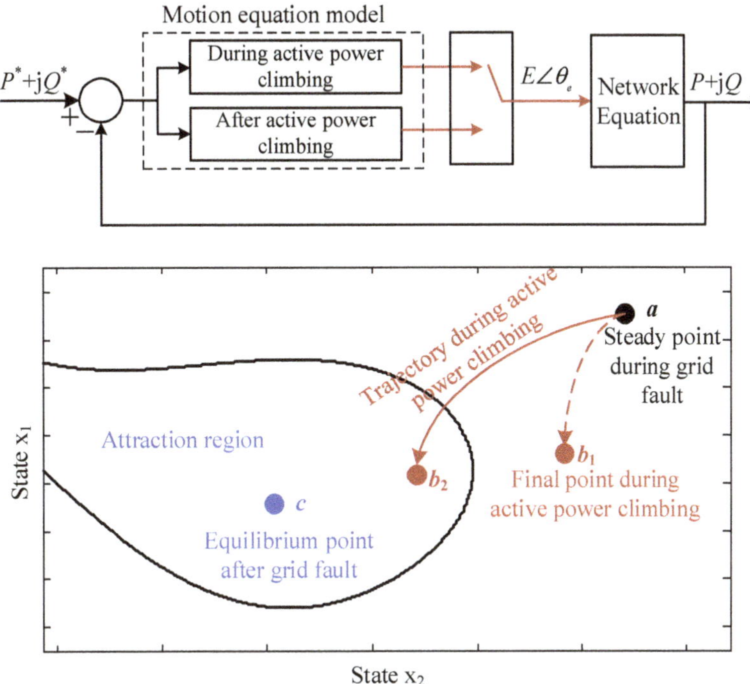

Figure 7. Transient analysis idea during fault recovery.

4.1. Transient Analysis in Stage of Active Power Climbing

In the stage of active power climbing, closed-loop dynamics of internal voltage can be investigated by combining the motion equation model and network model. Simplified network topology is shown in Figure 8, and the grid is represented by its Thevenin equivalent circuit. When considering transient behavior in the DC-link voltage control time scale, fast dynamics of the network are neglected, and an algebraic equation is used to calculate power through voltage vectors [8]. In Figure 9, it is shown that phase dynamics of internal voltage are induced by time-varying excitation from the active current order. In the model, the time-varying excitation can be further replaced by the integral calculus of constant k_{ramp}.

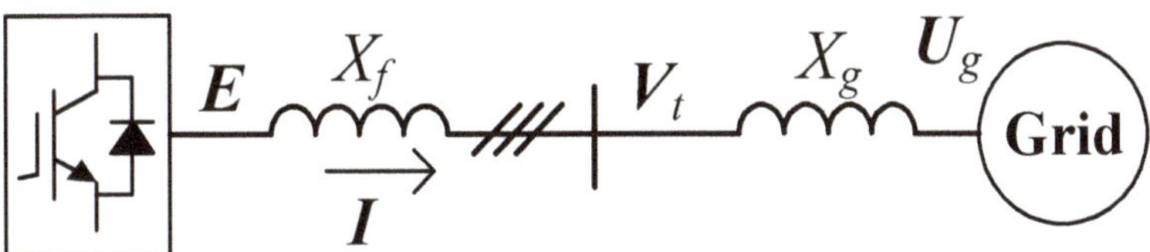

Figure 8. Simplified network topology.

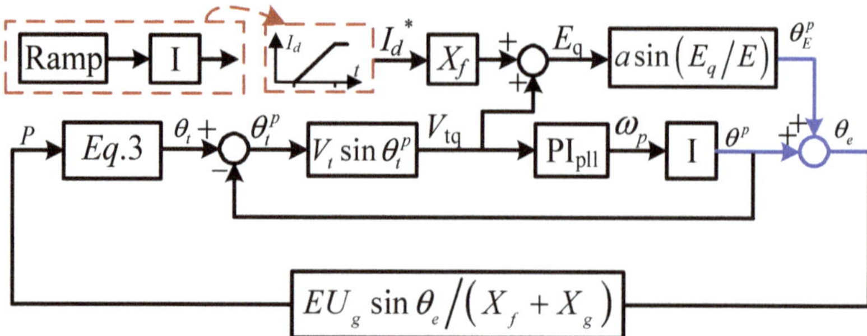

Figure 9. Phase dynamics of internal voltage in stage of active power climbing.

It is known that phase of internal voltage is composed of two parts: θ_p that is dominated by PLL and θ_E^p of internal voltage relative to PLL. Both of them are induced by active current order excitation. Since θ_p is directly related with V_{tq}, its dynamics can be investigated based on the relationship of V_{tq} and the active current order. Utilizing terminal voltage V_t and grid voltage U_g in Figure 8, active power can be represented by

$$P = \frac{V_t U_g \sin \theta_t}{X_g} \tag{10}$$

In addition to this, active power can also be calculated through *d*-axis and *q*-axis current, which is obtained by

$$P = U_g I_d^p \cos \theta_p - U_g I_q^p \sin \theta_p \tag{11}$$

Since the phase of terminal voltage is also composed of two parts: θ_p and θ_t^p of terminal voltage relative to PLL, $V_t \sin \theta_t$ in (10) has another form represented by

$$V_t \sin \theta_t = V_t \sin \theta_t^p \cos \theta_p + V_t \cos \theta_t^p \sin \theta_p \tag{12}$$

Combing (10)–(12), relationship of V_{tq} and active current order excitation is obtained by

$$V_{tq} = X_g \int k_{ramp} dt - U_g \sin \theta_p \tag{13}$$

Further, differentiating (13), the following expression can be obtained.

$$\frac{dV_{tq}}{dt} = k_{ramp} X_g - \omega_p U_g \cos \theta_p \tag{14}$$

Then the frequency dynamics of internal voltage dominated by PLL can be shown in Figure 10. It is a step response of a third-order nonlinear dynamical system, and excitation is related with X_g and k_{ramp}. Nonlinearity exists in red dashed line frame in Figure 10. In addition to these, the initial states in Figure 10 reflect the influence of states during grid fault on the step response, and the initial states of θ_p and V_{tq} are represented by (15) and (16), which is calculated based on the state-equation during grid fault [13]. Since it is assumed that steady states are achieved during grid fault, the initial integral state of PLL's PI controller is usually zero.

$$\theta_{p\_initial} = a \sin \left(\frac{X_{gf} I_{df0}}{U_{df}} \right) \tag{15}$$

$$V_{tq\_initial} = X_g I_{df0} - U_g \sin \left(\theta_{p\_initial} \right) \tag{16}$$

where U_{df} and X_{gf} are Thevenin equivalents of grid and I_{df0} is the active current order, which are all during grid fault.

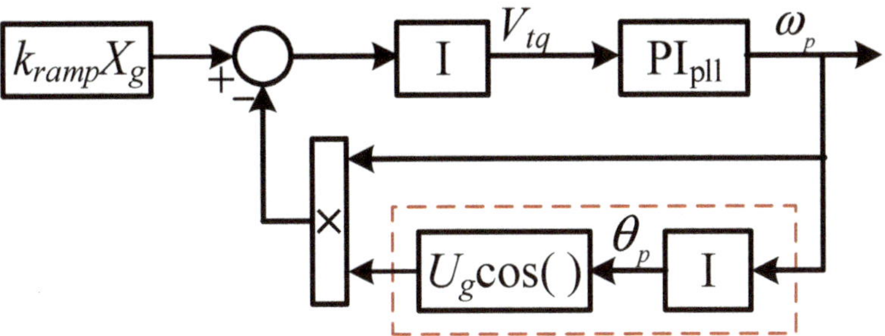

Figure 10. Frequency dynamics of internal voltage dominated by PLL.

For convenience, assume that deep voltage sag is considered and the active current order during grid fault is zero. Thus, V_{tq} will not jump and keep zero at the beginning of fault recovery. Since the initial state of θ_p is also zero, V_{tq} begins to increase driving by $k_{ramp}X_g - \omega_p U_g\cos\theta_p$, and then ω_p and θ_p both increase from the zero initial state. However, in a short period of time, θ_p is very small and $U_g\cos\theta_p$ is approximate to be constant U_g. Thus, the nonlinear part in the red dashed line frame can be replaced by a constant and ω_p is approximate to be the step response of the second-order system as represented by

$$\omega_{p\_approximation} = L^{-1}\left[\frac{k_{p\_pll}s + k_{i\_pll}}{s^2 + k_{p\_pll}s + k_{i\_pll}}\frac{k_{ramp}X_g}{s}\right] \tag{17}$$

As time prolongs, θ_p becomes large, and the influence of the nonlinear part should be considered. At this time, due to the fast response of PLL, dynamical regulation resulted in a large deviation of $k_{ramp}X_g$ and $\omega_p U_g\cos\theta_p$ can be thought to be finished and approximation of $k_{ramp}X_g \approx \omega_p U_g\cos\theta_p$ is reasonable. Thus, the dynamics of ω_p can be represented by the quasi-steady-state solution shown below.

$$\omega_{p\_quasi\_steady\_state} = \frac{k_{ramp}X_g}{U_g\cos(\int \omega_p)} \tag{18}$$

Based on the above, the dynamics of ω_p, in the whole stage of active power climbing, can be approximately represented by

$$\omega_p \approx \omega_{p\_approximation} + \omega_{p\_quasi\_steady\_state} - k_{ramp}X_g \tag{19}$$

In the initial stage, it can be depicted by the step response of the second-order system, and then the quasi-steady-state solution reflects the subsequent dynamics. Further, the quasi-steady-state solution also has an approximate relationship represented by

$$\int (k_{ramp}X_g)dt \approx \int (\omega_p U_g\cos\theta_p)dt \tag{20}$$

Then θ_p at the end of active power climbing can be estimated by

$$\theta_{1s} = a\sin\left(\frac{X_g I_{d0}}{U_g}\right) \tag{21}$$

ω_p reaches the maximum at this time, which is represented by

$$\omega_{p\_max} \approx \frac{(k_{ramp} X_g)}{(U_g \cos \theta_{1s})} \tag{22}$$

Simulated results verified the analysis is shown in Figure 11. Since θ_p is integral of ω_p, dynamics of θ_p is charactered by monotonous increase. Further analysis reveals that the influence of states of amplitude branch on the oscillation in the second stage is very small. Thus dynamics of the amplitude branch in the first stage will not be deeply investigated.

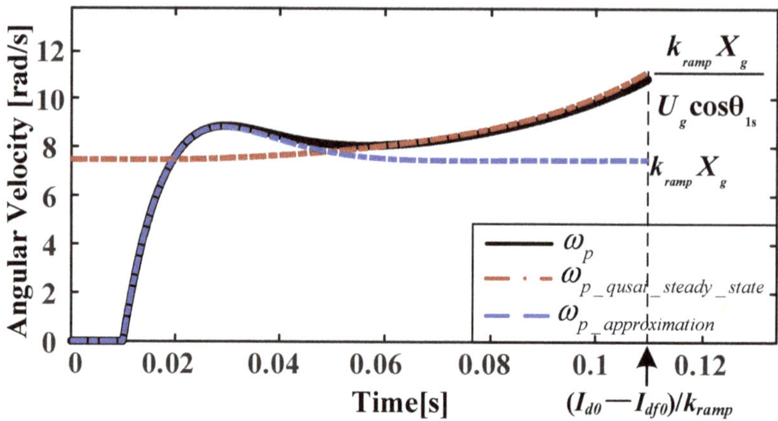

Figure 11. Frequency response of internal voltage dominated by PLL.

From Figure 10, active power climbing rate k_{ramp} has much influence on frequency response of internal voltage dominated by PLL. By numerical calculation, the frequency response at different active power climbing rates is shown in Figure 12. It is seen that the frequency offset tends to be large at the end of active power climbing with the increase in active power climbing rate. Since the final state in the stage determines the initial state after active power climbing, it is indicated that the initial state in the second stage will deviate from the equilibrium point far away with the increase in active power climbing rate, which will deteriorate the transient behavior and even bring transient instability issue after active power climbing.

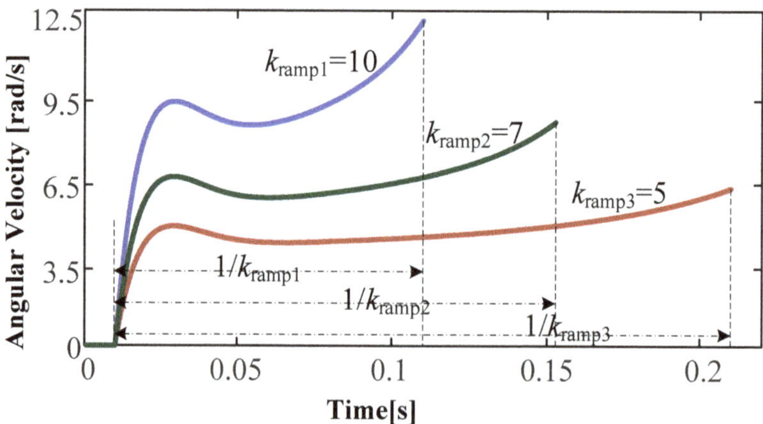

Figure 12. Influence of k_{ramp} on frequency response of internal voltage.

4.2. Nonlinear Oscillation Analysis after Active Power Climbing

Based on the motion equation model in Figure 5, it is known that the open-loop characteristics of a wind turbine are depicted by two input and two output nonlinear transfer functions. In order to qualitatively investigate the influence of nonlinearity on large-signal oscillation behavior, single input and single output dynamical equation are employed for convenience based on a hypothesis. Here open-loop phase dynamics induced by unbalanced active power are investigated, as shown in Figure 13.

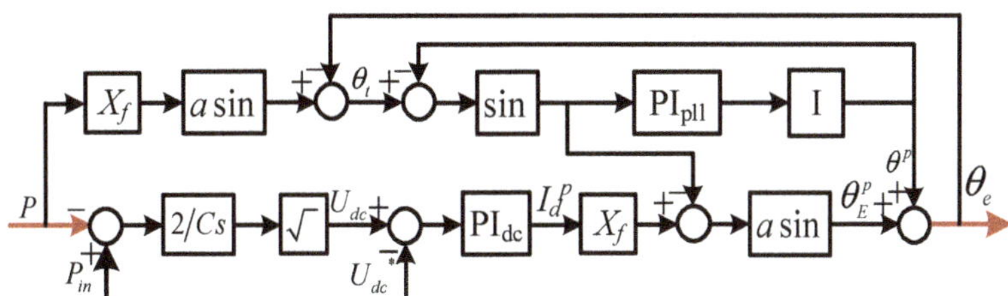

Figure 13. Open-loop phase dynamics with amplitude dynamics neglected.

The nonlinear function has three types: sin, asin, and square root. The influence of nonlinear function on open-loop characteristics can be reflected by (23). The small increase in the amount of output relative to the small increase in the amount of input at different operating points at a disturbed trajectory is different. The complexity induced by nonlinearity has resulted from this.

$$\sin(x_i + \Delta x) = \sin x_i + (\cos x_i)\Delta x \tag{23}$$

However, based on the geometric interpretation of the Euler integral, short time window dynamics around any operating point at a disturbed trajectory can be represented by a time-varying linear equation and constant excitation. The linear equation is obtained by linearization at the studied operating point, which usually is not the equilibrium point. Thus, open-loop characteristics in a short time window can be represented by (24).

$$\Delta \theta_e = G(s, X_i)\Delta P \tag{24}$$

Since $G(s, X_i)$ is related with operating point X_i, it is not constant and changes with time. When the disturbance is small, it means that X_i is very close to the equilibrium point X_e and the influence of change of X_i on $G(s, X_i)$ is so small that it can be neglected. As a result, $G(s, X_i)$ is fixed, and small-signal dynamics have constant oscillation modes. Amplitude attenuation and the oscillation frequency are fixed. However, when the disturbance is large, $G(s, X_i)$ changes a lot with time. It is known that short time window oscillation characteristics are related with $G(s, X_i)$. Thus large-signal oscillation characteristics may be very different from small-signal oscillation, and its amplitude attenuation and oscillation frequency are not fixed.

$$\sin(x_e + x) = \sin x_e + (\cos x_e)x - 0.5(\sin x_e)x^2 + O\left(x^2\right) \tag{25}$$

Further, the influence of nonlinear function on open-loop characteristics can be investigated from the viewpoint of Taylor's high-order expansion. From (25), it is known that the output of a nonlinear function has other frequency components even if the input is a single frequency sinusoidal signal, and as the amplitude of the input signal increases, other frequency components in the output signal tend to be large. Due to these characteristics, closed-loop oscillation behavior will be more complex.

In order to study closed-loop oscillation behavior, the motion equation model in Figure 5 is combined with the network model in Figure 8. Oscillatory modes of linearized system at equilibrium point are listed in Table 1. It is shown that the linearized system is poor damping and mode three dominates the small-signal oscillation. Comparative simulated results shown in Figure 14 reveals that large signal oscillation characteristics (amplitude attenuation and oscillation frequency) are different from that of linear oscillation. In order to further investigate the large-signal oscillation characteristics, time-frequency analysis based on the Hilbert transform [31] is employed.

Table 1. Oscillatory modes of equilibrium point after active power climbing.

Mode	Eigenvalue	Freq.(Hz)	Damping Ratio
1	$-100 \pm 99j$	15.8	71%
2	$-29.4 \pm 35j$	5.6	64%
3	$-0.6 \pm 67.9j$	10.8	1.3%

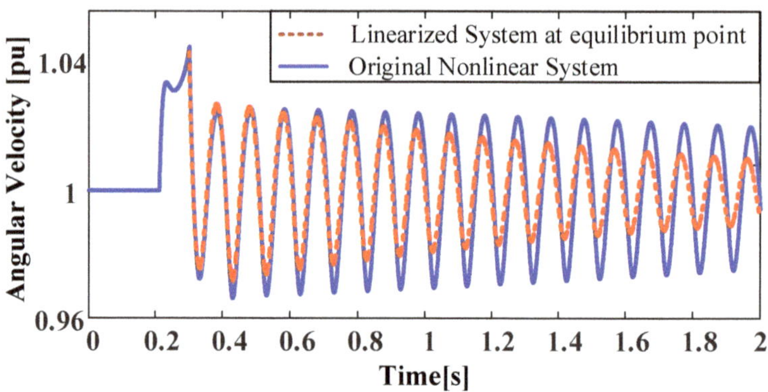

Figure 14. Comparative simulated results of internal voltage's angular velocity between original nonlinear system and linearized system at equilibrium point.

Since damping ratios of modes one and two are large, the oscillation component dominated by them will attenuate quickly, and the obtained simulated responses can be thought to be a single component from the viewpoint of a nonstationary signal. The single-component large-signal oscillation has the unified form represented by

$$x(t) = Ae^{\int \alpha(t)dt} \cos\left(\int \omega(t)dt\right) \tag{26}$$

Here $\alpha(t)$ and $\omega(t)$ are defined as instantaneous amplitude attenuation ratio and oscillation frequency, respectively. If the oscillation is linear, $\alpha(t)$ and $\omega(t)$ are constantly determined by mode three. However, due to nonlinearity, $\alpha(t)$ and $\omega(t)$ change with time. In order to obtain $\alpha(t)$ and $\omega(t)$, Hilbert transform of $x(t)$ is utilized and $y(t)$ can be attained represented by

$$y(t) = Ae^{\int \alpha(t)dt} \sin\left(\int \omega(t)dt\right) \tag{27}$$

Based on $x(t)$ and $y(t)$, the amplitude dynamics $A(t)$ and phase dynamics $\theta(t)$ can be represented by

$$A(t) = Ae^{\int \alpha(t)dt} = \sqrt{x(t)^2 + y(t)^2} \tag{28}$$

$$\theta(t) = \int \omega(t)dt = \arctan\left(\frac{y(t)}{x(t)}\right) \tag{29}$$

Then $\alpha(t)$ and $\omega(t)$ can be calculated by

$$\alpha(t) = \frac{dA(t)/dt}{A(t)} \tag{30}$$

$$\omega(t) = \frac{d\theta(t)/dt}{\theta(t)} \tag{31}$$

Based on the Hilbert transform, the obtained instantaneous amplitude attenuation ratio $\alpha(t)$ and oscillation frequency $\omega(t)$ are shown in Figure 15. It is known that $\alpha(t)$ and $\omega(t)$ are both not constant and change with time. Further, $\alpha(t)$ and $\omega(t)$ oscillates around mode 3. This also verifies the idea of piecewise linearization with short time window. Since the short time window open-loop characteristics are determined by $G(s, X_i)$ and states X_i oscillates around equilibrium point, instantaneous amplitude attenuation ratio, and oscillation frequency are inevitable to change around mode three with time. Figure 15 also shows that $\alpha(t)$ varies a lot around real part of mode three. This reveals that $\alpha(t)$ is very sensitive to change of states. Integral of $\alpha(t)$ reflects attenuation of amplitude. Figure 16 shows that $\int \alpha(t)dt$ tends to be larger than $\int \alpha_0 dt$ as time increases. This reveals that the nonlinearity deteriorates amplitude attenuation.

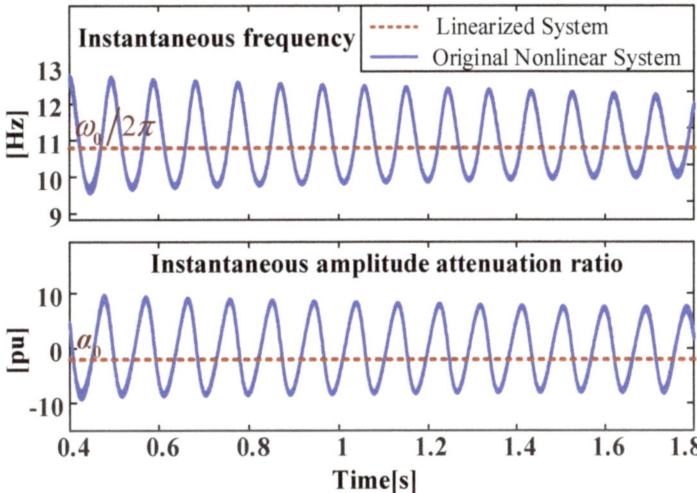

Figure 15. Instantaneous frequency and amplitude attenuation ratio characteristic.

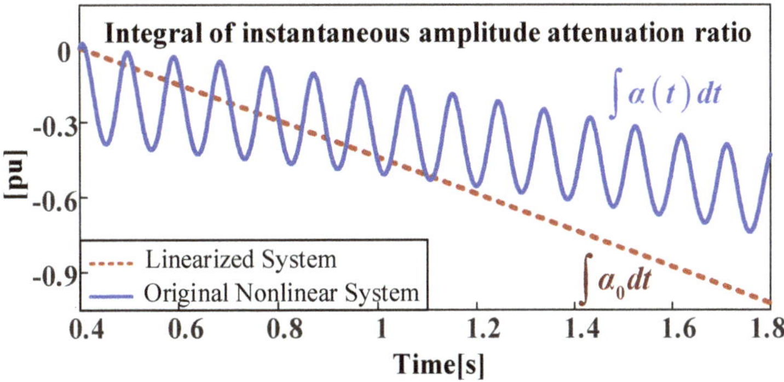

Figure 16. Integral of instantaneous amplitude attenuation ratio.

Time-frequency analysis digs deep into the nonlinear oscillation characteristics. In order to further carry out the mechanism explanation, Taylor's high-order expansion joint with the analysis idea of Normal Form is employed. From (25), it is known that Taylor's high-order expansion can achieve a good approximation of nonlinear function, and the order of high order term is related to the disturbance. Here second-order approximation is considered, and dynamics of the jth state can be represented by

$$\frac{dx_j}{dt} \approx f_j(X_e) + A_j(X - X_e) + 0.5(X - X_e)^T H^j (X - X_e) \tag{32}$$

Where A_j is the jth row of the Jacobian matrix $[\partial f / \partial X]$, and H^j is the Hessian matrix. Comparative simulated results among the original nonlinear system, first-order and second-order approximated systems are shown in Figure 17. It is known that second-order approximation almost achieves the same dynamical response as that of the original nonlinear system.

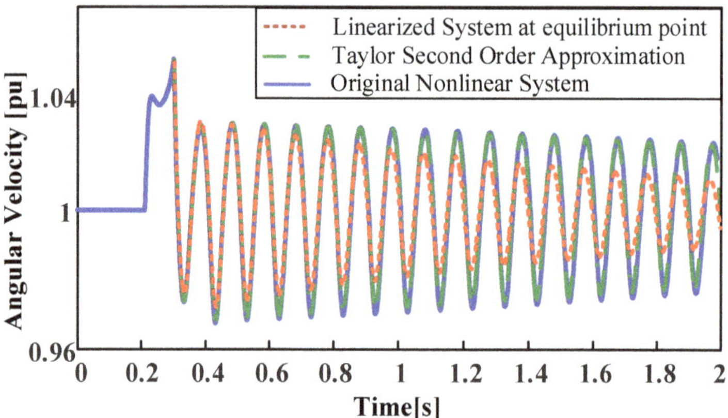

Figure 17. Comparative simulated results among original nonlinear system, first order and second order approximated system at equilibrium point.

Inspired by the analysis idea of Normal Form, analysis of the second-order approximated system can further explain nonlinear oscillation characteristics. The idea of Normal Form is that by the nonlinear transform of state variables, second-order terms in a state-space equation can be eliminated and an approximate solution of the nonlinear system is obtained, and the oscillation behavior is dominated by individual system modes, $\lambda_1, \lambda_2, \cdots, \lambda_n$ that are calculated by Jacobian matrix and second-order modes, $\lambda_1 + \lambda_1, \lambda_1 + \lambda_2, \cdots, \lambda_{n-1} + \lambda_n, \lambda_n + \lambda_n$. However, the base of the solution of Normal Form is still eigenvalues of the linearized system at the equilibrium point, and the obtaining of an approximate solution is under the condition that the influence of higher-order terms is neglected. These are reasons for the approximate solution's error. Since the approximate solution is not the target here and just an analysis idea is employed, the base of the solution can be selected in aid of Fourier and prony analysis, and oscillation behavior is still dominated by individual system modes and second-order modes. Fourier spectra in Figure 18 reveal that the second-order mode exists, and its frequency is almost twice the base dominant mode's. The second-order mode results from the nonlinear modal interaction of the base dominant mode. Due to the existence of the second-order mode component, the instantaneous oscillation frequency changes with time can be explained, which can also be understood from (33) and Figure 19. Further, prony analysis results in Table 2 show that the nonlinear oscillation is dominated by two modes: poor damping base mode and second-order mode, which is the combination of the poor damping base mode.

The real part of the base mode is smaller than that of the linear system. This further verifies that nonlinearity deteriorates amplitude attenuation.

$$x(t) = Ae^{\int \alpha(t)dt} \cos(\int \omega(t)dt)$$
$$\approx A_1 e^{-\alpha_1 t} \cos(\omega_1 t + \theta_1) + A_2 e^{-2\alpha_1 t} \cos(2\omega_1 t + \theta_2) \tag{33}$$

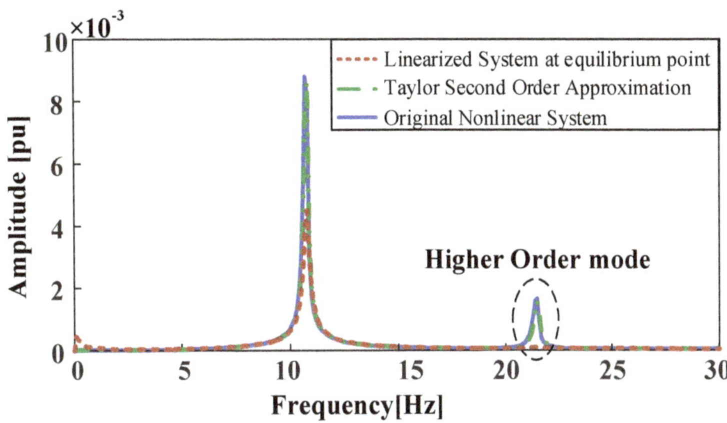

Figure 18. Fourier spectra of simulated results.

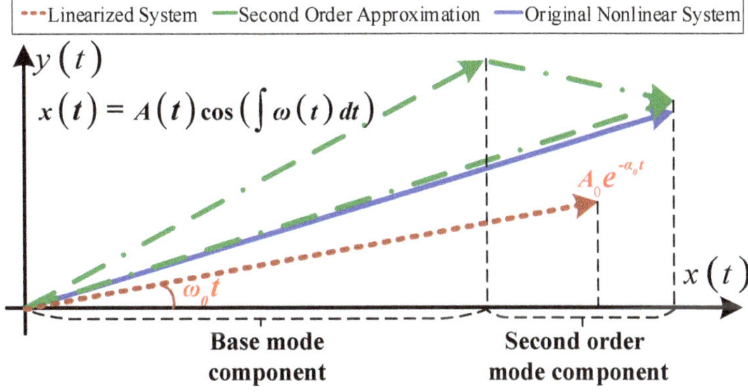

Figure 19. Mechanism explanation of instantaneous oscillation frequency changing with time.

Table 2. Prony analysis results of nonlinear oscillation response.

Mode	Eigenvalue	Freq.(Hz)	Damping Ratio
3	$-0.32 \pm 60\mathrm{j}$	10	0.53%
3,3	$-0.75 \pm 127\mathrm{j}$	21	0.59%

4.3. Influence of Ramp Rate Limit in First Stage on Oscillation Behavior in Second Stage

Based on the above analysis, it is known that the large signal oscillation behavior after active power climbing is very different from linear oscillation. Due to the influence of nonlinearity, its instantaneous amplitude attenuation ratio and oscillation frequency change with time, and the size of fluctuation is related to the state at the end of active power climbing. The above analysis further reveals that the comprehensive effect of the time-varying amplitude attenuation ratio is to deteriorate amplitude attenuation. When the initial state is far away from the equilibrium point in the second stage, oscillation with increasing amplitude may occur.

Since the initial state after active power climbing is dependent on the final state during active power climbing, transient behavior during the stage of active power climbing influences the subsequent oscillation. Based on the analysis in Section 4.1, it is known that initial states after active power climbing tend to be far away from the equilibrium point when k_{ramp} increases. As a result, when k_{ramp} is large, the influence of nonlinearity on large-signal oscillation behavior is strong. Since the comprehensive effect of nonlinearity is to deteriorate amplitude attenuation based on the analysis in Section 4.2, it is indicated that the nonlinear oscillation after active power climbing decays slowly and even diverges with the increase in active power climbing rate. Comparative simulated results at different ramp rate limit based on MATLAB® is shown in Figure 20. It is seen that the frequency offset at the end of active power climbing tends to be large, and then the subsequent oscillation after active power climbing tends to decay slowly and even diverges with the increase of active power climbing rate, which verifies the above theoretical analysis.

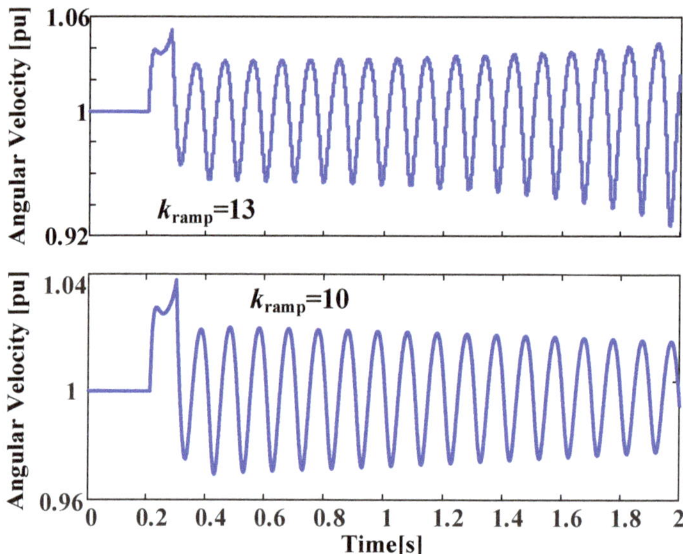

Figure 20. Comparative simulated results at different ramp rate limit.

5. Conclusions

In this paper, transient stability of direct drive wind turbine in DC-link voltage control timescale during LVRT is studied, with practice transient switch control considered. A novel two-stage motion equation model that depicts the phase and amplitude dynamics of internal voltage driven by unbalanced active and reactive power is developed firstly to physically study the transient characteristics of a direct drive wind turbine. Then the transient behavior during fault recovery is explored based on the developed model. When considering discontinuity resulting from the transient switch control, the whole transient process during fault recovery is divided into two stages: during and after active power climbing. In the first stage, frequency dynamics of a direct drive wind turbine at the excitation of active power climbing are studied. A novel approximate analytical expression is proposed to clearly reveal the transient frequency response and the influence of the active power climbing rate on it. After active power climbing, a novel analysis idea combining time-frequency analysis based on the Hilbert transform and high order modes is employed to investigate and reveal the nonlinear oscillation. The influence of transient behavior in the stage of active power climbing on the nonlinear oscillation after active power climbing is also explored. The conclusions and key findings are as follows.

(1) During active power climbing, an approximate monotonic increase in the wind turbine's angular frequency is identified at the excitation of active power climbing. With the increase in active power climbing rate, the frequency offset at the end of active power climbing tends to be large.

(2) After active power climbing, nonlinear oscillation characterized by time-varying oscillation frequency and amplitude attenuation ratio is revealed. It is found that the comprehensive effect of the time-varying amplitude attenuation ratio is to deteriorate amplitude attenuation. When the initial state tends to be far away from the equilibrium point in this stage, the nonlinear oscillation tends to decay slowly and even diverge, bringing in transient instability.

(3) The final state during active power climbing determines the initial state after active power climbing. With the increase in active power climbing rate, the final state during active power climbing will deviate from the equilibrium point after active power climbing far away. Then the amplitude attenuation of the nonlinear oscillation deteriorates, and oscillation with increasing amplitude is easier to occur.

Author Contributions: Conception and analysis, Q.H.; simulation and prony analysis, C.L.; high order modal analysis, G.W.; writing—original draft preparation, Q.H.; writing—review and editing, Y.X. and C.L.; literature research, Y.M. All authors have read and agreed to the published version of the manuscript.

Funding: This work was supported by National Natural Science Foundation of China under Grant 52107137.

Institutional Review Board Statement: Not applicable.

Informed Consent Statement: Not applicable.

Data Availability Statement: Not applicable.

Conflicts of Interest: The authors declare no conflict of interest.

Nomenclature

Symbol	Explanation
E	GSC internal voltage vector
V_t	Terminal voltage vector
U_g	Equivalent grid voltage vector
I	Current vector across filter inductor
P,Q	Active and reactive power output of GSC
X_g	Equivalent grid inductor
X_f	Grid-side filter inductor
θ_p	PLL output angle relative to grid voltage
ω_p	Angular velocity of PLL relative to grid voltage
k_{ramp}	Ramp rate limit
T_{rv}, k_v	Parameters of AVC's controller
k_{p_dc}, k_{i_dc}	Parameters of DVC's PI controller
k_{p_pll}, k_{i_pll}	Parameters of PLL's PI controller
Subscripts: dq	components in PLL reference frame

References

1. Wang, X.; Taul, M.G.; Wu, H.; Liao, Y.; Blaabjerg, F.; Harnefors, L. Grid-Synchronization Stability of Converter-Based Resources—An Overview. *IEEE Open J. Ind. Appl.* **2020**, *1*, 115–134. [CrossRef]
2. Hu, Q.; Fu, L.; Ma, F.; Ji, F.; Zhang, Y.; Wang, G. Small Signal Synchronizing Stability Analysis of PLL-based VSC Connected to Weak AC Grid. *Proc. CSEE* **2021**, *41*, 98–108. (In Chinese) [CrossRef]
3. Zhao, M.; Yuan, X.; Hu, J.; Yan, Y. Voltage Dynamics of Current Control Time-Scale in a VSC-Connected Weak Grid. *IEEE Trans. Power Syst.* **2016**, *31*, 2925–2937. [CrossRef]
4. Hu, J.; Hu, Q.; Wang, B.; Tang, H.; Chi, Y. Small Signal Instability of PLL-Synchronized Type-4 Wind Turbines Connected to High-Impedance AC Grid During LVRT. *IEEE Trans. Energy Convers.* **2016**, *31*, 1676–1687. [CrossRef]

5. Harnefors, L.; Bongiorno, M.; Lundberg, S. Input-Admittance Calculation and Shaping for Controlled Voltage-Source Converters. *IEEE Trans. Ind. Electron.* **2007**, *54*, 3323–3334. [CrossRef]
6. Chen, X.; Du, W.; Wang, H. Analysis on Wide-range-Frequency Oscillations of Power Systems Integrated with PMSGs Under the Condition of Open-loop Modal Resonance. *Proc. CSEE* **2019**, *39*, 2625–2635. (In Chinese) [CrossRef]
7. Wu, W.; Pu, T.; Chen, Y.; Luo, A.; Zhou, L.; Zhou, X.; Yang, L.; He, Z. Megawatt Wide-bandwidth Impedance Measurement Device Design and its Control Method. *Proc. CSEE* **2018**, *38*, 4096–4106. (In Chinese) [CrossRef]
8. Hu, J.; Yuan, H.; Yuan, X. Modeling of DFIG-Based WTs for Small-Signal Stability Analysis in DVC Timescale in Power Electronized Power Systems. *IEEE Trans. Energy Convers.* **2017**, *32*, 1151–1165. [CrossRef]
9. Gao, Z.; Liu, X. An Overview on Fault Diagnosis, Prognosis and Resilient Control for Wind Turbine Systems. *Processes* **2021**, *9*, 300. [CrossRef]
10. Fu, Y.; Gao, Z.; Liu, Y.; Zhang, A.; Yin, X. Actuator and Sensor Fault Classification for Wind Turbine Systems Based on Fast Fourier Transform and Uncorrelated Multi-Linear Principal Component Analysis Techniques. *Processes* **2020**, *8*, 1066. [CrossRef]
11. Ying, J.; Yuan, X.; Hu, J. Inertia Characteristic of DFIG-Based WT Under Transient Control and Its Impact on the First-Swing Stability of SGs. *IEEE Trans. Energy Convers.* **2017**, *32*, 1502–1511. [CrossRef]
12. Taul, M.G.; Wang, X.; Davari, P.; Blaabjerg, F. An Overview of Assessment Methods for Synchronization Stability of Grid-Connected Converters Under Severe Symmetrical Grid Faults. *IEEE Trans. Power Electron.* **2019**, *34*, 9655–9670. [CrossRef]
13. Hu, Q.; Fu, L.; Ma, F.; Ji, F. Large Signal Synchronizing Instability of PLL-Based VSC Connected to Weak AC Grid. *IEEE Trans. Power Syst.* **2019**, *34*, 3220–3229. [CrossRef]
14. Hu, Q.; Fu, L.; Ma, F.; Ji, F.; Zhang, Y. Analogized Synchronous-Generator Model of PLL-Based VSC and Transient Synchronizing Stability of Converter Dominated Power System. *IEEE Trans. Sustain. Energy* **2021**, *12*, 1174–1185. [CrossRef]
15. Ma, S.; Geng, H.; Liu, L.; Yang, G.; Pal, B.C. Grid-Synchronization Stability Improvement of Large Scale Wind Farm During Severe Grid Fault. *IEEE Trans. Power Syst.* **2018**, *33*, 216–226. [CrossRef]
16. He, X.; Geng, H.; Li, R.; Pal, B.C. Transient Stability Analysis and Enhancement of Renewable Energy Conversion System During LVRT. *IEEE Trans. Sustain. Energy* **2020**, *11*, 1612–1623. [CrossRef]
17. Wu, H.; Wang, X. Design-Oriented Transient Stability Analysis of PLL-Synchronized Voltage-Source Converters. *IEEE Trans. Power Electron.* **2020**, *35*, 3573–3589. [CrossRef]
18. Sun, J. Small-Signal Methods for AC Distributed Power Systems–A Review. *IEEE Trans. Power Electron.* **2009**, *24*, 2545–2554. [CrossRef]
19. Zhou, J.Z.; Ding, H.; Fan, S.; Zhang, Y.; Gole, A.M. Impact of Short-Circuit Ratio and Phase-Locked-Loop Parameters on the Small-Signal Behavior of a VSC-HVDC Converter. *IEEE Trans. Power Deliv.* **2014**, *29*, 2287–2296. [CrossRef]
20. Zhao, M.; Yuan, X.; Hu, J. Modeling of DFIG Wind Turbine Based on Internal Voltage Motion Equation in Power Systems Phase-Amplitude Dynamics Analysis. *IEEE Trans. Power Syst.* **2018**, *33*, 1484–1495. [CrossRef]
21. Lu, J.; Yuan, X.; Hu, J.; Zhang, M.; Yuan, H. Motion Equation Modeling of LCC-HVDC Stations for Analyzing DC and AC Network Interactions. *IEEE Trans. Power Deliv.* **2020**, *35*, 1563–1574. [CrossRef]
22. Wang, X.; Harnefors, L.; Blaabjerg, F. Unified Impedance Model of Grid-Connected Voltage-Source Converters. *IEEE Trans. Power Electron.* **2018**, *33*, 1775–1787. [CrossRef]
23. Wen, B.; Boroyevich, D.; Burgos, R.; Mattavelli, P.; Shen, Z. Analysis of D-Q Small-Signal Impedance of Grid-Tied Inverters. *IEEE Trans. Power Electron.* **2016**, *31*, 675–687. [CrossRef]
24. Liao, Y.; Wang, X. Impedance-Based Stability Analysis for Interconnected Converter Systems with Open-Loop RHP Poles. *IEEE Trans. Power Electron.* **2020**, *35*, 4388–4397. [CrossRef]
25. Tavoosi, J.; Mohammadzadeh, A.; Pahlevanzadeh, B.; Kasmani, M.B.; Band, S.S.; Safdar, R.; Mosavi, A.H. A Machine Learning Approach for Active/Reactive Power Control of Grid-Connected Doubly-Fed Induction Generators. *Ain Shams Eng. J.* **2022**, *13*, 101564. [CrossRef]
26. Cao, H.; Yu, T.; Zhang, X.; Yang, B.; Wu, Y. Reactive Power Optimization of Large-Scale Power Systems: A Transfer Bees Optimizer Application. *Processes* **2019**, *7*, 321. [CrossRef]
27. Mohammadi Moghadam, H.; Mohammadzadeh, A.; Hadjiaghaie Vafaie, R.; Tavoosi, J.; Khooban, M.-H. A Type-2 Fuzzy Control for Active/Reactive Power Control and Energy Storage Management. *Trans. Inst. Meas. Control* **2022**, *44*, 1014–1028. [CrossRef]
28. Messina, A.R.; Vittal, V. Nonlinear, Non-Stationary Analysis of Interarea Oscillations via Hilbert Spectral Analysis. *IEEE Trans. Power Syst.* **2006**, *21*, 1234–1241. [CrossRef]
29. Andrade, M.A.; Messina, A.R.; Rivera, C.A.; Olguin, D. Identification of Instantaneous Attributes of Torsional Shaft Signals Using the Hilbert Transform. *IEEE Trans. Power Syst.* **2004**, *19*, 1422–1429. [CrossRef]
30. Sanchez-Gasca, J.J.; Vittal, V.; Gibbard, M.J.; Messina, A.R.; Vowles, D.J.; Liu, S.; Annakkage, U.D. Inclusion of Higher Order Terms for Small-Signal (Modal) Analysis: Committee Report-Task Force on Assessing the Need to Include Higher Order Terms for Small-Signal (Modal) Analysis. *IEEE Trans. Power Syst.* **2005**, *20*, 1886–1904. [CrossRef]
31. Lin, C.-M.; Vittal, V.; Kliemann, W.; Fouad, A.A. Investigation of Modal Interaction and Its Effects on Control Performance in Stressed Power Systems Using Normal Forms of Vector Fields. *IEEE Trans. Power Syst.* **1996**, *11*, 781–787. [CrossRef]
32. Liu, S.; Messina, A.R.; Vittal, V. Characterization of Nonlinear Modal Interaction Using Normal Forms and Hilbert Analysis. In Proceedings of the IEEE PES Power Systems Conference and Exposition, New York, NY, USA, 10–13 October 2004; Volume 2, pp. 1113–1118. [CrossRef]

33. Tsili, M.; Papathanassiou, S. A Review of Grid Code Technical Requirements for Wind Farms. *IET Renew. Power Gener.* **2009**, *3*, 308–332. [CrossRef]
34. Western Electricity Coordinating Council. WECC Second Generation Wind Turbine Models. Available online: https://www.wecc.org/Reliability/WECC%20Second%20Generation%20Wind%20Turbine%20Models%20012314.pdf (accessed on 3 June 2021).

Article

Multi-Objective Optimal Scheduling for Multi-Renewable Energy Power System Considering Flexibility Constraints

Lei Yang [1], Wei Huang [2], Cheng Guo [3], Dan Zhang [2], Chuan Xiang [3], Longjie Yang [1] and Qianggang Wang [1,*]

[1] State Key Laboratory of Power Transmission Equipment and System Security and New Technology, Chongqing University, Chongqing 400044, China; 15911577929@139.com (L.Y.); 20191101333@cqu.edu.cn (L.Y.)

[2] Electric Power Dispatching and Control Center of Yunnan Power Grid Co., Ltd., Kunming 650200, China; haxwell@163.com (W.H.); 6950704@foxmail.com (D.Z.)

[3] Electric Power Research Institute of Yunnan Power Grid Co., Ltd., Kunming 650217, China; gc325@126.com (C.G.); 1091930966@foxmail.com (C.X.)

* Correspondence: qianggang1987@cqu.edu.cn; Tel.: +86-136-4055-8474

Abstract: As renewable energy penetration increases, the lack of flexibility in a multi-renewable power system can seriously affect its own economics and reliability. To address this issue, three objectives are considered in this study: power fluctuations on tie-line, operating cost, and curtailment rate of renewable energy. Presented also is an optimal day-ahead scheduling model based on the MREPS for distributed generations with flexibility constraints. The multi-objective particle swarm optimization (MOPSO) algorithm can be applied to obtain a set of Pareto non-dominated solutions for the day-ahead scheduling strategy with the proposed model. By using fuzzy comprehensive evaluation, the optimal compromise solution is determined in the set. The presented method sacrifices a small amount of economy and power fluctuation, but it can reduce the deviation between forecast and realized power fluctuations on the tie-line, while improving the utilization of renewable energy.

Keywords: flexibility constraints; fuzzy comprehensive evaluation method; MOPSO; MREPS; optimal day-ahead scheduling

Citation: Yang, L.; Huang, W.; Guo, C.; Zhang, D.; Xiang, C.; Yang, L.; Wang, Q. Multi-Objective Optimal Scheduling for Multi-Renewable Energy Power System Considering Flexibility Constraints. *Processes* **2022**, *10*, 1401. https://doi.org/10.3390/pr10071401

Academic Editor: Jie Zhang

Received: 28 June 2022
Accepted: 15 July 2022
Published: 18 July 2022

Publisher's Note: MDPI stays neutral with regard to jurisdictional claims in published maps and institutional affiliations.

1. Introduction

In modern power systems, the scarcity of fossil fuels and increasing pollution of the environment contribute to the development of renewable energy sources, such as solar and wind. Despite this, the stochastic nature of renewable energy generation is likely to have significant effects on system reliability and economy [1–3].

When operating a multi-renewable energy power system (MREPS), it is necessary to develop an optimal schedule to cope with the stochasticity of renewable energy generation [4,5]. MREPS scheduling is divided into two categories: day-ahead and real-time scheduling. There are several strategies for achieving various operational objectives using day-ahead scheduling. Reference [6] used a two-stage stochastic optimization model for an MREPS for minimizing the short-term operation cost, which introduces uncertainty in renewable generation, and showed that stochastic scheduling can provide significant reliability benefits to multi-energy supply systems. Reference [7] presented a day-ahead scheduling model that considers the seasonal uncertainty of renewable energy for a micro-grid equipped with multi-renewable energy units. An improved optimization algorithm was proposed to solve optimization problems that focus on minimizing the operation cost. The results indicate that day-ahead scheduling based on the proposed algorithm can provide an efficient solution for managing MREPS energy. An MREPS is highly influenced by both economic indicators and the rate at which renewable energy is utilized, as illustrated in [8–10]. Furthermore, power fluctuations on the tie-line serve as relevant indicators for the main grid connected to an MREPS [11,12]. In the current MREPS scheduling process, the impact of these power fluctuations on the main grid is rarely considered. Although operating costs, curtailment rates of renewable energy, and power fluctuations on the tie-lines are all commonly considered in current studies on optimal dispatch [4–15], few studies consider these three indicators simultaneously. This can lead to a situation where, while some of these indicators are optimal, the other indicators may be poor. As a result, the present study proposes a day-ahead scheduling strategy based on operating costs, curtailment rates of renewable energy, and power fluctuations on the tie-lines.

In the case of real-time scheduling, the schedule is amended if the actual renewable energy output differs from the forecasted output. An improved particle swarm-optimization (PSO)-based strategy for managing energy over a two-time scale was presented in [16]. In day-ahead scheduling, one of the primary objectives is to achieve the most cost-effective schedule, and in real-time scheduling, the primary goal is to track day-ahead scheduling, compensate for power fluctuations, and maintain system stability. According to the experimental results, the proposed method could minimize the cost of generated electricity and maximize the efficiency of renewable energy systems. Real-time scheduling, however, relies on day-ahead scheduling. It is possible to increase the utilization rate of renewable energy if uncertainties associated with the generation of renewable energy can be properly incorporated into day-ahead scheduling. To address uncertainties in power systems [17–19], flexibility has been proposed. In [20], an optimal scheduling model for flexible resources was presented from both the generation and load sides. According to the results, a dynamic line rating model that incorporates optimal scheduling can maximize the utilization of flexible resources without curtailing wind power and minimize dispatch costs. Consequently, we propose a day-ahead scheduling strategy for an MREPS that accounts for flexibility constraints and concentrates on operation costs, renewable energy curtailment rates, and power fluctuations on the tie-line.

In this study, we attempt to solve a multi-objective optimization problem. It is possible to solve a multi-objective optimization problem in several ways. Among them, the linear weighted sum method and intelligent algorithms are frequently adopted. The linear weighted sum method is simple and fast, but it usually gives unclear physical results. In addition, the resulting error is typically large because several targets are of different dimensions and orders of magnitude. These factors remarkably impact the results and conclusions of practical problems [21]. Intelligent algorithms, which have clearer physical meaning, outperform the linear weighted sum method in accuracy, flexibility, and effectiveness in solving multi-objective problems. A number of intelligent algorithms have been successfully applied to engineering optimization, such as multi-objective particle swarm optimization (MOPSO) due to its unique search mechanism, excellent convergence performance, and convenient calculation capabilities [22,23]. As a result, MOPSO is adopted in this study to identify the Pareto non-dominated set of objective functions. Following the determination of the Pareto non-dominated solution set, a fuzzy comprehensive evaluation method [24] is adopted to determine the optimal compromise solution. The optimal day-ahead scheduling strategy can be determined based on the optimum compromise solution.

The contributions of this paper are summarized as follows:

1. Considering the operation cost, renewable energy curtailment rates, and power fluctuations on the tie-line, a day-ahead scheduling model for the MREPS is established.
2. MOPSO and a fuzzy comprehensive evaluation method are used to evaluate the day-ahead scheduling model, and a day-ahead scheduling strategy for the MREPS considering flexibility is proposed.

Following is the remainder of this paper. The MREPS presents the day-ahead optimal scheduling model in Section 2, along with its constraints, taking flexibility into account. In Section 3, MOPSO and fuzzy comprehensive evaluation are discussed. An analysis of the experimental results is presented in Section 4, which simulates an actual MREPS. In Section 5, the conclusions are summarized.

2. Model for Multi-Objective Optimal Scheduling

The MREPS, consisting of wind turbines (WTs), photovoltaic (PV) arrays, diesel generators (DGs), energy storage systems (ESSs), and loads, only purchases electricity from the main grid. Figure 1 illustrates the details of the MREPS.

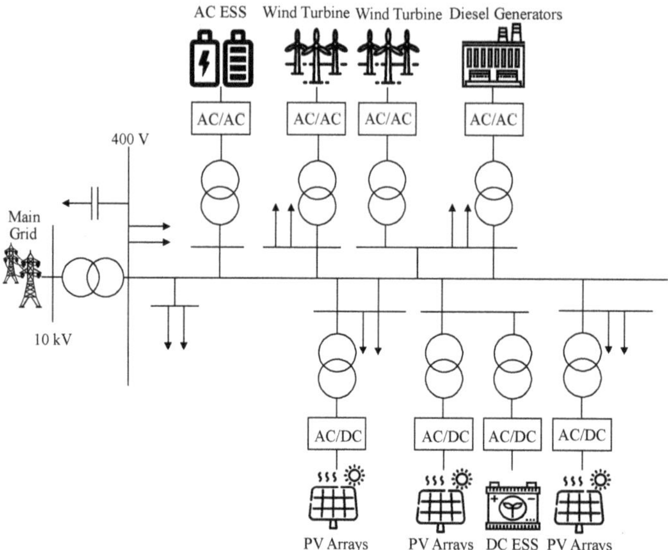

Figure 1. MREPS model.

2.1. Objective Function

2.1.1. Operation Cost

A major consideration in the MREPS's day-ahead scheduling is the operation cost. The operation cost mainly includes fuel cost, operation and maintenance, pollutant control, and purchased power cost. The daily operation cost of the MREPS can be expressed as follows:

$$F_1 = C_{\text{fuel}} + C_{\text{OM}} + C_{\text{PC}} + C_{\text{PP}} \tag{1}$$

where C_{fuel}, C_{OM}, and C_{PC} are the total fuel, operation and maintenance (OM), and pollutant control (PC) costs associated with each distributed power generation in one day and C_{PP} is the total cost of the purchased power (PP) from the main grid in one day. The formulas for C_{fuel}, C_{OM}, C_{PC}, and C_{PP} are as follows:

$$C_{\text{fuel}} = \sum_{n=1}^{N} \sum_{t=1}^{T} \xi_{\text{f.DG}.n} \cdot P_{\text{DG}.n}(t) \cdot \Delta t \tag{2}$$

$$C_{\text{OM}} = \sum_{m=1}^{M} \sum_{t=1}^{T} \xi_{\text{OM}.m} \cdot P_m(t) \cdot \Delta t \tag{3}$$

$$C_{\text{PC}} = \sum_{m=1}^{M} \sum_{j=1}^{J} \sum_{t=1}^{T} \xi_{m.j} \cdot P_m(t) \cdot \Delta t \tag{4}$$

$$C_{\text{PP}} = \sum_{t=1}^{T} \xi_{\text{PP}}(t) \cdot P_{\text{PP}}(t) \cdot \Delta t \tag{5}$$

where N is the total number of DGs; T is the total scheduling time; $\xi_{\text{f.DG}.n}$ is the consumption cost per kW·h of the nth DG; $P_{\text{DG}.n}(t)$ is the output power of the nth DG in period t; Δt is the scheduling interval; M is the total number of distributed generation types; $\zeta_{OM.m}$ is the cost of operation and maintenance per kW·h of the mth type of distributed generation variety; $P_m(t)$ is the output power of the mth type of distributed generation during period t; J is the total number of pollutant types; $\zeta_{m.j}$ is the emission cost factor for the jth type of pollutants generated by the mth type of distributed power generation; $\zeta_{PP}(t)$ and $P_{PP}(t)$ are the power purchased by the MREPS from the main grid for the period t, respectively.

2.1.2. Renewable Energy Curtailment Rate

The renewable energy curtailment rate is the ratio of the power curtailed by all renewable energy units to the total power that can be generated by renewable energy units during the dispatching period. The lower the value, the higher the utilization rate of renewable energy is. Utilization of the renewable energy is ideally increased while reducing the operation cost of the MREPS [15]. Thus, this study considers the curtailment rate of renewable energy. Following is the daily curtailment rate of the MREPS's renewable energy:

$$F_2 = \sum_{t=1}^{T}\sum_{h=1}^{H} P_{h.c}(t) \Big/ \sum_{t=1}^{T}\sum_{h=1}^{H} P_{h.all}(t) \tag{6}$$

where H represents the total number of the renewable energy types; $P_{h.c}(t)$ and $P_{h.all}(t)$ reflect the power curtailment and the available power generation of the hth type of renewable energy in period t, respectively.

2.1.3. Tie-Line Power Fluctuations

The frequent power exchange between the MREPS and the main grid can be attributed to the stochastic nature of renewable energy. In general, the unit commitment and economic load dispatch of the main grid have no impact on the MREPS. This is because the MREPS can be self-sufficient. Only when the MREPS is not self-sufficient, it is necessary for the main grid to help the MREPS to achieve power balance through unit commitment or economic load dispatch. As a consequence, the significant fluctuation in power on the tie-line has a negative impact on the main grid. The MREPS day-ahead scheduling must be designed to take into account power fluctuations across the tie-line.

As the square root of the variance, the standard deviation is a metric used to measure how dispersed a dataset is relative to its mean. Consequently, the standard deviation is typically used to describe the fluctuation of a data series. However, the standard deviation lacks comparability for different objects or samples with varying means of the same object. Hence, the coefficient of variation is used in this study to avoid these problems.

As a measure of data dispersion around the mean, the coefficient of variation represents the ratio of the standard deviation to the mean for a series of data points. The degree of variation from one data series to another can be compared, although the means are remarkably different from one another. A small variation indicates a small fluctuation degree [11]. On the tie-line, the power fluctuation can be defined as follows:

$$F_3 = \sqrt{\frac{1}{T} \cdot \sum_{t=1}^{T} \left(P_{TL}(t) - \mu_{TL}\right)^2 / \mu_{TL}} \tag{7}$$

where $P_{TL}(t)$ represents the transmission power of the tie-line during period t and μ_{TL} represents the average transmission power of the tie-line during one day, which can be expressed as follows:

$$\mu_{TL} = \frac{1}{T} \cdot \sum_{t=1}^{T} P_{TL}(t) \tag{8}$$

Based on the above three objectives, the total objective function for the multi-objective optimal scheduling model is as follows:

$$\min F = [F_1, F_2, F_3] \tag{9}$$

2.2. Constraint Conditions

2.2.1. Constraints on the Power Balance

During the operation of the MREPS, at the end of each period, the sum of the output power of the DGs, the charging and discharging power of the ESS, and the purchased power from the main grid should be equal to the net load (NL) power of the system. As a result of this relationship, we may state:

$$P_{DG}(t) + P_{ESS}(t) + P_{TL}(t) = P_L(t) - P_{WT}(t) - P_{PV}(t) = P_{NL}(t) \tag{10}$$

where $P_{DG}(t)$ is the sum of the output power of the DGs in period t; $P_{ESS}(t)$ is the charge and discharge power of the ESS in period t and becomes negative when the ESS is being charged; $P_{TL}(t)$ is the transmission power of the tie-line in period t; $P_L(t)$, $P_{WT}(t)$, $P_{PV}(t)$, and $P_{NL}(t)$ are the load power, output of WTs, output of PV arrays, and net load power in period t, respectively.

2.2.2. Constraints of Output Power for DGs Considering Flexibility

In light of the uncertainty associated with renewable energy generation, it is imperative that the MREPS be flexible. The DG is a commonly used flexible resource in MREPSs because of its satisfactory controllability and fast response speed. Therefore, flexibility should be considered and exploited in the output power constraints of DGs.

The uncertainty of renewable energy generation leads to prediction errors in net load power and the MREPS's flexibility requirements. These flexibility requirements are provided by DGs in the MREPS. DGs are typically restricted in their output power range as follows [7]:

$$\begin{cases} P_{DG}(t) \geq P_{DG.min} \\ P_{DG}(t) \leq P_{DG.max} \end{cases} \tag{11}$$

where $P_{DG.min}$ and $P_{DG.max}$ represent the minimum and maximum output power of DGs, respectively.

In terms of conventional constraints, DGs are only considered for their minimum and maximum output. Due to these constraints, the output power of DGs may be inflexible during operation. When large prediction errors occur, DGs output power fails to meet the flexibility requirements. Therefore, new constraints that consider flexibility based on conventional constraints can be expressed as follows:

$$\begin{cases} P_{DG}(t) \geq (P_{DG.min} + F_{NL.D}(t)) \\ P_{DG}(t) \leq (P_{DG.max} - F_{NL.U}(t)) \end{cases} \tag{12}$$

where $F_{NL.U}(t)$ and $F_{NL.D}(t)$ represent the maximum upward and downward flexibility requirements of the MREPS in each period t.

Because of the error in predicting net load power, the MREPS is required to be flexible. Variables such as wind, solar, and load power are among the factors that affect the prediction errors of the net load power. As a result, wind, solar, and load power prediction errors are assumed to follow a normal distribution with a zero mean [25]. $\sigma_{PV}(t)$, $\sigma_{WT}(t)$, and $\sigma_L(t)$ have the following standard deviations:

$$\begin{cases} \sigma_{PV}(t) = 0.2W_{PV.F}(t) + 0.02W_{PV.C} \\ \sigma_{WT}(t) = 0.2W_{WT.F}(t) + 0.02W_{WT.C} \\ \sigma_L(t) = 0.02W_{L.F}(t) \end{cases} \tag{13}$$

where $W_{\text{PV.F}}(t)$, $W_{\text{WT.F}}(t)$, and $W_{\text{L.F}}(t)$ are the predicted wind, solar, and load power in period t, respectively, and $W_{\text{PV.C}}$ and $W_{\text{WT.C}}$ are the total installed capacity of PV arrays and WTs, respectively.

There are no correlations among wind, solar, and load power prediction errors. According to the basic characteristics of the normal distribution, the sum of two independent normally distributed random variables is normal, with the sum of the two means as the mean and the sum of the two variances as the variance. Hence, prediction errors of the net load power also follow the normal distribution with mean zero and standard deviation $\sigma_{\text{NL}}(t)$, as shown below:

$$\sigma_{\text{NL}}(t) = \sqrt{\sigma_{\text{PV}}^2(t) + \sigma_{\text{WT}}^2(t) + \sigma_{\text{L}}^2(t)} \tag{14}$$

The MREPS must offer flexibility equal to the forecast errors of net load power. However, in the prediction of net load power, there is a very small probability of maximum errors. Hence, the scenario where the flexibility requirement and forecast error are equal is uneconomical and wasteful. As a result, we need to consider the confidence intervals of these prediction errors. If the confidence level is $1 - \alpha$, then $F_{\text{NL.U}}(t)$ and $F_{\text{NL.D}}(t)$ can be expressed as:

$$\begin{cases} F_{\text{NL.U}}(t) = u_{1-\alpha/2} \cdot \sigma_{\text{NL}}(t) \\ F_{\text{NL.D}}(t) = -u_{1-\alpha/2} \cdot \sigma_{\text{NL}}(t) \end{cases} \tag{15}$$

where $u_{1-\alpha/2}$ is the upper $(1 - \alpha/2)$ critical value for the standard normal distribution.

Conventional ramping-rate constraints of DGs are expressed as follows:

$$\begin{cases} P_{\text{DG}}(t) - P_{\text{DG}}(t - \Delta t) \geq -R_{\text{DG.max}}^{\text{D}} \cdot \Delta t \\ P_{\text{DG}}(t) - P_{\text{DG}}(t - \Delta t) \leq R_{\text{DG.max}}^{\text{U}} \cdot \Delta t \end{cases} \tag{16}$$

where $R_{\text{DG.max}}^{\text{D}}$ and $R_{\text{DG.max}}^{\text{U}}$ are the maximum downward and upward ramping rates of DGs, respectively, with both values positive, and Δt is the time interval.

Conventional constraints only consider the minimum and maximum ramping rates. The maximum ramping rate of DGs may occur in some periods under these constraints. The output power of DGs fails to meet the fluctuation of the net load power in this situation when DGs lack the ramping rate. Consequently, ramping-rate constraints should take flexibility into consideration.

As shown in Figure 2, A_0, B_0, and C_0 are the prediction output power values of DGs at different times. DGs must reduce their output power to A_2 if the maximum downward prediction errors of the net load power occur at time $t - \Delta t$. According to Equation (15), the reduced output power should be equal to $F_{\text{NL.D}}(t - \Delta t)$. It is imperative that the DGs increase their output power to B_1 if they experience maximum upward prediction errors at time t. According to Equation (15), the increased output power should be equal to $F_{\text{NL.U}}(t)$. Therefore, the margin of the ramping rate that DGs should reserve is the sum of $F_{\text{NL.D}}(t - \Delta t)$ and $F_{\text{NL.U}}(t)$ from time $t - \Delta t$ to time t. Similarly, the margin of the ramping rate that DGs should reserve is the sum of $F_{\text{NL.U}}(t)$ and $F_{\text{NL.D}}(t + \Delta t)$ from time t to time $t + \Delta t$. As a result, new constraints that consider flexibility can be expressed as follows:

$$\begin{cases} P_{\text{DG}}(t) - P_{\text{DG}}(t - \Delta t) \geq -R_{\text{DG.max}}^{\text{D}} \cdot \Delta t + (F_{\text{NL.D}}(t) + F_{\text{NL.U}}(t - \Delta t)) \\ P_{\text{DG}}(t) - P_{\text{DG}}(t - \Delta t) \leq R_{\text{DG.max}}^{\text{U}} \cdot \Delta t - (F_{\text{NL.U}}(t) + F_{\text{NL.D}}(t - \Delta t)) \end{cases} \tag{17}$$

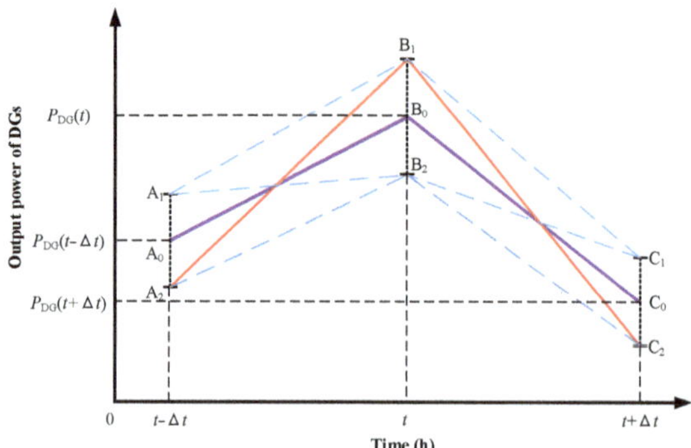

Figure 2. Output power of DGs.

2.2.3. Constraints of the ESS Considering Flexibility in Charging and Discharging

In this study, the main function of the ESS is to provide assistance to DGs and ensure that DGs have adequate flexibility. Thus, the ESS and its conventional constraints are similar in terms of charging and discharging power:

$$\begin{cases} S_{\text{ESS}}(t+\Delta t) = (1-\delta) \cdot S_{\text{ESS}}(t) - \dfrac{\Delta t \cdot \alpha_{\text{ESS}}(t) \cdot P_{\text{ESS}}(t)}{E_{\text{ESS}}} \\ S_{\text{ESS.min}} \le S_{\text{ESS}}(t) \le S_{\text{ESS.max}} \\ |P_{\text{ESS}}(t)| \cdot \Delta t' \le \eta_{\text{ESS}} \cdot E_{\text{ESS}} \end{cases} \tag{18}$$

where $S_{\text{ESS}}(t)$ represents the state of charge (SOC) of the ESS in period t; δ represents the self-discharge efficiency of the ESS; E_{ESS} represents the maximum capacity of the ESS; $S_{\text{ESS.min}}$ and $S_{\text{ESS.max}}$ represent the minimum and maximum SOC of the ESS, respectively; $\Delta t'$ represents an hour; η_{ESS} represents the percentage of the maximum charge–discharge capacity per hour to the maximum capacity of the ESS; $\alpha_{\text{ESS}}(t)$ represents the charge–discharge efficiency of the ESS in period t. The following formula can be used to calculate $\alpha_{\text{ESS}}(t)$:

$$\alpha_{\text{ESS}}(t) = \begin{cases} \alpha_{\text{c}}, \forall P_{\text{ESS}}(t) < 0 \\ 1/\alpha_{\text{d}}, \forall P_{\text{ESS}}(t) > 0 \end{cases} \tag{19}$$

where α_{c} and α_{d} are the charge and discharge efficiencies of the ESS, respectively.

DGs, however, fail to achieve the maximum prediction error of the net load power. Because of this, we should take into account flexibility in the power constraints of the ESS when charging and discharging. This is because the secondary function of the ESS is to meet these maximum prediction errors. Here are the revised constraints:

$$\begin{cases} S_{\text{ESS}}(t+\Delta t) = (1-\delta) \cdot S_{\text{ESS}}(t) - \dfrac{\Delta t \cdot \alpha_{\text{ESS}}(t) \cdot P_{\text{ESS}}(t)}{E_{\text{ESS}}} \\ (S_{\text{ESS.min}} + F_{\text{ESS.U}}) \le S_{\text{ESS}}(t) \le (S_{\text{ESS.max}} - F_{\text{ESS.D}}) \\ |P_{\text{ESS}}(t)| \cdot \Delta t' \le \eta_{\text{F.ESS}} \cdot E_{\text{ESS}} \end{cases} \tag{20}$$

where $F_{\text{ESS.U}}$ and $F_{\text{ESS.D}}$ are the reserved SOC of the ESS and $\eta_{\text{F.ESS}}$ is the percentage of the maximum charge–discharge capacity per hour to the maximum capacity of the ESS after considering flexibility.

If the maximum prediction error of the net load power occurs, DGs are capable of meeting most of these prediction errors. There is only a small portion of these prediction errors that must be met by the ESS. As a result, the constraints of $F_{\text{ESS.U}}$, $F_{\text{ESS.D}}$, and $\eta_{\text{F.ESS}}$ are as follows:

$$\begin{cases} 0 \le F_{\text{ESS.U}} \le 0.1 \\ 0 \le F_{\text{ESS.D}} \le 0.1 \\ 0.8 \cdot \eta_{\text{ESS}} \le \eta_{\text{F.ESS}} \le \eta_{\text{ESS}} \end{cases} \tag{21}$$

Since the ESS operation is periodic, the initial SOC should be equal to the final SOC in a day and expressed as follows:

$$S_{\mathrm{ESS}}(t = 0) = S_{\mathrm{ESS}}(t = T) \tag{22}$$

where T represents the total scheduling time.

2.2.4. Tie-Line Transmission Power Constraints in Consideration of Flexibility

In the MREPS, DGs and the ESS can sometimes fail. In this case, the tie-line must be flexible to facilitate the normal functioning of the MREPS. Therefore, the power limitation for tie-line transmission considering flexibility is expressed as follows:

$$P_{\mathrm{TL.min}} \leq P_{\mathrm{TL}}(t) \leq P_{\mathrm{TL.max}} - F_{\mathrm{TL.U}}(t) \tag{23}$$

where $P_{\mathrm{TL.min}}$ and $P_{\mathrm{TL.max}}$ represent the minimum and maximum transmission power of the tie line, respectively, and $F_{\mathrm{TL.U}}(t)$ represents the reserved transmission power of the tie-line that can be increased in period t.

Load shedding is prohibited in the operation of the MREPS. In the event that DGs and the ESS fail, it is necessary to reserve the transmission power on the tie-line, which may be increased in case of a breakdown. However, if the load power decreases, then we may be able to achieve power balance by reducing the output power of DGs or curtailing the use of renewable energy sources. As a consequence, it is not necessary to reserve the transmission power of the tie-line.

3. Algorithm for the Solution of the Multiobjective Optimization Model

3.1. MOPSO

The multi-objective optimization problem in the presented model is solved via MOPSO in this study. Although MOPSO is one of the commonly used intelligent optimization algorithms [22,23], it is briefly introduced since it is the core of the algorithm in this paper. MOPSO updates the position and the velocity of each particle in every iteration and searches for the local and global best positions of particles. The velocity is modified and updated as follows:

$$v_i^{k+1} = \omega v_i^k + c_1 r_1 (p_l^k - x_i) + c_2 r_2 (p_g^k - x_i) \tag{24}$$

where v_i^k and x_i are the velocity and position of the ith particle in the kth iteration; p_l^k is the local optimal position of the ith particle; p_g^k is the global optimal position of particles; ω is the inertia weight that influences the local and global exploitation abilities for MOPSO; c_1 and c_2 are the cognitive and social learning factors that maintain the movement of particles to the local and global optimal positions, respectively; r_1 and r_2 are two uniform random functions in the range [0,1].

Constraints of the velocity are expressed as follows:

$$\begin{cases} v_i^{k+1} = v_{\max}, & \forall v_i^{k+1} > v_{\max} \\ v_i^{k+1} = v_{\min}, & \forall v_i^{k+1} < v_{\min} \end{cases} \tag{25}$$

where $v_{\min}$ and $v_{\max}$ are the minimum and maximum velocities of particles, respectively.

The position of particles is indicated as follows:

$$x_i^{k+1} = x_i^k + v_i^k \tag{26}$$

MOPSO obtains the non-dominated solution set, unlike the single-objective PSO algorithm. Therefore, an external file is needed to store the set. The flowchart of MOPSO is shown in Figure 3.

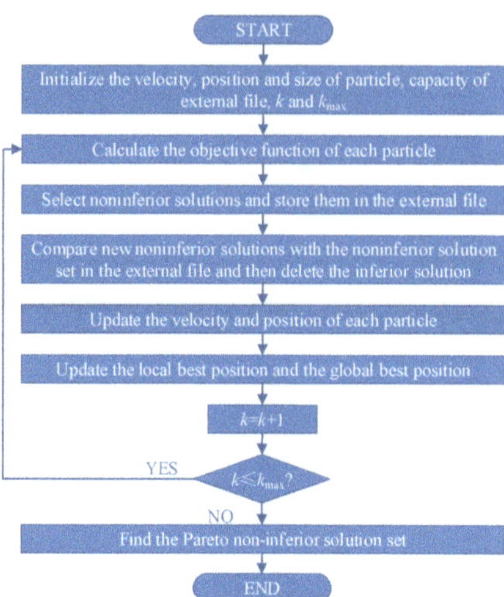

Figure 3. Flowchart of MOPSO.

The SOC of the ESS, the output power of DGs, and the tie-line transmission power are unknown in this study. Therefore, the SOC of the ESS and the output power of DGs in each period are considered decision variables. The tie-line power of transmission in each period can be obtained using Equation (10). Due to the fact that the set of Pareto non-dominated solutions is determined by these decision variables, each solution in the set corresponds to an operation state of the MREPS. Therefore, if we wish to determine the optimal day-ahead scheduling strategy, then we need to first identify the optimum compromise solution from the set of non-dominated Pareto solutions.

3.2. Fuzzy Comprehensive Analysis Methodology

This study employs the fuzzy comprehensive evaluation method to determine the most appropriate compromise solution. An effective and widely adopted method for evaluating hierarchical and integrated problems is a fuzzy comprehensive evaluation. The top two critical steps of this method are determining the weight vector of each evaluation objective appropriately and selecting the appropriate fuzzy membership function. Figure 4 illustrates the fuzzy comprehensive evaluation method flowchart.

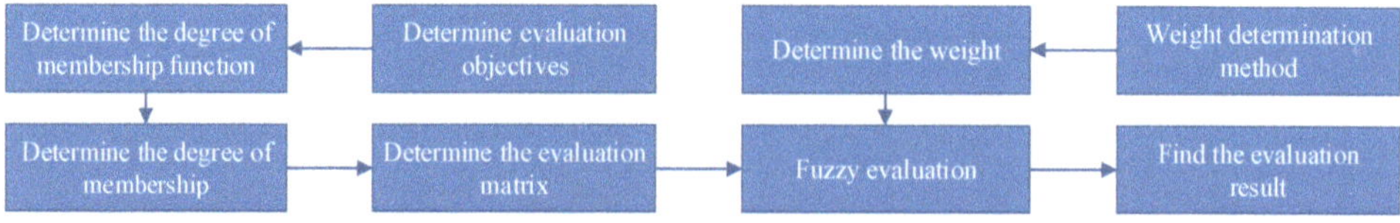

Figure 4. Flowchart of the fuzzy comprehensive evaluation method.

Assume that X is the Pareto non-dominated solution set, which is expressed as follows:

$$X = \begin{bmatrix} x_{11} & \cdots & x_{1j} & \cdots & x_{1n} \\ \vdots & \ddots & \vdots & \ddots & \vdots \\ x_{i1} & \cdots & x_{ij} & \cdots & x_{in} \\ \vdots & \ddots & \vdots & \ddots & \vdots \\ x_{m1} & \cdots & x_{mj} & \cdots & x_{mn} \end{bmatrix} \tag{27}$$

where m is the total number of objectives, n is the total number of the Pareto non-dominated solutions, and x_{ij} is the value of the ith objective of the jth Pareto non-dominated solution. To determine the most effective compromise solution from the Pareto non-dominated solution set, follow the steps outlined below:

(1) For determining the membership degree of each objective in each Pareto non-dominated solution, a single-factor fuzzy evaluation is adopted. The fuzzy relation matrix can then be obtained as follows:

$$
R = \begin{bmatrix}
r_{11} & \cdots & r_{1j} & \cdots & r_{1n} \\
\vdots & \ddots & \vdots & \ddots & \vdots \\
r_{i1} & \cdots & r_{ij} & \cdots & r_{in} \\
\vdots & \ddots & \vdots & \ddots & \vdots \\
r_{m1} & \cdots & r_{mj} & \cdots & r_{mn}
\end{bmatrix}
\tag{28}
$$

where r_{ij} is the membership degree of the ith objective in the jth Pareto non-dominated solution. r_{ij} can be calculated as follows:

$$
r_{ij} = \begin{cases}
1 & , \ x_{ij} \leq \underline{X}_i \\
\frac{\overline{X}_i - x_{ij}}{\overline{X}_i - \underline{X}_i} & , \underline{X}_i < x_{ij} < \overline{X}_i \\
0 & , \ x_{ij} \geq \overline{X}_i
\end{cases}
\tag{29}
$$

where $\underline{X}_i$ and $\overline{X}_i$ are the minimum and maximum expectations of the decision-maker for the ith objective, respectively.

(2) The analytic hierarchy process (AHP)-entropy weight method (EWM) can be employed to determine the comprehensive weight vector for each objective. Assume that the comprehensive weight vector is:

$$
A = [\omega_1, \omega_2, \cdots, \omega_i, \cdots, \omega_m]^{\mathrm{T}}
\tag{30}
$$

where ω_i is the comprehensive weight vector of the ith objective. A can then be calculated using the following equation:

$$
\omega_i = \frac{\omega_i' \cdot \omega_i''}{\sum\limits_{i=1}^{m} (\omega_i' \cdot \omega_i'')}
\tag{31}
$$

where ω_i' and ω_i'' are the objective weights based on AHP and EWM, respectively. The discussion on AHP [26] and EWM [27] is excluded from this paper given that both have been extensively analyzed in the literature.

(3) The comprehensive fuzzy evaluation vector B can be calculated as follows:

$$
\begin{cases}
B = A \odot R = [b_1, b_2, \cdots, b_j, \cdots, b_m] \\
b_j = \sum\limits_{i=1}^{m} (\omega_i \cdot r_{ij})
\end{cases}
\tag{32}
$$

where b_j is the membership degree of the jth Pareto non-dominated solution.

When b_j is close to 1, we can evaluate the jth Pareto non-dominated solution comprehensively. Therefore, from the set of Pareto non-dominated solutions, the optimal compromise solution corresponds to the maximum b_j.

4. Case Study

Practical MREPS data are presented in this study within a simulation and analysis environment. The parameters of each power supply unit in the MREPS are listed in Table 1.

The cost of each distributed generation is presented in Table 2. Table 3 presents the pollutant emission coefficients generated by each distributed generation. Table 4 shows the electricity purchase price of the MREPS from the main grid [15].

Table 1. Parameters of each power supply unit in the MREPS.

Power Supply Unit Type	Parameter Type	Parameter Value
Photovoltaic array	Power rating	100 kW
Wind turbine	Power rating	33 kW
Lead-acid battery	Rated capacity	100 kW·h
	Range of SOC	0.2–1
	Maximum charge and discharge power	25 kW
Diesel generator	Power rating	200 kW
	Maximum upward ramping rate	120 kW/h
	Maximum downward ramping rate	120 kW/h
Tie line	Maximum transmission power	90 kW

Table 2. Cost coefficient of each distributed generation.

Generation Unit Type	Fuel Cost (CNY·(kW·h)$^{-1}$)	Operation Management Coefficient (CNY·(kW·h)$^{-1}$)
Photovoltaic array	—	0.0096
Wind turbine	—	0.0296
Lead-acid battery	—	0.0322
Diesel generator	0.81	0.0880

Table 3. Pollutant emission coefficient generated by each distributed generation.

Pollutant Type		CO_2	SO_2
Handling Expense (CNY·kg^{-1})		0.21	14.842
Pollutant emission coefficient (g·(kW·h)$^{-1}$)	Photovoltaic power generation	0	0
	Wind power generation	0	0
	Diesel power generation	649	0.206

Table 4. Electricity purchase price of the MREPS from the main grid.

Type of Period	Period (h)	Purchase Price (CYN)
Peak period	8:00–11:00 13:00–15:00 18:00–21:00	1.25
Ordinary period	6:00–8:00 11:00–13:00 15:00–18:00 21:00–22:00	0.80
Valley period	0:00–6:00 22:00–0:00	0.40

This study adopts a simulation of the load and output date of a typical day in the MREPS for WTs and PV arrays. To determine an optimal scheduling strategy, the forecast values of these data are used. Utilizing the realized values of these data, the optimal scheduling strategy is evaluated. Figures 5 and 6 present the predicted and realized curves for the load and output from WTs and PV arrays in the MREPS for a typical day, respectively.

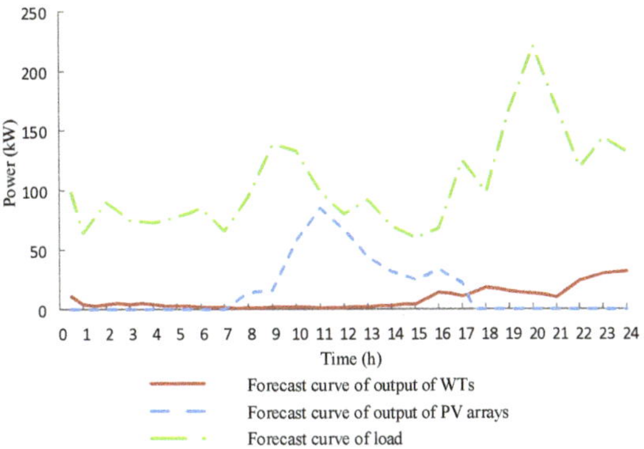

Figure 5. Forecast curves of the load and output of WTs and PV arrays.

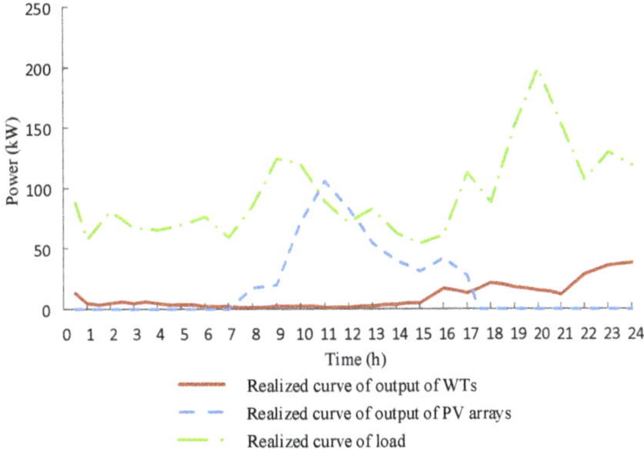

Figure 6. Realized curves of the load and output of WTs and PV arrays.

As shown in Section 4, it is possible to determine the optimal scheduling strategy of a typical day after obtaining the non-dominated solution set. On the basis of the models, constraints, data, and decision variables of this study, the non-dominated solution set of objectives can therefore be obtained using MOPSO. Figures 7 and 8 illustrate non-dominated solutions based on the conventional and proposed strategies, respectively. From Figures 7 and 8, we can see that the three objectives cannot reach the optimal solution at the same time for the non-dominated solution set obtained by MOPSO. The distribution of the non-dominated solution set shows a narrow shape at both ends and is wide in the middle.

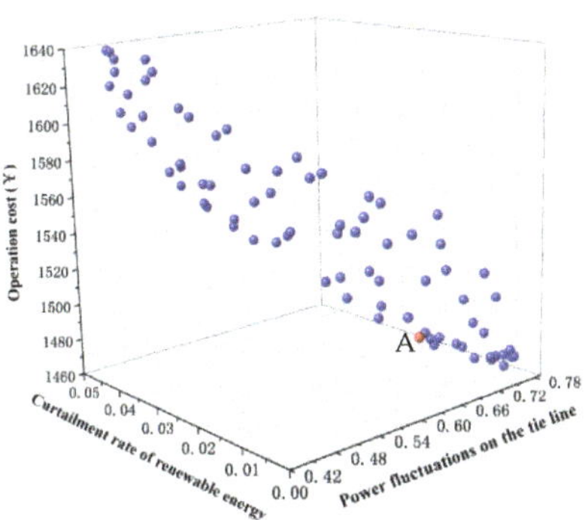

Figure 7. Three-objective non-dominated solution set based on the conventional strategy.

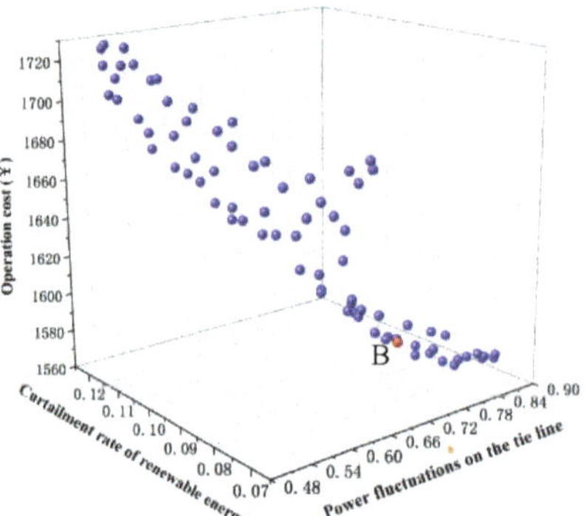

Figure 8. Three-objective non-dominated solution set based on the proposed strategy.

Furthermore, Figures 7 and 8 indicate that the non-dominated solution set of the proposed strategy is larger than that of the conventional strategy. According to this finding, after evaluating the flexibility constraints, the MREPS will sacrifice a portion of the economy, the utilization rate of renewable energy, and power fluctuations.

We must then determine the optimal compromise solution from the Pareto non-dominated solution set to find the optimal day-ahead scheduling strategy for a typical day. The fuzzy comprehensive evaluation method can be used to obtain optimal compromise solutions from non-dominated Pareto solution sets, based on the conventional and proposed strategies, as discussed in Section 4. By the fuzzy comprehensive evaluation method, Point A and Point B are the solutions with the highest evaluation scores in the respective non-dominated solution sets of Figures 7 and 8, respectively. Therefore, Points A and B represent optimal compromise solutions in Figures 7 and 8, respectively.

Finally, based on the relevant decision variables, the optimal day-ahead scheduling strategy for a typical day according to every optimal compromise solution can be determined. These are illustrated in Figures 9 and 10.

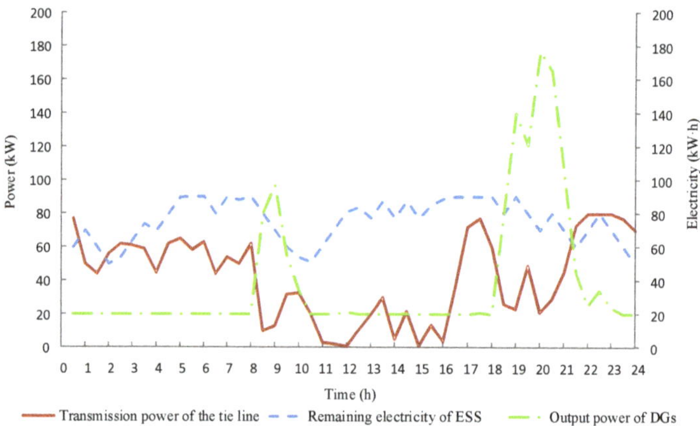

Figure 9. Optimal scheduling strategy of the typical day based on the conventional strategy.

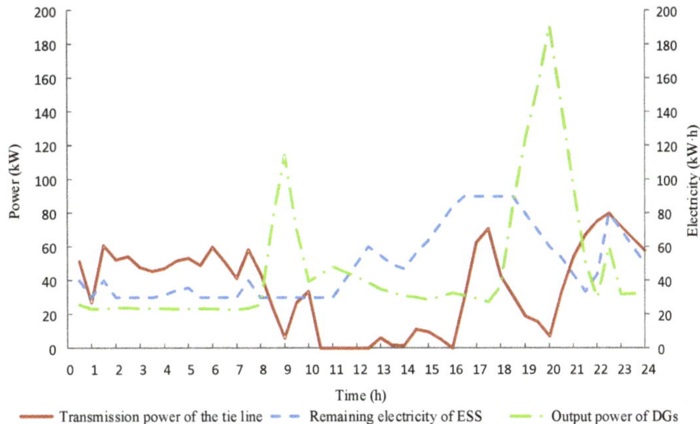

Figure 10. Optimal scheduling strategy of the typical day based on the proposed strategy.

Using the conventional and proposed strategies, Strategies A and B provide an optimal scheduling strategy for the typical day based on what has been denoted, respectively, for ease of use. Figures 9 and 10 illustrate Strategy A's preference for purchasing electricity from the main grid. It was found that the tie-line transmission power of Strategy A is close to its average. The consequence is a relatively low level of power fluctuations at the tie-line under Strategy A, whereas Strategy B prefers to generate electricity from DGs. It is more expensive to generate electricity through DGs than to purchase electricity from the main grid; therefore, Strategy B has a higher operation cost.

To better compare Strategies A and B, they were both applied in the context of a realistic scenario based on a typical day. Figure 5 shows such a situation. In Figures 11 and 12, the realized operation results under Strategies A and B are presented, respectively.

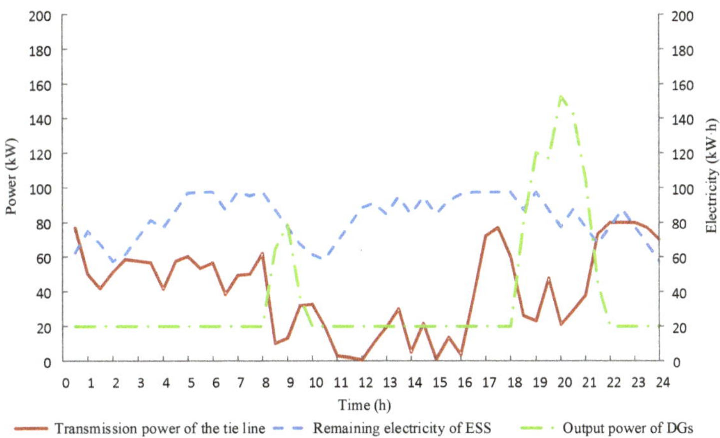

Figure 11. Realized operation results under Strategy A.

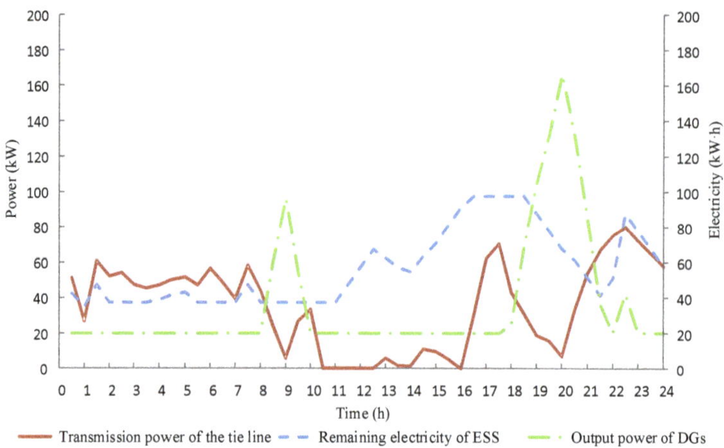

Figure 12. Realized operation results under Strategy B.

The realized output of WTs and PV arrays is larger than the predicted value, as depicted in Figures 5 and 6. Consequently, the realized output power of DGs in both Figures 11 and 12 is smaller than the predicted value. However, DGs based on Strategy A generally have a minimum output power. DGs fail to reduce the output power in this case. Hence, curtailing renewable energy is the most economical way to maintain power balance after the ESS absorbs the excess power to the extent possible. By comparison, DGs based on this strategy are able to reduce the output power because Strategy B takes flexibility constraints into account. Therefore, the MREPS can decrease the output power of DGs to increase the utilization rate of renewable energy. In addition, Strategy B reduces the power purchased from the main grid during peak load periods from 18:00 to 22:00 in comparison to Strategy A. This phenomenon significantly eases the demand for power on the main grid during peak load periods.

The simulation results of Strategies A and B are listed in Table 5. It can be seen that Strategy A has a lower forecast operation cost, forecast curtailment rate of renewable energy, and forecast power fluctuations on the tie-line. All three indicators are higher for Strategy B. However, the IFR is 31.47% higher, the FSR is 45.83% lower, and the AIF is 13.36 kW/h higher for Strategy A compared to Strategy B. This indicates that Strategy A is less capable of coping with uncertainties during the real-time operation of the MREPS compared to Strategy B. This is corroborated by the real-time operation results of the two strategies.

Table 5. Simulation results of Strategies A and B.

Parameters and Units	Strategy A	Strategy B
Insufficient flexibility rate of the day-ahead scheme (IFR) (%) [15]	46.21	14.74
Flexibility sufficiency rate (FSR) (%) [28]	4.17	50.00
Average insufficiency of flexibility (AIF) (kW·h) [28]	28.458	15.098
Forecast operation cost (CNY)	1503.96	1598.37
Realized operation cost (CNY)	1756.52	1794.48
Forecast curtailment rate of renewable energy (%)	0	6.96
Realized curtailment rate of renewable energy (%)	37.58	19.55
Forecast power fluctuations on the tie-line (%)	59.71	69.65
Realized power fluctuations on the tie-line (%)	60.18	69.70
Deviation rate of forecast and realized power fluctuations on the tie-line (%)	29.17	14.58

What is more, the MREPS is operating at a higher cost under Strategy B, and the tie-line power fluctuations are 9.52% higher than they are under Strategy A. However, the rate of utilization of renewable energy for the MREPS under Strategy B is 18.03% higher than that under Strategy A. In addition, the deviation rate of forecast and realized power fluctuations of power under Strategy B is 14.59% lower than that under Strategy A. These findings indicate that Strategy B can reduce the deviation between forecast and realized power fluctuations on the tie-line and increase the utilization of renewable energy at the expense of a small amount of economy and power fluctuations at the same time. Moreover, the realized values of the three objectives under Strategy A increased by 16.79%, 37.58%, and 0.47% compared with their forecast values. In comparison with their forecast values, the realized values of the three objectives under Strategy B increased by 12.27%, 12.59%, and 0.05%. Based on this finding, the results formulated by Strategy B are more consistent with the realized operation results.

5. Conclusions

In this study, a day-ahead scheduling strategy designed to account for flexibility constraints was presented. A multi-objective problem was solved using MOPSO, and a set of the non-dominated solutions was derived for the three objectives. Using the fuzzy comprehensive evaluation method, the optimal compromise solution of the non-dominated solution set was determined. In addition, the optimal strategy proposed in this study was the day-ahead scheduling strategy corresponding to the optimal compromise solution. During peak load periods, the simulation results showed that the proposed strategy was effective in relieving the main grid's pressure on power supply. Moreover, compared with those formulated using the conventional strategy, the results obtained from day-ahead scheduling formulated using the proposed strategy were closer to the results obtained from the MREPS. Although the economy and power fluctuations on the tie-line were slightly higher under the revised strategy, renewable energy usage was significantly higher, and the differences between forecast and realized power fluctuations on the tie-line were relatively small. This finding showed that the revised strategy has the potential to significantly improve the flexibility and reliability of the MREPS's operation at the cost of a small amount of economy and fluctuations in power supply.

Author Contributions: Conceptualization, L.Y. (Lei Yang); and Q.W.; methodology, L.Y. (Lei Yang); software, C.G. and D.Z.; validation, W.H., C.X., and L.Y. (Longjie Yang). All authors have read and agreed to the published version of the manuscript.

Funding: This research received no external funding.

Institutional Review Board Statement: Not applicable.

Informed Consent Statement: Not applicable.

Data Availability Statement: The data presented in this study are available on request from the corresponding author. The data are not publicly available due to data confidentiality requirements.

Conflicts of Interest: The authors declare no conflict of interest.

References

1. Hong, Z.; Hao, S.; Zhang, Q.; Kong, G. Microgrid Spinning Reserve Optimization with Improved Information Gap Decision Theory. *Energies* **2018**, *11*, 2347.
2. Cz, A.; Hc, A.; Lu, L.; Zhang, H.; Zhang, X.; Li, G. Coordination planning of wind farm, energy storage and transmission network with high-penetration renewable energy. *Int. J. Electr. Power Energy Syst.* **2020**, *120*, 105944.
3. Li, J.; Liu, J.; Yan, P.; Li, X.; Zhou, G.; Yu, D. Operation Optimization of Integrated Energy System under a Renewable Energy Dominated Future Scene Considering Both Independence and Benefit: A Review. *Energies* **2021**, *14*, 1103. [CrossRef]
4. Menezes, R.; Soriano, G.D.; Aquino, R. Locational Marginal Pricing and Daily Operation Scheduling of a Hydro-Thermal-Wind-Photovoltaic Power System Using BESS to Reduce Wind Power Curtailment. *Energies* **2021**, *14*, 1441. [CrossRef]
5. Wang, L.; Li, Q.; Ding, R.; Sun, M.; Wang, G. Integrated scheduling of energy supply and demand in microgrids under uncertainty: A robust multi-objective optimization approach. *Energy* **2017**, *130*, 1–14. [CrossRef]
6. Shams, M.H.; Shahabi, M.; Khodayar, M.E. Stochastic Day-ahead Scheduling of Multiple Energy Carrier Microgrids with Demand Response. *Energy* **2018**, *155*, 326–338. [CrossRef]

7. Chen, W.; Shao, Z.; Wakil, K.; Aljojo, N.; Samad, S.; Rezvani, A. An efficient day-ahead cost-based generation scheduling of a multi-supply microgrid using a modified krill herd algorithm. *J. Clean. Prod.* **2020**, *272*, 122364. [CrossRef]
8. Han, S.; Yin, H.; Alsabbagh, A.; Ma, C. A flexible distributed approach to energy management of an isolated microgrid. In Proceedings of the 26th International Symposium on Industrial Electronics (ISIE), Edinburgh, UK, 19–21 June 2017; pp. 2063–2068.
9. Kumar, K.P.; Saravanan, B. Day ahead scheduling of generation and storage in a microgrid considering demand Side management. *J. Energy Storage* **2019**, *21*, 78–86. [CrossRef]
10. Ebrahimi, M.R.; Amjady, N. Adaptive robust optimization framework for day-ahead microgrid scheduling. *Int. J. Electr. Power Energy Syst.* **2019**, *107*, 213–223. [CrossRef]
11. Shi, J.; Huang, W.; Tai, N.; Zhu, Q.; Liu, D. Strategy to smooth tie-line power of microgrid by considering group control of heat pumps. *J. Eng.* **2017**, *2017*, 2417–2422. [CrossRef]
12. Yao, Y.; Zhang, P. Transactive control of air conditioning loads for mitigating microgrid tie-line power fluctuations. In Proceedings of the 2017 IEEE Power & Energy Society General Meeting, Chicago, IL, USA, 16–20 July 2017; pp. 1–5.
13. Hosseini, S.E.; Najafi, M.; Akhavein, A.; Shahparasti, M. Day-Ahead Scheduling for Economic Dispatch of Combined Heat and Power with Uncertain Demand Response. *IEEE Access* **2022**, *10*, 42441–42458. [CrossRef]
14. Shan, X.; Xue, F. A Day-Ahead Economic Dispatch Scheme for Transmission System with High Penetration of Renewable Energy. *IEEE Access* **2022**, *10*, 11159–11172. [CrossRef]
15. Yang, L.; Li, H.; Yu, X.; Zhang, L.; Pang, B.; Yi, R.; Gai, P.; Xin, C. Multi-Objective Day-Ahead Optimal Scheduling of Isolated Microgrid Considering Flexibility. *Power Syst. Technol.* **2017**, *5*, 1432–1440.
16. Yi, W.; Jiang, H.; Xing, P. Improved PSO-based energy management of Stand-Alone Micro-Grid under two-time scale. In Proceedings of the 2016 IEEE International Conference on Mechatronics and Automation, Harbin, China, 7–10 August 2016; pp. 2128–2133.
17. Varghese, S.; Dalvi, S.; Narula, A.; Webster, M. The Impacts of Distinct Flexibility Enhancements on the Value and Dynamics of Natural Gas Power Plant Operations. *IEEE Trans. Power Syst.* **2021**, *36*, 5803–5813. [CrossRef]
18. Li, H.; Lu, Z.; Qiao, Y.; Zhang, B.; Lin, Y. The Flexibility Test System for Studies of Variable Renewable Energy Resources. *IEEE Trans. Power Syst.* **2021**, *36*, 1526–1536. [CrossRef]
19. Eltohamy, M.S.; Moteleb, M.S.A.; Talaat, H.; Mekhemer, S.F.; Omran, W. Technical investigation for power system flexibility. In Proceedings of the 2019 6th International Conference on Advanced Control Circuits and Systems (ACCS) & 2019 5th International Conference on New Paradigms in Electronics & information Technology (PEIT), Hurghada, Egypt, 17–20 November 2019; pp. 299–309.
20. Song, C.; Chu, X. Optimal Scheduling of Flexibility Resources Incorporating Dynamic Line Rating. In Proceedings of the 2017 IEEE Power & Energy Society General Meeting, Chicago, IL, USA, 16–20 July 2017; pp. 1–5.
21. Lu, W.; Hang, N. A Multiobjective Evaluation Method for Short-term Hydrothermal Scheduling. *IEEJ Trans. Electr. Electron. Eng.* **2017**, *12*, 31–37. [CrossRef]
22. Elgammal, A.; El-Naggar, M. Energy management in smart grids for the integration of hybrid wind–PV–FC–battery renewable energy resources using multi-objective particle swarm optimisation (MOPSO). *J. Eng.* **2018**, *11*, 1806–1816. [CrossRef]
23. Liu, X.; Zhang, P.; Fang, H.; Zhou, Y. Multi-Objective Reactive Power Optimization Based on Improved Particle Swarm Optimization With ε-Greedy Strategy and Pareto Archive Algorithm. *IEEE Access* **2021**, *9*, 65650–65659. [CrossRef]
24. Li, C.; Yang, J.; Xu, Y.; Wu, Y.; Wei, P. Classification of voltage sag disturbance sources using fuzzy comprehensive evaluation method. *CIRED Open Access Proc. J.* **2017**, *2017*, 544–548. [CrossRef]
25. Sang, Y.; Zheng, Y. Reserve scheduling in the congested transmission network considering wind energy forecast errors. In Proceedings of the 2020 52nd North American Power Symposium (NAPS), Tempe, AZ, USA, 11–13 October 2021; Volume 1, pp. 1–6.
26. Saaty, T.L.; Vargas, L.G. Models, Methods, Concepts & Applications of the Analytic Hierarchy Process. *International* **2017**, *7*, 9–172.
27. Huang, X. Time-series analysis model based on data visualization and entropy weight method. In Proceedings of the 4th International Conference on Automation, Electronics and Electrical Engineering (AUTEEE), Shenyang, China, 19–21 November 2021; pp. 501–503.
28. Liu, W.; Li, H.; Zhang, H.; Xiao, Y. Expansion Planning of Transmission Grid Based on Coordination of Flexible Power Supply and Demand. *Autom. Electr. Power Syst.* **2018**, *42*, 56–63.

Article

An Improved Model for Five-Phase Induction Motor Based on Magnetic Noise Reduction Part II: Pole-Slot Scheme

Hansi Chen, Jinghong Zhao, Yiyong Xiong *, Sinian Yan and Hao Xu

School of Electrical Engineering, Naval University of Engineering, Jiefang Road 717, Wuhan 430033, China; chs8000@163.com (H.C.); zhaojinghong@163.com (J.Z.); ysnian0504@126.com (S.Y.); xuhao10141205@163.com (H.X.)
* Correspondence: xiongyiyong1989@163.com

Abstract: For the reduction in the electromagnetic noise level of three-phase induction motors, many empirical rules and analytical models have been established to select the matching scheme of pole and slot, but they are not fully applicable to five-phase squirrel cage induction motors (FSCIM). In this paper, combined with the slot-number phase diagram (SNPD), and the electromagnetic force-vibration-acoustic analytical model deduced in Part I, the influence of pole-slot schemes, including five-phase regular-size phase-belt and fractional-slot winding, on magnetic noise is analyzed. The feasibility of electromagnetic noise prediction is verified by finite element simulation and experiments. Taking a 4 kW FSCIM as a prototype, noise prediction is carried out for all the slot-number matching schemes with pole pairs not exceeding three. For two noise reduction targets, which reduce the maximum single-frequency noise in the steady-state operation and the average noise during startup, the pole-slot numbers matching rule of FSCIM is given. This improved model is also applicable in different power ranges for the noise reduction design of five-phase motors.

Keywords: five-phase induction motor; magnetic noise; pole-slot numbers; electromagnetic vibration

Citation: Chen, H.; Zhao, J.; Xiong, Y.; Yan, S.; Xu, H. An Improved Model for Five-Phase Induction Motor Based on Magnetic Noise Reduction Part II: Pole-Slot Scheme. *Processes* **2022**, *10*, 1430. https://doi.org/10.3390/pr10081430

Academic Editors: Haoming Liu, Jingrui Zhang and Jian Wang

Received: 27 June 2022
Accepted: 20 July 2022
Published: 22 July 2022

Publisher's Note: MDPI stays neutral with regard to jurisdictional claims in published maps and institutional affiliations.

1. Introduction

With the promotion and popularization of new energy vehicles, in addition to power density and fault tolerance performance, the electromagnetic vibration noise of electric motors, which is closely related to the noise, vibration, and harshness (NVH) characteristics of electric vehicles, has also received more and more attention. Similar to the three-phase motor, the pole-slot coordination is an important factor affecting the electromagnetic vibration characteristics of the motor. Some of the original rules are in [1], and numerous empirical rules have been developed in the ensuing decades. An exhaustive list of these laws can be found in Timar's book [2]. However, these results mainly focused on three-phase motors, without the consideration of the natural frequency and modal characteristics of the motor structure. Additionally, some of them are based on electromagnetic torque pulsation as the limiting condition, which may not be suitable for reducing audible magnetic noise [3]. Due to the difference in the number of phases, the air-gap magnetic density and electromagnetic force harmonic of the five-phase squirrel cage induction motors (FSCIM) are various.

To design an FSCIM for driving, it is necessary to enhance NVH characteristics. In almost all induction machines, noise originates from aerodynamic, mechanical, and electromagnetic problems [4–6]. Aerodynamic and mechanical problems, which mainly result from turbines and assembly errors, can be ignored in the design stage. However, electromagnetic vibration levels can directly affect the NVH characteristics and cause failures, such as bearing failure and insulation breakdown [7]. Magnetic pull force [8], torque ripple [9], cogging torque [9], and unbalanced magnetic force [8] are the main electromagnetic sources of NVH characteristics. Improvement of the overall force characteristics is required for improving the NVH characteristics. Additionally, optimizing the individual force characteristics while improving the overall electromagnetic force characteristics is a better choice to improve the NVH characteristics. The electromagnetic vibration characteristics of the five-phase motors are similar to those of the three-phase motors. The pole-slot number scheme has a decisive effect on the magnetic noise [10–13]. Obviously, it is not suitable to directly apply the pole-slot matching of the three-phase motor to FSCIM.

A 40/30 slot, four poles, and an FSCIM are used to study the influence of saturation on the air-gap flux density waveform [14]. Wang Dong et al. selected three five-phase motors with 60/38 slots to form a fifteen-phase induction motor, and analyzed the air-gap magnetic potential of the induction motor under non-sinusoidal power supply conditions [15]. A comparative analysis of the operating characteristics between the three-phase and five-phase induction motors under the same structural size, that is, 30/44 slots and 2-poles induction motor, is discussed in [16]. Pereira LA et al. analyzed the mathematical model of the five-phase induction motor, deduced the self-inductance and mutual inductance of the stator and rotor, and calculated the time and space harmonics of the air-gap flux density [17,18]. The effect of applying stator shifting to five-phase winding to suppress the effect of the slot harmonics by doubling the number of slots is investigated [19]. Based on the measured sample data, a new radial vibration model is proposed, consisting of a vibration acceleration impulse model and a natural oscillation model [20]. The vibration and noise levels in a permanent magnet synchronous motor with different slot-pole combinations are discussed, mainly through the FEM to establish the analytical model [21]. An analytical model of the acoustic behavior of pulse-width modulation (PWM) controlled induction machines is applied to a three-phase fractional-slot winding machine. However, the acoustic radiation is simplified as a 2D cylindrical shell model [22,23]. Reference [24] established the electromagnetic vibration and noise model of the three-phase induction motor based on the acoustic model of the infinitely long ring, and carried out the optimization study of the slot-number scheme, but did not consider the influence of the pole-pair-number and the axial modes of the stator frame in an acoustic radiator. Reference [25] aimed at reducing the resonance noise of an evaporative cooling motor induced by an electromagnetic and two-phase flow based on the fluid-structure coupling theory. The subdomain method is used to optimize the noise, vibration, and harshness (NVH) characteristics of a permanent magnet synchronous motor. The predecessors mainly analyzed the operating characteristics and control strategies of the existing five-phase motor or those created by re-embedding the stator winding of the three-phase motor. However, there rarely are selection basis and parameter optimization processes of pole-slot number schemes for five-phase induction motors.

This paper aims at the optimization design of the electromagnetic noise of an FSCIM, as shown in Figure 1. Firstly, the slot-number assignment of the five-phase regular-size phase-belt winding is detailed via the slot-number phase diagram (SNPD) [26,27]. Then, the expression of stator magnetomotive force (MMF) can be given by the superposition of a single conductor or a coil. Next, an improved analytical model of the induction machine's electromagnetic force-vibration-acoustic radiation is proposed. Based on these mathematical models, the influence of time and space harmonics of electromagnetic forces on vibro-noise will be considered as a whole. The accuracy of this model for predicting magnetic noise phenomena is validated by a prototype at different stages (natural frequency, vibration, and sound power level) by numerical methods and tests. Finally, focused on two different low electromagnetic noise targets, the magnetic noise level of every pole-slot number scheme of five-phase induction motors with less than four pole pairs is simulated. Taking a 4 kW FSCIM with an outer diameter of 175 mm as a prototype allows for recommending schemes for low-noise slot matching.

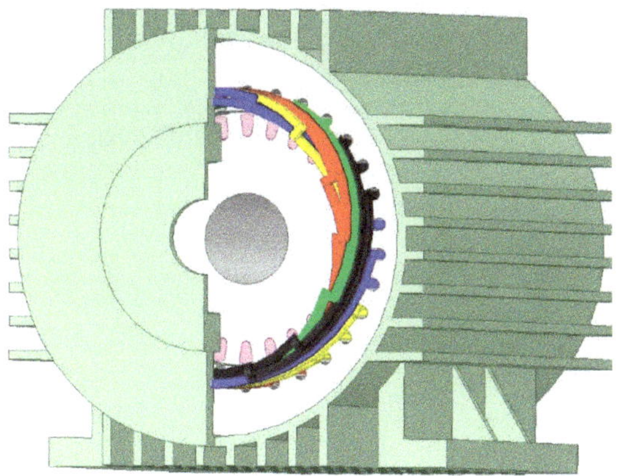

Figure 1. Model of five-phase cage induction motor.

2. Electromagnetic Noise Analytical Prediction Model

2.1. Subordination of Slot-Numbers and Phase

SNPD is a useful method for expressing the winding MMF that quickly obtains the multi-phase symmetrical winding distribution and is convenient for nested programming. The tabular form (including Z_1 positive slots and Z_1 negative slots) presents the unit space vector distribution of the MMF generated in each slot. For the five-phase winding, the normal connection winding can be divided into 36° phase-belt winding, regular-size phase band winding, and 72° phase-belt winding. As shown in Figure 2, the positive phase-belt is expanded by L slots, as the negative phase-belt reduces L slots. When $L = 0$, the common 36° phase-belt winding is given.

360° — 72° (A Phase) | 72° (B Phase) | 72° (C Phase) | 72° (D Phase) | 72° (E Phase)

A Phase						B Phase						C Phase						D Phase						E Phase					
1		2		3		4		5		6		7		8		9		10		11		12		13		14		15	
16		17		18		19		20		21		22		23		24		25		26		27		28		29		30	
	-24		-25		-26		-27		-28		-29		-30		-1		-2		-3		-4		-5		-6		-7		-8
	-9		-10		-11		-12		-13		-14		-15		-16		-17		-18		-19		-20		-21		-22		-23

(a)

360° — 72° (A Phase) | 72° (B Phase) | 72° (C Phase) | 72° (D Phase) | 72° (E Phase)

A Phase					B Phase						C Phase					D Phase						E Phase				
1		2		3		4		5		6		7		8		9		10		11		12		13		14
	15		16		17		18		19		20		21		22		23		24		25		26		27	
28		29		30		31		32		33		34		35		36		37		38		39		40		
	-35		-36		-37		-38		-39		-40		-1		-2		-3		-4		-5		-6		-7	
	-8		-9		-10		-11		-12		-13		-14		-15		-16		-17		-18		-19		-20	
-21		-22		-23		-24		-25		-26		-27		-28		-29		-30		-31		-32		-33		-34

(b)

Figure 2. SNPD of five−phase fractional slot size phase−belt winding. (**a**) M1 motor; (**b**) M2 motor.

Assuming that the slot-number J is located at the R-th cell of the phase diagram when the phase of the slot-number is represented by R, the following formula should be satisfied:

$$R^*(J) = D(|J| - 1) + 1 + m_1 N \frac{1 - \mathrm{sgn}(J)}{2} \tag{1}$$

$$R(J) = \mathrm{mod}(R^*(J), 2m_1 N) \tag{2}$$

where $q = \frac{N}{D}$, N, and D are co-prime, mod( ) indicates the remainder, sgn( ) is the signum function, and sign of J only indicates the current flow in the slot.

In terms of Figure 2, the phase of the slot can be determined according to the phase R of the slot-number. If the number of cells occupied by the positive and negative phase-belt are $N_p + L$ and $N_p - L$, respectively, the slots-number that belong to the β-th phase (1–5 corresponding to the A–E phase, respectively) J, can be assigned according to the following rules:

$$2(\beta - 1)N + L\frac{1 - \operatorname{sgn}(J)}{2} \le R(\beta, J) \le (2\beta - 1)N + L\frac{1 + \operatorname{sgn}(J)}{2} \tag{3}$$

According to the aforementioned, the expression of the winding coefficient can be deduced for each pole-slot match. Taking the two motors shown in Table 1 as an example, the phase-belt distribution can be obtained as shown in Figure 2 by (3).

Table 1. Generic five-phase fractional slot motors M1 and M2.

Symbol	M1	M2
Number of stator slots Z_1	30	40
Number of rotor slots Z_2	26	32
Number of pole pairs p	2	3
Number of slots moved L	1	2

It can be seen from Figure 2 (a) that for the M2 motor, the subordinations of the slot-numbers of $L = 0$ and $L = 1$ are consistent. So far, by combining (1), (2), and (3), the symmetrical distribution of the five-phase normal windings and the affiliation of each slot-number can be determined only via Z_1 and p. If the center line of the A-phase winding is taken as the origin, the winding function of the five-phase normal winding can be obtained from the slot-number distribution $R(m)$ and (4) by the superposition principle.

$$WF_s(\theta_s) = \begin{cases} \sum\limits_{v=1}^{\infty} \frac{k_{sv}}{\pi v} \sin v\theta_s, \text{single conductor} \\ \sum\limits_{v=1}^{\infty} \frac{2N_c}{\pi v} k_{sv} k_{yv} \cos v\theta_s, \text{single coil} \end{cases} \tag{4}$$

where N_c is the number of turns of the coil; k_{sv} and k_{yv} are the slotting coefficient and short distance coefficient of v pair poles harmonics, respectively.

2.2. Analytical Calculation of Electromagnetic Force

For induction motors, the MMF generated by the rotor is often much smaller than the stator in the case of no load, not resulting in new electromagnetic resonance phenomena [28]. Therefore, the radial component B_r of the magnetic flux density can be expressed as [29,30]:

$$B_r(t, \theta_s) = f_{mmf}^s(t, \theta_s)\Lambda(t, \theta_s) \tag{5}$$

$$\Lambda(t, \theta_s) = \Lambda_0 + \sum_{k_1}\lambda_{k_1} + \sum_{k_2}\lambda_{k_2} + \sum_{k_1}\sum_{k_2}\lambda_{k_1}\lambda_{k_2} \tag{6}$$

The specific expressions of each part of Λ have been given in detail in [29,31]. Noticeably, the effects of both sides slotting in Λ are proportional to the width of the slot opening. Both λ_{k_1} and λ_{k_2} are inversely proportional to the harmonic order of the permeance k_1 and k_2. The MMF of the stator is obtained when the sequential five-phase currents pass through the windings. Substituting (3), (4) and (5) into Maxwell's radial force expression (7), the radial electromagnetic forces generated by all radial magnetic flux harmonics can be determined.

$$P_{eR}(t, \theta_s) = B_r^2(t, \theta_s)/2\mu_0 \tag{7}$$

Considering only the major force waves in magnetic noise, that is, the low-order force waves P_{belt} sourced from the interaction between the harmonics of phase-belt of stator MMF and the tooth harmonics of air-gap permeability. The main force characteristics are given in Table 2.

Table 2. Characteristics of phase belt force wave P_{belt}.

Symbol	Frequency f_P	Force Wave Mode m	Remark
P_{belt}	$f_n[k_2Z_2(1- - s)/p + l]$	$k_2Z_2- - 2m_1pk_1 + lp$	$l = 0, \pm2$

It can be seen from Table 2 that the pole-slot matching scheme has a decisive influence on the force wave characteristics. The number of stator and rotor slots together affects the force wave order, while the frequency is mainly affected by the Z_2, p, s, and supply frequency f_1. Furthermore, five-phase motors are mostly powered by PWM with rich harmonics, so the frequency range of Maxwell forces is likely to cover the natural frequency of lower modes of the stator system.

2.3. Electromagnetic Vibration-Acoustic Radiation Model

The stator system mainly consists of an iron core, windings–tooth, and housing, which can be regarded as equivalent rings, respectively. The equivalent model can be divided into finitely and infinitely cylindrical shells for analytical calculation of vibration and noise based on Table 3.

Table 3. Criteria for type of analytical model.

Condition	Type of Equivalent Shells
$ml \geq 10\,aR$	Infinitely cylindrical shell
$ml < 10\,aR$	Finitely cylindrical shell

For an infinite cylindrical shell, the analytical expressions for vibration and acoustics are given in [32,33]. In another case, a more detailed description of the finite shell can be found in the submitted paper *Part I*.

Due to the large stiffness of the motor shaft, the natural frequencies of the same circumferential mode are relatively close, and the influence of the axial mode on the natural frequency is smaller than that of the circumferential mode. Therefore, the sound power of different axial modes under the same circumferential mode excited by the m-order force wave can be considered uniformly. The noise of the low axial orders of the same circumferential mode is summed up by using the principle of modal superposition:

$$L_W(n, m) = 10\,\log_{10}\left(\sum_{a=1}^{3} W_m(n, a, m)/W_0\right) \tag{8}$$

$$L_{WA}(n, m) = 10\,\log_{10}\left(\sum_{n} 10^{0.1(L_w(n,m)+\Delta L_A(n))}\right) \tag{9}$$

where $\Delta L_A(n)$ is sound frequency dependent A-weighting factor.

2.4. Model Validation

The model has been validated at different stages (natural frequency, vibration, and sound power level) by numerical methods and tests in the submitted paper *Part I* and [22,28,34].

Furthermore, in this paper, the experimental measurement and analysis of the vibration-noise of the prototype in the transient process are carried out, and the results further verify the accuracy of the analytical model. The layout of the vibration acceleration sensors (#A1–#A5) and sound sensors (#S1–#S3) is shown in Figure 3, and an overall layout of the test platform is given in Figure 4.

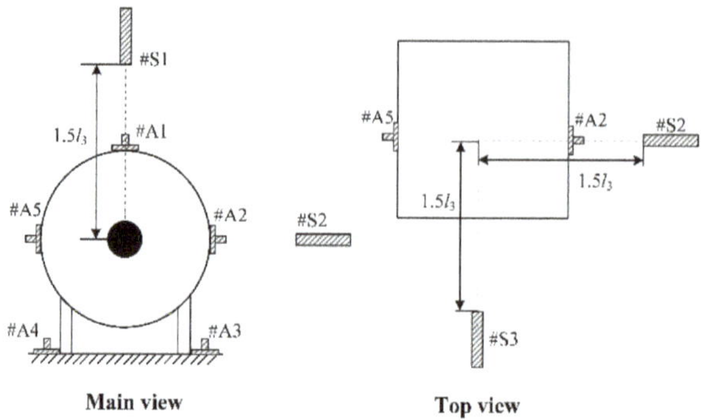

Figure 3. Schematic diagram of sensors layout.

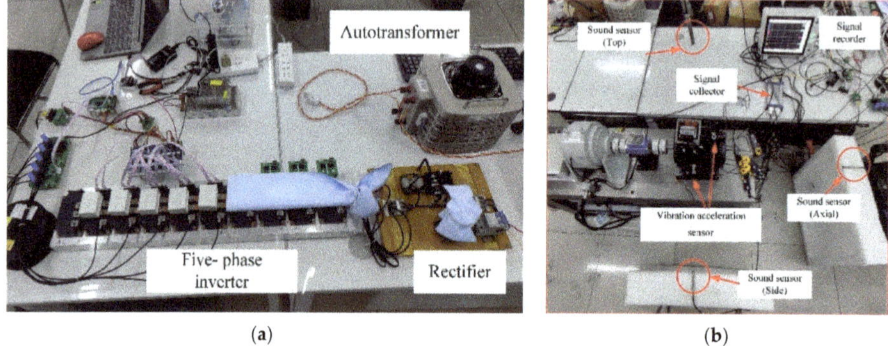

Figure 4. FSCIM drive control and noise measurement platform. (**a**) Five-phase PWM drive system; (**b**) vibration-noise measurement system.

According to the principle of frequency conversion speed regulation, the starting process is divided into 10 equal processes for vibration and noise measurements. The five-phase currents and speed curves over time are shown in Figure 5.

Theoretically, the stable speed of 10 stages should be proportional to the power supply frequency, but due to the idling loss, the speed is not strictly proportional to the frequency. The stator current shows a trend of increasing-decreasing-increasing. Meanwhile, the vibration and noise results obtained by sensors placed as in Figure 3 are shown in Figures 6–9.

As shown in Figure 6, acceleration amplitudes and trends between the top and bottom are basically consistent. However, the side vibration amplitude is almost half of the top, and the change trend with the supply frequency is different, but consistent with the trend of the current.

The natural frequency of the stator system is given in the submitted paper *Part I*, so this paper will use it as known data. Through spectrum analysis of the acceleration at each stage in Figure 7, it is found that the main vibration frequency is concentrated at approximately 5000 Hz and 1000 Hz, which corresponds to the modes (3, 1), (3, 2), (1, 4), (2, 4) and (3, 4) of the stator system. For vibration acceleration, it is mainly reflected in the circumferential modem, that is, m = 1, 2, and 4. In terms of Figure 7j, the analytical model is used to calculate the vibration acceleration of the test motor in main modes. From Table 4, it is obvious that mode (3, 2) contributes the most to the vibration, while mode (3, 1) is the opposite, even though the natural frequencies of both are very close. Furthermore, all three modes with m = 4 have contributions to the vibration, hence the main modes are confirmed to m = 2 and 4 for the test motor, which is consistent with Figure 10.

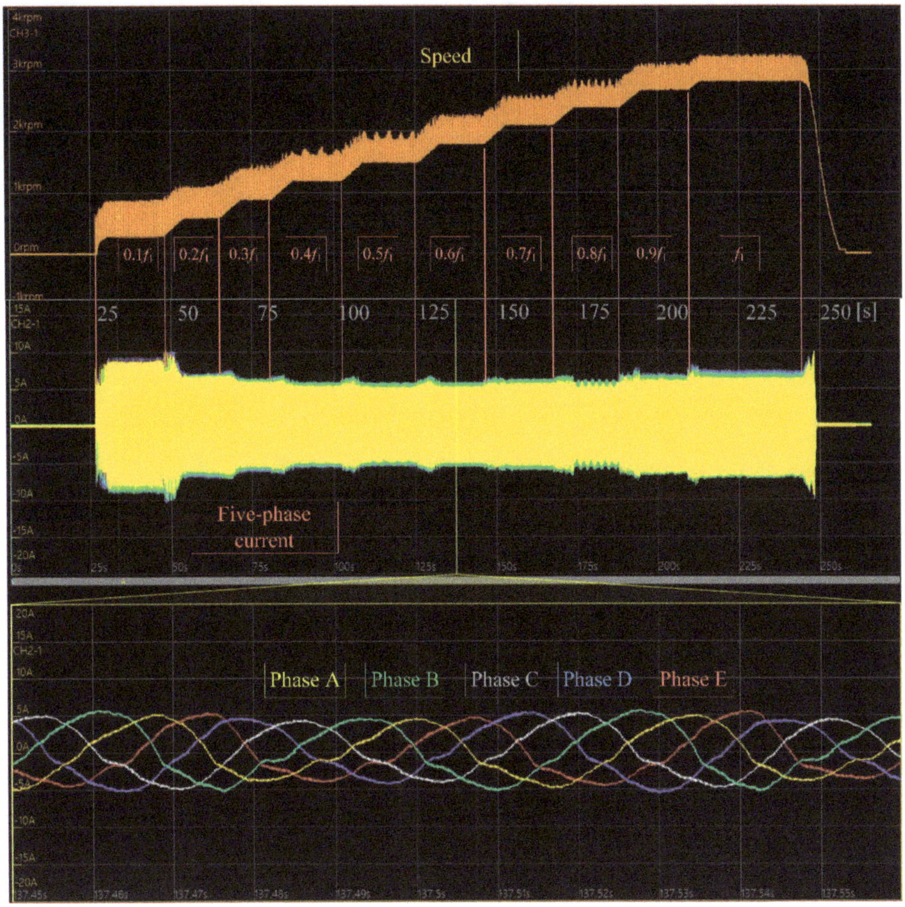

Figure 5. Five−phase current and speed curves from start−up to steady-state.

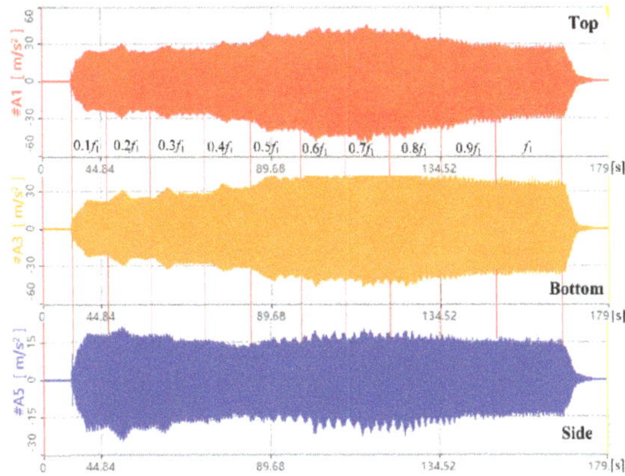

Figure 6. Vibration acceleration of three measuring points on the frame.

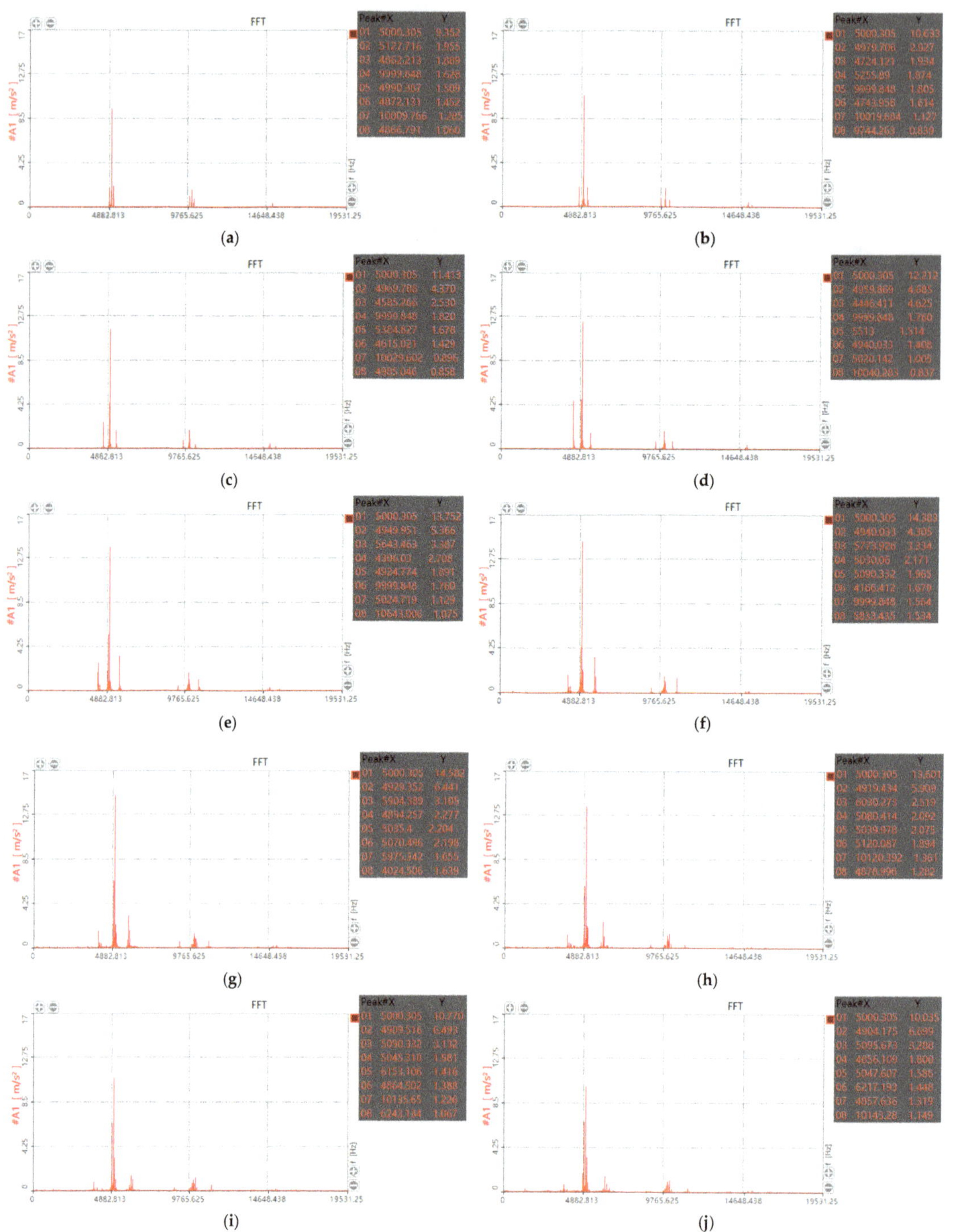

Figure 7. The vibration acceleration spectrum of the top point #A1 in 10 stages. (**a**) Stage 1, $f = 0.1 f_1$; (**b**) Stage 2, $f = 0.2 f_1$; (**c**) Stage 3, $f = 0.3 f_1$; (**d**) Stage 4, $f = 0.4 f_1$; (**e**) Stage 5, $f = 0.5 f_1$; (**f**) Stage 6, $f = 0.6 f_1$; (**g**) Stage 7, $f = 0.7 f_1$; (**h**) Stage 8, $f = 0.8 f_1$; (**i**) Stage 9, $f = 0.9 f_1$; (**j**) Stage 10, $f = f_1$.

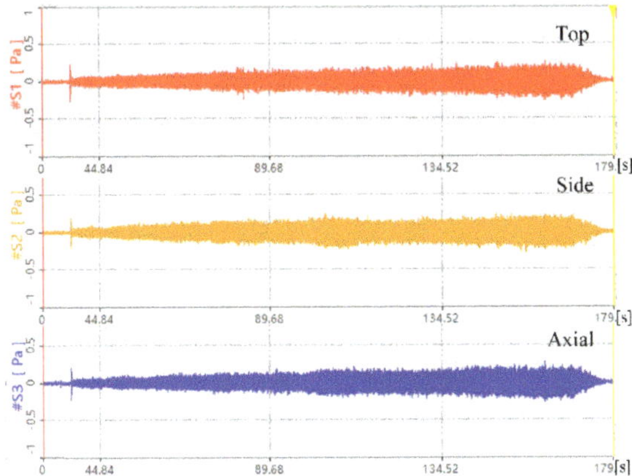

Figure 8. Vibration acceleration of three measuring points on the frame.

Table 4. Frame surface vibration acceleration $a_{a,m}$ (unit is m/s^2) by improved analytical method.

$a_{a,m}$ (m/s^2)	Mode a		
Mode m	1	2	3
0	1.88	1.11	0.41
1	0	0	0
2	3.65	3.49	13.96
3	0	0	0
4	4.69	2.10	3.57

From Figure 8, it is found that the sound pressure intensities and trends in the top, side, and axial directions are more consistent than they are in vibration. Focused on side sensor #S2, 10 stages of spectrum analysis are performed in Figure 9.

With the increase in the power supply frequency and rotation speed, the noise in the low-frequency region (500–2000 Hz) gradually increases and fluctuates. This frequency band that does not appear in the frequency spectrum of vibration acceleration is related to the supply frequency and speed. The mechanical noise corresponding to speed is mainly generated by the bearing connection, and the part related to supply frequency is the result of the resonance of the PWM module and the heat sink. Additionally, the magnetic noise of the test motor also fluctuates with the operation condition, and the amplitude spectrum also agrees with the vibration in both the analytical model and experiment.

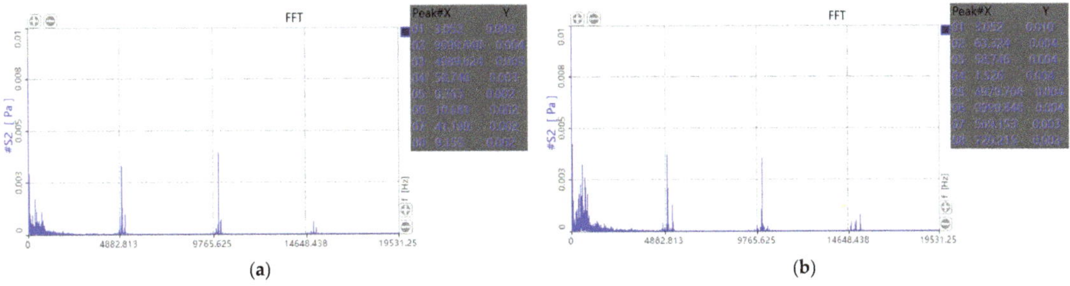

(a) **(b)**

Figure 9. *Cont.*

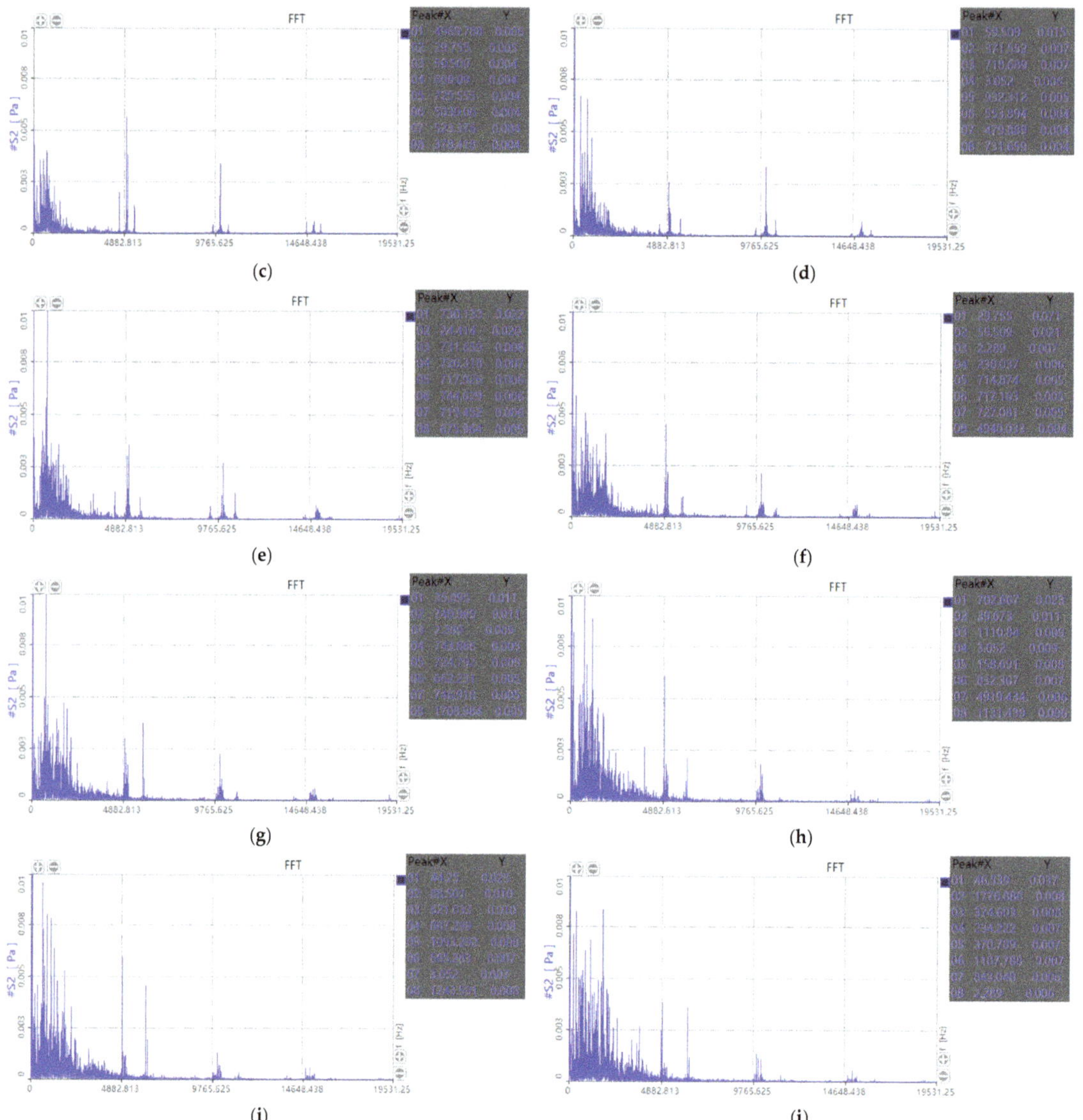

Figure 9. The sound pressure spectrum of the top point #S2 in 10 stages. (**a**) Stage 1, $f = 0.1\,f_1$; (**b**) Stage 2, $f = 0.2\,f_1$; (**c**) Stage 3, $f = 0.3\,f_1$; (**d**) Stage 4, $f = 0.4\,f_1$; (**e**) Stage 5, $f = 0.5\,f_1$; (**f**) Stage 6, $f = 0.6\,f_1$; (**g**) Stage 7, $f = 0.7\,f_1$; (**h**) Stage 8, $f = 0.8\,f_1$; (**i**) Stage 9, $f = 0.9\,f_1$; (**j**) Stage 10, $f = f_1$.

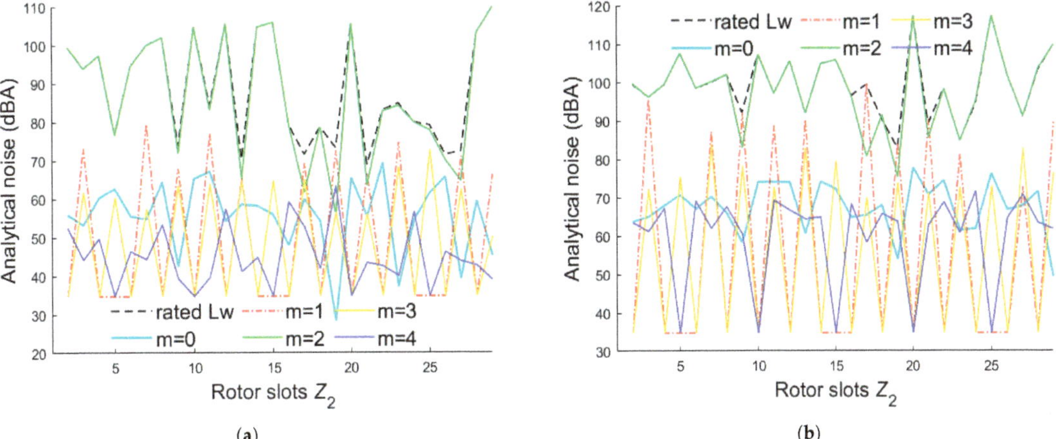

Figure 10. A $Z_1 = 30$ FSCIM electromagnetic sound power level in function Z_2 for both operation conditions. (**a**) Rated condition; (**b**) start-up condition.

3. Low-Noise Pole-Slot Optimal Combination

3.1. Low-Noise Targets

At present, the five-phase motor mainly adopts PWM variable frequency speed control. During the process from start-up to stable operation, the power supply frequency rises linearly from 0 to f_1, during which a rich force wave will be generated. In order to avoid the resonance between the low-order force wave P_{belt} shown in Table 2 and the circumferential mode of the stator, the following conditions must be fulfilled:

$$f_1[k_2Z_2(1-s)/p+l] < f_m, m = |k_2Z_2 - k_1Z_1 + lp| \tag{10}$$

According to the noise reduction requirements in different environments, two noise optimization goals are set, respectively:

(1) The minimum total sound power level of each frequency in rated condition:

$$\min(\sum_m L_{WA}(f_m)) \tag{11}$$

(2) The sound power amplitude of the maximum noise generated during the start-up is the smallest:

$$\min(\max(L_{WA}(f_m))) \tag{12}$$

For the pole-slot matching scheme, the above two optimization objectives should be considered, and it should be determined according to the application environment of the motor.

3.2. Pole-Slot Matching Strategy

When the power demand of a certain type of motor is determined, its outer diameter is also determined according to the national standard. On the basis of the aforementioned analytical model, all possible slot matching schemes are traversed, and the noise radiation in the process of frequency conversion and speed regulation is analyzed and calculated. The low-noise slot fit solution can be screened out. This paper takes the test motor shown in Table 3 as the object, and comprehensively considers the outer diameter of the motor, the principle of fewer slots and near-slots, and the symmetry conditions of multi-phase motor windings [10]. All cases of symmetrical distribution of stator windings in the range of $Z_1 \in [5, 50]$, $Z_2 \in [2, Z_1]$, and $p \in [1, 3]$, and the following assumptions are made:

(1) The slip ratio changes linearly.

(2) In each case, five-phase sequential currents of the same amplitude are passed through the stator windings.

(3) The magnitude of the air-gap MMF generated by the stator winding remains unchanged, so the $N_1 k_{w1}$ under various schemes remains approximately unchanged. It should be noted that the weight of the winding and the stiffness will change.

$$N_1 k_{w1} = const \tag{13}$$

(4) In order to control the variables, the influence of slot opening and slot type is not considered when changing the number of stator and rotor slots. It can be known from the amplitude of air-gap permeability [29,31]. It is necessary to keep $Z_1 b_{01}$ and $Z_2 b_{02}$ unchanged.

$$Z_1 b_{01} = const, Z_2 b_{02} = const \tag{14}$$

(5) Excluding the influence of magnetic saturation, the air-gap permeance does not contain the harmonic content caused by saturation.

(6) The slot number Z_1 is greater than Z_2.

The electromagnetic noise results of the slot fitting scheme are obtained by calculation, which include the maximum and average noise levels radiated by the motor during operation under each scheme, and the noise radiation under each modal resonance. The predicted results of the test motor are shown in Figure 10.

As can be seen from Figure 10, the sound power level in (b) during the starting process is higher than that in (a) under steady-state. This is because the frequency of force waves is much more abundant during the starting process, which is more likely to result in stator system resonance. Firstly, consistent with the previous experiment's results, the magnetic noise is mainly generated from modes m = 2 and 4 on the test motor with 30/26 slots. In addition, both the average magnetic noise level in the starting process and the maximum noise level in the steady-state show discrete changes with the increase in the number of rotor slots. As expected, each stator mode has different effects on the overall noise level. For the five-phase motor with $Z_1 = 30$ and $p = 1$, the resonance effect at $m = 2$ (that is, the elliptical mode) is the strongest and dominant in both operating conditions. It is worth noting that for steady-state operation, the effects of modes $m = 2$ and 4 tend to have the same trend, which is also consistent with the previous conclusions of the test motor, but the effects of the two modes are quite the opposite during start-up. In summary, the initial design should focus on avoiding the slot matches that produce such modal resonance.

3.3. Optimal Pole-Slot Scheme

By analyzing the noise prediction results of all schemes, for the two different optimization objectives in Section 3.1, the quietest combinations between Z_2 and p in each five-phase symmetrical stator winding are screened out, as shown in Tables 5–7. According to the noise level, the rotor slots number are arranged in increasing order from left to right. From the above three tables, it can be found that no matter what the parity of Z_1 and Z_2 are, there will always be some low-noise combinations. For inter-slot motors (underlined in the table), the quiet slot combinations will be significantly different from the fractional-slot situation due to the change in the flux density harmonics content. Generally speaking, the quietest rotor slot number of a fractional-slot winding will be less than that of the integer-slot case. If Z_1 and Z_2 are both a multiple of $2p$, the expression of force wave order in Table 2 can be transformed into (15), hence the main force orders should be a multiple of $2p$ which is in agreeance with Figure 10.

$$m = 2p(k_2 q_2 - k_1 q_1 + 0.5l) \tag{15}$$

Table 5. Low-noise pole-slot numbers scheme for $p = 1$.

Z_1	Z_2 of Target (10)	Z_2 of Target (11)
5	4, 2, 3	3, 2, 4
10	5, 9, 3, 6, 4	7, 3, 8, 4, 2
15	5, 14, 6, 9, 10	5, 10, 4, 12, 13
20	16, 13, 2, 17, 9	13, 19, 18, 9, 3
25	24, 5, 21, 12, 19	23, 7, 18, 14, 9
30	21, 13, 17, 26, 27	19, 23, 21, 18, 27
35	34, 32, 5, 26, 4	34, 28, 22, 25, 5
40	32, 38, 24, 2, 37	31, 38, 23, 36, 39

Table 6. Low-noise pole-slot numbers scheme for $p = 2$.

Z_1	Z_2 of Target (10)	Z_2 of Target (11)
5	2, 3, 4	2, 3, 4
10	2, 5, 7, 3, 8	8, 2, 4, 6, 7
15	2, 8, 13, 4, 6	10, 9, 2, 4, 11
20	12, 4, 18, 10, 8	12, 4, 16, 8, 10
25	23, 15, 20, 18, 10	24, 20, 23, 10, 8
30	20, 10, 24, 23, 15	10, 20, 23, 29, 18
35	23, 24, 30, 32, 7	23, 30, 16, 32, 10
40	36, 32, 30, 23, 12	32, 36, 12, 4, 16

Table 7. Low-noise pole-slot numbers scheme for $p = 3$.

Z_1	Z_2 of Target (10)	Z_2 of Target (11)
5	2, 4, 3	3, 2, 4
10	6, 2, 8, 4, 3	2, 5, 6, 4, 8
15	12, 4, 2, 14, 8	5, 3, 10, 2, 11
20	19, 10, 18, 6, 12	10, 12, 13, 5, 11
25	23, 19, 18, 24, 6	13, 3, 6, 11, 2
30	24, 19, 10, 25, 23	10, 24, 20, 22, 3
35	19, 25, 29, 34, 17	3, 13, 34, 26, 6
40	19, 30, 25, 38, 10	30, 10, 12, 24, 20

It should be noted that the several limitations of the pole-slot combination mentioned above are: it only considers the symmetrical winding with a number of pole pairs less than four, and it calculates the magnetic noise level without consideration of the effects of saturation, slot shape, and PWM harmonics. However, there is no doubt that this prediction model can be applied to five-phase induction motors with other power, size, and power supply conditions, in the initial design process to avoid strong magnetic noise and vibrations due to slotting harmonics.

4. Conclusions

Aiming at the relationship of the FSCIM between the electromagnetic noise characteristics and the pole-slot numbers match, this paper deduces and improves the radial electromagnetic force-vibration-noise radiation model based on the analysis of the five-phase symmetrical windings and the acoustic model in finite cylindrical shells. This model can comprehensively consider the natural frequency of the system composed of the three parts of the stator superimposed, as well as the acoustic radiation characteristics of the circumferential and axial modes of the finite cylindrical shell. Therefore, the results are more in line with the real situation of the electromagnetic noise of the motor. The accuracy of the improved model was verified by simulation and experiments at multiple stages under the conditions of a given power, supply frequency, and slot type. The noise prediction of as many slots as possible was carried out, and some quiet pole-slots for the FSCIM test were selected by different noise optimization objectives. The results reflect the effects and rules of pole-slot matching for noise reduction in five-phase induction motors.

Although the matching results in this paper have some limitations, which are that the results only reflect the effect of pole-slot numbers without the consideration of the slot geometries and PWM harmonics. However, they can still be applied to motors with similar natural frequencies, and the analytical noise prediction model can be applied to FSCIM with other requirements. To avoid severe resonance and noise caused by electromagnetic force waves, the model is still beneficial for the selection of pole-slot numbers during motor initial design.

Future works should focus on the multi-objective optimization algorithms for low-noise optimization design, comprehensively considering the pole-slot scheme, and slot geometries.

Author Contributions: Writing—original draft preparation, H.C.; validation, Y.X.; software, H.X.; supervision, J.Z.; data curation, S.Y. All authors have read and agreed to the published version of the manuscript.

Funding: This research was funded by the National Natural Science Foundation of China under the grant number 51507813.

Institutional Review Board Statement: Not applicable.

Informed Consent Statement: Not applicable.

Data Availability Statement: Not applicable.

Conflicts of Interest: The authors declare no conflict of interest.

Nomenclature

Electrical notations:

B_r	radial air-gap magnetic density
θ_s	angular position on the circumference
μ_0	air magnetic permeability
f_n	frequency of electromagnetic force
m	force wave order, circumferential mode
f_1	fundamental supply frequency
f^s_{mmf}	stator magnetomotive force
WF_s	stator winding function
P_{belt}	major electromagnetic force

Mechanical and acoustic notations:

Z_1	stator slots number
Z_2	rotor slots number
D_1	stator core outer diameter
D_{i1}	stator core inner diameter
p	pole pair number
v	harmonic pole pairs
k_{sv}	slotting coefficient
k_{yv}	short distance coefficient
s	slip
δ	air-gap length
m_1	phase number, m1 = 5
q	number of slots per pole per phase
b_{01}	stator slots opening width
b_{02}	rotor slots opening width
N_c	number of turns
a	axial mode
l	length of the housing
R	radius of the housing
W_m	sound power
W_0	reference sound power
L_W	sound power level

References

1. Kron, G. Induction Motor Slot Combinations Rules to Predetermine Crawling, Vibration, Noise and Hooks in the Speed-Torque Curve. *Am. Inst. Electr. Eng. Trans.* **1931**, *50*, 757–767. [CrossRef]
2. Timár, P.L. *Noise and Vibration of Electrical Machines*; Elsevier: Amsterdam, The Netherlands, 1989.
3. Besnerais, J.L.; Lanfranchi, V.; Hecquet, M.; Romary, R.; Brochet, P. Optimal Slot Opening Width for Magnetic Noise Reduction in Induction Motors. *IEEE Trans. Energy Convers.* **2009**, *24*, 869–874. [CrossRef]
4. Frosini, L. Novel Diagnostic Techniques for Rotating Electrical Machines—A Review. *Energies* **2020**, *13*, 5066. [CrossRef]
5. Choi, S.; Haque, M.S.; Tarek, M.T.B.; Mulpuri, V.; Duan, Y.; Das, S.; Garg, V.; Ionel, D.M.; Masrur, M.A.; Mirafzal, B.; et al. Fault Diagnosis Techniques for Permanent Magnet AC Machine and Drives—A Review of Current State of the Art. *IEEE Trans. Transp. Electrif.* **2018**, *4*, 444–463. [CrossRef]
6. Zhang, S.; Zhang, S.; Wang, B.; Habetler, T.G. Deep Learning Algorithms for Bearing Fault Diagnostics—A Comprehensive Review. *IEEE Access* **2020**, *8*, 29857–29881. [CrossRef]
7. Shin, K.-H.; Bang, T.-K.; Kim, K.-H.; Hong, K.; Choi, J.-Y. Electromagnetic Analysis and Experimental Study to Optimize Force Characteristics of Permanent Magnet Synchronous Generator for Wave Energy Converter Using Subdomain Method. *Processes* **2021**, *9*, 1825. [CrossRef]
8. Wang, Y.; Zhu, Z.-Q.; Feng, J.; Guo, S.; Li, Y.; Wang, Y. Investigation of Unbalanced Magnetic Force in Fractional-Slot Permanent Magnet Machines Having an Odd Number of Stator Slots. *IEEE Trans. Energy Convers.* **2020**, *35*, 1954–1963. [CrossRef]
9. Ocak, O.; Aydin, M. An Innovative Semi-FEA Based, Variable Magnet-Step-Skew to Minimize Cogging Torque and Torque Pulsations in Permanent Magnet Synchronous Motors. *IEEE Access* **2020**, *8*, 210775–210783. [CrossRef]
10. Chen, S. *Motor Design*; Machinery Industry Press: Beijing, China, 2004; pp. 244–247.
11. Akhtar, M.J.; Behera, R.K. Optimal Design of Stator and Rotor Slot of Induction Motor for Electric Vehicle Applications. *IET Electr. Syst. Transp.* **2019**, *9*, 35–43. [CrossRef]
12. Qiu, H.; Zhang, Y.; Yang, C.; Yi, R. The Influence of Stator–Rotor Slot Combination on Performance of High-Voltage Asynchronous Motor. *J. Control. Autom. Electr. Syst.* **2019**, *30*, 1126–1134. [CrossRef]
13. Wang, D.; Wu, X.; Ma, W.; Wang, X.; Guo, Y. Air-gap MMF Analysis of Fifteen-phase Induction Motor with Non-sinusoidal Supply. *Proc. CSEE* **2009**, *29*, 88–94.
14. Pereira, L.A.; Scharlau, C.C.; Pereira, L.; Haffner, S. Influence of Saturation on the Airgap Induction Waveform of Five-Phase Induction Machines. *IEEE Trans. Energy Convers.* **2012**, *27*, 29–41. [CrossRef]
15. Pile, R.; Le Menach, Y.; Le Besnerais, J.; Parent, G. Study of the Combined Effects of the Air-Gap Transfer for Maxwell Tensor and the Tooth Mechanical Modulation in Electrical Machines. *IEEE Trans. Magn.* **2020**, *56*, 1–4. [CrossRef]
16. Rosa, R.; Pereira, L.A.; Pereira, L. Comparison of Operating Curves of Five-Phase and Three-Phase Induction Machines of Same Size. In Proceedings of the Haffner IECON 2014—40th Annual Conference of the IEEE Industrial Electronics Society, Dallas, TX, USA, 29 October–1 November 2014.
17. Scharlau, C.C.; Pereira, L.F.A. Model of a Five-Phase Induction Machine Allowing for Harmonics in the Air-Gap Field. Part I. Parameter Determination and General Equations. In Proceedings of the Haffner 2004. IECON 2004. 30th Annual Conference of IEEE, Busan, Korea, 23–24 October 2004.
18. Scharlau, C.C.; Pereira, L.A.; Pereira, L.F.A.; Haffner, J.F. Model of a Five-Phase Induction Machine Allowing for Harmonics in the Air-Gap Field. Part II: Transformation of Co-Ordinates and d-q Models. In Proceedings of the 30th Annual Conference of IEEE Industrial Electronics Society, 2004. IECON 2004, Busan, Korea, 2–6 November 2004.
19. Abdel-Khalik, A.S.; Ahmed, S.; Massoud, A. Application of Stator Shifting to Five-Phase Fractional-Slot Concentrated Winding Interior Permanent Magnet Synchronous Machine. *IET Electr. Power Appl.* **2016**, *10*, 681–690. [CrossRef]
20. Zhong, R.; Guo, X.; Yin, J.; Zhao, L.; Sun, W. Method for Radial Vibration Modelling in Switched Reluctance Motor. *IET Electr. Power Appl.* **2016**, *10*, 834–842.
21. Lin, F.; Zuo, S.; Wu, X. Electromagnetic Vibration and Noise Analysis of Permanent Magnet Synchronous Motor with Different Slot-Pole Combinations. *IET Electr. Power Appl.* **2016**, *10*, 900–908. [CrossRef]
22. Le Besnerais, J.; Lanfranchi, V.; Hecquet, M.; Brochet, P.; Friedrich, G. Acoustic Noise of Electromagnetic Origin in a Fractional-Slot Induction Machine. *COMPEL Int. J. Comput. Maths. Electr. Electron. Eng.* **2008**, *27*, 1033–1052. [CrossRef]
23. Besnerais, J.L.; Lanfranchi, V.; Hecquet, M.; Brochet, P.; Friedrich, G. Prediction of Audible Magnetic Noise Radiated by Adjustable-Speed Drive Induction Machines. *IEEE Trans. Ind. Appl.* **2010**, *46*, 1367–1373. [CrossRef]
24. Leissa, A.W.; Nordgren, R.P. Vibration of Shells. *J. Appl. Mech.* **1993**, *41*, 544. [CrossRef]
25. Cheng, Z.; Ruan, L.; Huang, S.; Yang, J. Research on Noise Reduction of 3.6 MW Evaporative Cooling Wind Motor Induced by Electromagnetic and Two-Phase Flow Resonance Based on Stator Optimization. *Processes* **2021**, *9*, 669. [CrossRef]
26. Xu, S.; Motor, A.C. *Winding Theory*; Machinery Industry Press: Beijing, China, 1985; pp. 3–25.
27. Yu, K.; Xu, S. Twin pole pairs slot-number phase diagram and its application in the design of pole chaning windings. *Trans. China Electrotech. Soc.* **1993**, *6*, 6–8.
28. Le Besnerais, J. Reduction of Magnetic Noise in PWM-Supplied Induction Machines—Low-Noise Design Rules and Multi-Objective Optimisation. Ph.D. Thesis, Ecole Centrale de Lille, Lille, France, 2008.
29. Chen, Y. *Analysis and Control of Motor Noise*; Zhejiang University Press: Hangzhou, China, 1987; pp. 5–23.

problems due to the uncertainty of the parameters during the operation of the motor. In order to overcome these uncertainties, several methods based on model estimation, such as a model reference adaptive system, full-order observer, extended Luenberger observer, sliding mode observer, and the extended Kalman filter, have become the main research focuses [10–13]. Different from the deterministic method used in observer designs in model reference adaptive system technology, the extended Kalman filter takes the system error and measurement error into account in the estimation process. The ability to adjust the Kalman filter according to the noise characteristics of measurement and initial disturbance highlights the advantages of the stochastic method over the deterministic method [14,15].

Different from the three-phase induction motor, the vector control of the five-phase induction motor with centralized winding should consider both the fundamental space and the third-harmonic space, and it generally adopts the control method of third-harmonic current suppression [16]. Therefore, when designing an EKF observer, the order of the state equation of the five-phase induction motor is higher than that of the three-phase induction motor. Considering the difference in the state equation of the five-phase induction motor, if the fundamental space and the third-harmonic state-space variables of the five-phase induction motor are controlled at the same time, the state equation of the system will reach the ninth order [17]. If other state variables are introduced, it will be higher and increase the complexity of the system. Therefore, a double EKF structure is proposed in this paper, which can decompose the original ninth-order state equation into a fourth-order and a fifth-order state equation when the third-harmonic current is small. The EKF observer based on this structure can simultaneously observe the rotor angular velocity, fundamental space rotor flux linkage, and third-harmonic space rotor flux linkage.

The rest of this paper is structured as follows: in Section 2, the linear and discrete state equations of the fundamental spatial components of the five-phase squirrel cage induction motor are derived; Section 3 introduces the EKF algorithm and the double EKF structure; then, the results of speed prediction and flux linkage prediction are discussed in Section 4; finally, Section 5 summarizes the research of this paper.

2. Linear, Discrete State-Space Model for Five-Phase Squirrel Cage Induction Motors

Similarly to the three-phase induction motor, the five-phase induction motor can also transform the five-phase voltage and current into the two-phase stationary coordinate system and the two-phase synchronous rotating coordinate system through coordinate transformation. The Clark conversion formula from the five-phase rotating coordinate system to the two-phase stationary coordinate system can be written as [18]:

$$
\begin{bmatrix} x_{\alpha 1} \\ x_{\beta 1} \\ x_{\alpha 3} \\ x_{\beta 3} \\ x_0 \end{bmatrix} = \sqrt{\frac{2}{5}}
\begin{bmatrix}
1 & \cos\left(\frac{2\pi}{5}\right) & \cos\left(\frac{4\pi}{5}\right) & \cos\left(\frac{6\pi}{5}\right) & \cos\left(\frac{8\pi}{5}\right) \\
0 & \sin\left(\frac{2\pi}{5}\right) & \sin\left(\frac{4\pi}{5}\right) & \sin\left(\frac{6\pi}{5}\right) & \sin\left(\frac{8\pi}{5}\right) \\
1 & \cos\left(\frac{4\pi}{5}\right) & \cos\left(\frac{8\pi}{5}\right) & \cos\left(\frac{2\pi}{5}\right) & \cos\left(\frac{6\pi}{5}\right) \\
0 & \sin\left(\frac{4\pi}{5}\right) & \sin\left(\frac{8\pi}{5}\right) & \sin\left(\frac{2\pi}{5}\right) & \sin\left(\frac{6\pi}{5}\right) \\
\sqrt{\frac{1}{2}} & \sqrt{\frac{1}{2}} & \sqrt{\frac{1}{2}} & \sqrt{\frac{1}{2}} & \sqrt{\frac{1}{2}}
\end{bmatrix}
\begin{bmatrix} x_a \\ x_b \\ x_c \\ x_d \\ x_e \end{bmatrix}
\tag{1}
$$

The Park conversion equation from a two-phase stationary coordinate system to a two-phase simultaneous rotational coordinate system can be written as [19]:

$$
\begin{bmatrix} x_d \\ x_q \end{bmatrix} =
\begin{bmatrix} \cos\theta & \sin\theta \\ -\sin\theta & \cos\theta \end{bmatrix}
\begin{bmatrix} x_\alpha \\ x_\beta \end{bmatrix}, \quad x = u, i, \psi
\tag{2}
$$

where x is voltage, current, or flux linkage. The fundamental space-stator current and rotor flux linkage in the two-phase stationary coordinate system are selected as state variables, and the corresponding state equations are as follows [20]:

$$
\begin{aligned}
\frac{di_{s\alpha 1}}{dt} &= \frac{L_r^2 R_s + L_m^2 R_r}{L_r(L_m^2 - L_s L_r)} i_{s\alpha 1} - \frac{L_m R_r}{L_r(L_m^2 - L_s L_r)}\psi_{r\alpha 1} - \frac{L_m \omega_r}{(L_m^2 - L_s L_r)}\psi_{r\beta 1} - \frac{L_r}{(L_m^2 - L_s L_r)}u_{s\alpha 1} \\
\frac{di_{s\beta 1}}{dt} &= \frac{L_r^2 R_s + L_m^2 R_r}{L_r(L_m^2 - L_s L_r)} i_{s\beta 1} + \frac{L_m \omega_r}{L_r(L_m^2 - L_s L_r)}\psi_{r\alpha 1} - \frac{L_m R_r}{(L_m^2 - L_s L_r)}\psi_{r\beta 1} - \frac{L_r}{(L_m^2 - L_s L_r)}u_{s\beta 1} \\
\frac{d\psi_{r\alpha 1}}{dt} &= \frac{L_m R_r}{L_r} i_{s\alpha 1} - \frac{R_r}{L_r}\psi_{r\alpha 1} - \omega_r \psi_{r\beta 1} \\
\frac{d\psi_{r\beta 1}}{dt} &= \frac{L_m R_r}{L_r} i_{s\beta 1} + \omega_r \psi_{r\alpha 1} - \frac{R_r}{L_r}\psi_{r\beta 1}
\end{aligned}
\tag{3}
$$

where L_m, L_s, and L_r are the equivalent mutual inductance, stator-side equivalent self-inductance, and rotor-side equivalent self-inductance, respectively, whose values are 2.5 times the mutual inductance, stator-side self-inductance, and rotor-side self-inductance in the fundamental space of the five-phase squirrel cage induction motor.

The dynamic state-space model of the fundamental space of the five-phase induction motor has four state variables. In order to realize vector control without a speed sensor, the rotor angular velocity is added as a state variable so that the expanded-order system equations of the five-phase induction motor are obtained. Since the Kalman filter algorithm is applicable to linear systems, the digitalization of the algorithm requires the discretization of the algorithm, while the state-space equations of the five-phase squirrel cage induction motor are nonlinear and continuous. Therefore, the state-space equations of the five-phase motor should be linearized and discretized. Consider the linear system shown in Figure 1, whose state and measurement equations can be written as:

$$
\begin{aligned}
\frac{dx(t)}{dt} &= f[x(t)] + Bu(t) + w(t) \\
y(t) &= h[x(t)] + v(t)
\end{aligned}
\tag{4}
$$

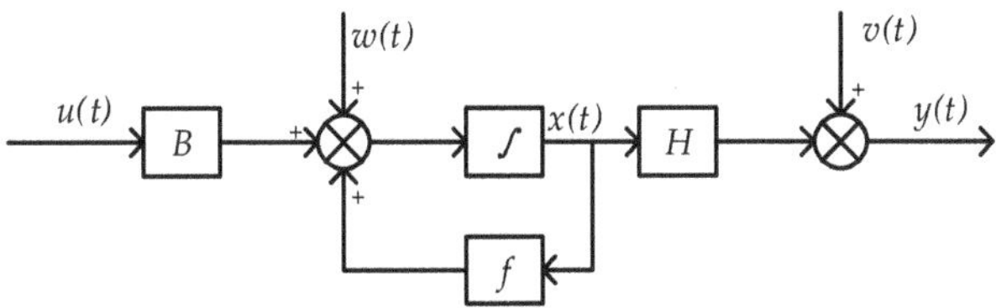

Figure 1. Linear system block diagram.

The Taylor expansion of Equation (4) at $\hat{x}(t)$, retaining the constant and primary terms and rounding off the higher terms, yields:

$$
\begin{aligned}
\frac{dx(t)}{dt} &= f[\hat{x}(t)] + \frac{\partial f[x(t)]}{\partial x(t)}\Big|_{x(t)=\hat{x}(t)}\Delta x + Bu(t) + w(t) \\
y(t) &= h[\hat{x}(t)] + \frac{\partial h[x(t)]}{\partial x(t)}\Big|_{x(t)=\hat{x}(t)}\Delta x + v(t)
\end{aligned}
\tag{5}
$$

where $\Delta x = x(t) - \hat{x}(t)$. $w(t)$ and $v(t)$ represent system noise and measurement noise, respectively, which are Gaussian white noise with the average expectation of 0.

The linearized state equation is discretized, and assuming that the sampling period T_s is sufficiently small, the following approximation can be made at the moment k:

$$
\frac{dx_k}{dt} \approx \frac{x_k - x_{k-1}}{T_s}
\tag{6}
$$

From Equation (4) to (6), the state equations of the discretized five-phase induction motor base-wave space linear system can be obtained as:

$$
\begin{cases}
x_k = x_{k-1} + [A_k x_{k-1} + Bu]T_s + w_{k-1} \\
y_k = Hx_k + v_k
\end{cases}
\tag{7}
$$

where:

$$
A = \begin{bmatrix} -\frac{R'}{\sigma} & 0 & \frac{L'}{T_r} & L'\omega_r & L'\psi_{r\beta 1} \\ 0 & -\frac{R'}{\sigma} & -L'\omega_r & \frac{L'}{T_r} & -L'\psi_{r\alpha 1} \\ \frac{L_m}{T_r} & 0 & -\frac{1}{T_r} & -\omega_r & -\psi_{r\beta 1} \\ 0 & \frac{L_m}{T_r} & \omega_r & -\frac{1}{T_r} & \psi_{r\alpha 1} \\ 0 & 0 & 0 & 0 & 0 \end{bmatrix}; \, B = \begin{bmatrix} \frac{1}{\sigma L_s} & 0 \\ 0 & \frac{1}{\sigma L_s} \\ 0 & 0 \\ 0 & 0 \\ 0 & 0 \end{bmatrix};
$$

$$
H = \begin{bmatrix} 1 & 0 & 0 & 0 & 0 \\ 0 & 1 & 0 & 0 & 0 \end{bmatrix};
$$

$$
x = \begin{bmatrix} i_{s\alpha 1} & i_{s\beta 1} & \psi_{r\alpha 1} & \psi_{r\beta 1} & \omega_r \end{bmatrix}^T ; y = \begin{bmatrix} i_{s\alpha 1} & i_{s\beta 1} \end{bmatrix}^T ; u = \begin{bmatrix} u_{s\alpha 1} & u_{s\beta 1} \end{bmatrix}^T
$$

$R' = \frac{L_r^2 R_s + L_m^2 R_r}{L_s L_r^2}; L' = \frac{L_m}{\sigma L_s L_r}; \sigma = 1 - \frac{L_m^2}{L_s L_r}$ is the magnetic flux leakage coefficient and $T_r = \frac{L_r}{R_r}$ is the rotor time constant. Define the system noise matrix $w(k)$ as a 5×1 order matrix, and the covariance matrix $Q = \mathrm{cov}(w) = E\{w, w^T\}$ as a 5×5 order matrix. The measurement noise $v(k)$ is a matrix of order 2×1, and its covariance matrix $R = \mathrm{cov}(v) = E\{v, v^T\}$ is a matrix of order 2×2.

3. Extended Kalman Filtering Algorithm

The application of the extended Kalman filter in the state estimation of the five-phase induction motor is shown in Figure 2. The red dotted box is the prediction part of the extended Kalman filter algorithm, and the blue dotted box is derived from the state-space equation of the motor. The function of the prediction part is to cause the error between the estimated value and the real value to be close to zero through a large number of calculations, so as to achieve the purpose of real-time tracking.

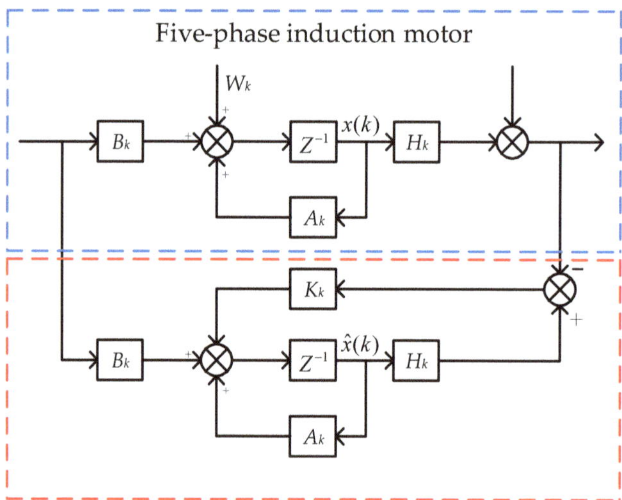

Figure 2. EKF observer.

Assuming that the state estimate $\hat{x}_{k-1}$ at the moment $k - 1$ is a known quantity, the calculation steps of the extended Kalman filter can be written as follows:

A priori prediction step: the state estimate at moment k is predicted based on the state estimate at moment $k - 1$, which is called a priori estimation and has the following expression:

$$
\begin{cases} \hat{x}_{k|k-1} = \hat{x}_{k-1|k-1} + \left(A_{k|k-1} + B_k u_k \right) T_s \\ \hat{y}_{k|k-1} = H_k \hat{x}_{k|k-1} \end{cases} \tag{8}
$$

Calculate the predicted covariance matrix:

$$P_{k|k-1} = F_{k|k-1}P_{k-1|k-1}F_{k|k-1}^T + Q \tag{9}$$

where:

$$F = I + T_s A = \begin{bmatrix} 1 - \frac{T_s R'}{\sigma} & 0 & \frac{T_s L'_s}{T_r} & T_s L'\omega_r & T_s L'\psi_{r\beta 1} \\ 0 & 1 - \frac{T_s R'}{\sigma} & -T_s L'\omega_r & \frac{T_s L'}{T_r} & -T_s L'\psi_{r\alpha 1} \\ \frac{T_s L_m}{T_r} & 0 & 1 - \frac{T_s}{T_r} & -T_s\omega_r & -T_s\psi_{r\beta 1} \\ 0 & \frac{T_s L_m}{T_r} & T_s\omega_r & 1 - \frac{T_s}{T_r} & T_s\psi_{r\alpha 1} \\ 0 & 0 & 0 & 0 & 1 \end{bmatrix} ;$$

Calculate the Kalman gain $K_{k|k-1}$:

$$K_{k|k-1} = P_{k|k-1}H^T \left(HP_{k|k-1}H^T + R \right)^{-1} \tag{10}$$

A posteriori correction step: The optimal solution obtained $\hat{x}_{k|k}$ from the gain $K_{k|k-1}$ calculated by Equation (6) is called the posterior value, and its expression is as follows:

$$\hat{x}_{k|k} = \hat{x}_{k|k-1} + K_{k|k-1}\left(y_k - \hat{y}_{k|k-1} \right) \tag{11}$$

Calculate the covariance matrix $P_{k|k}$ for the next moment:

$$P_{k|k} = \left(I - K_{k|k-1}H \right)P_{k|k-1} \tag{12}$$

If the fundamental magnetic chain, the third-harmonic magnetic chain, and the rotor angular velocity are to be observed simultaneously, the equation of state of the system as a whole is a matrix of order 9, and a general microcontroller or DSP cannot handle such a huge amount of data. In fact, after suppressing the current in the third-harmonic space, its electromagnetic torque generated in the third-harmonic space can be approximated as 0, and the effect on the rotor angular velocity can be negligible. Based on this, two EKF observers are used to observe the fundamental magnetic chain, the third-harmonic magnetic chain, and the rotor angular velocity. The first EKF observer is used to observe the fundamental magnetic chain and rotor angular velocity, and the second EKF observer is used to observe the third-harmonic magnetic chain. The predicted rotor angular velocity of the first EKF observer is used as the input to the second EKF observer, which is structured as follows.

As shown in Figure 3, EKF1 observes the rotor magnetic chain in the fundamental space and the rotor angular velocity, and EKF2 observes the rotor magnetic chain in the third-harmonic space. The equation of state in the third-harmonic space can be written as:

$$\begin{aligned} \frac{di_{s\alpha 3}}{dt} &= \frac{L_{r3}^2 R_s + L_{m3}^2 R_{r3}}{L_{r3}(L_{m3}^2 - L_{s3}L_{r3})}i_{s\alpha} - \frac{L_{m3}R_{r3}}{L_{r3}(L_{m3}^2 - L_{s3}L_{r3})}\psi_{r\alpha} - \frac{3L_{m3}\omega_r}{(L_{m3}^2 - L_{s3}L_{r3})}\psi_{r\beta} - \frac{L_{r3}}{(L_{m3}^2 - L_{s3}L_{r3})}u_{s\alpha 3} \\ \frac{di_{s\beta 3}}{dt} &= \frac{L_{r3}^2 R_s + L_{m3}^2 R_{r3}}{L_{r3}(L_{m3}^2 - L_{s3}L_{r3})}i_{s\beta 3} + \frac{3L_{m3}\omega_r}{L_{r3}(L_{m3}^2 - L_{s3}L_{r3})}\psi_{r\alpha 3} - \frac{L_{m3}R_{r3}}{(L_{m3}^2 - L_{s3}L_{r3})}\psi_{r\beta} - \frac{L_{r3}}{(L_{m3}^2 - L_{s3}L_{r3})}u_{s\beta 3} \\ \frac{d\psi_{r\alpha 3}}{dt} &= \frac{L_{m3}R_{r3}}{L_{r3}}i_{s\alpha 3} - \frac{R_{r3}}{L_{r3}}\psi_{r\alpha 3} - 3\omega_r\psi_{r\beta 3} \\ \frac{d\psi_{r\beta 3}}{dt} &= \frac{L_{m3}R_{r3}}{L_{r3}}i_{s\beta 3} + 3\omega_r\psi_{r\alpha 3} - \frac{R_{r3}}{L_{r3}}\psi_{r\beta 3} \end{aligned} \tag{13}$$

where $\hat{\psi}_{r\alpha 1}$ and $\hat{\psi}_{r\beta 1}$ are the predicted flux linkage in the fundamental space in the two-phase stationary coordinate system. $\hat{\omega}_r$ is the predicted angular velocity. $\hat{\psi}_{r\alpha 3}$ and $\hat{\psi}_{r\beta 3}$ are the predicted flux linkage of the third-harmonic space in the two-phase stationary coordinate system.

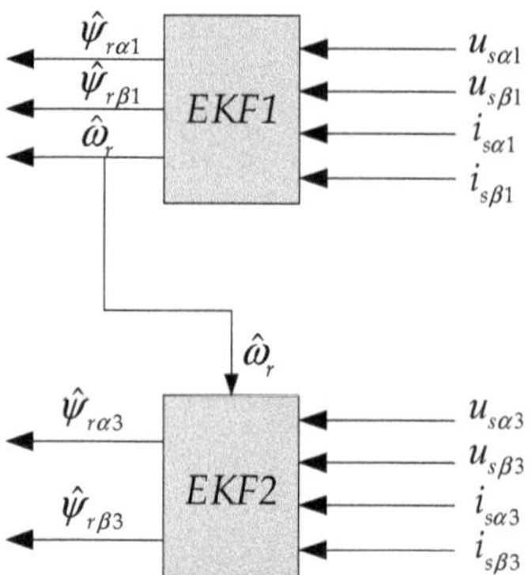

Figure 3. Double EKF observer.

Similarly, the equation of state in the third-harmonic space after linearizing and discretizing Equation (13) can be written as:

$$\begin{cases} x_k{}' = x_{k-1}{}' + [A_k{}'x_{k-1}{}' + B'u']T_s + w_{k-1}{}' \\ y_k{}' = H'x_k{}' + v_k{}' \end{cases} \tag{14}$$

where:

$$A' = \begin{bmatrix} -\frac{R_3{}'}{\sigma_3} & 0 & \frac{3L_3{}'}{T_{r3}} & L_3{}'\omega_r \\ 0 & -\frac{R_3{}'}{\sigma_3} & -L_3{}'\omega_r & \frac{3L_3{}'}{T_{r3}} \\ \frac{L_{m3}}{T_{r3}} & 0 & -\frac{1}{T_{r3}} & -3\omega_r \\ 0 & \frac{L_{m3}}{T_{r3}} & 3\omega_r & -\frac{1}{T_{r3}} \end{bmatrix}; B' = \begin{bmatrix} \frac{1}{\sigma_3 L_{s3}} & 0 \\ 0 & \frac{1}{\sigma_3 L_{s3}} \\ 0 & 0 \\ 0 & 0 \end{bmatrix}; H' = \begin{bmatrix} 1 & 0 & 0 & 0 \\ 0 & 1 & 0 & 0 \end{bmatrix};$$

$$x' = \begin{bmatrix} i_{s\alpha3} & i_{s\beta3} & \psi_{r\alpha3} & \psi_{r\beta3} \end{bmatrix}^T; y' = \begin{bmatrix} i_{s\alpha3} & i_{s\beta3} \end{bmatrix}^T; u' = \begin{bmatrix} u_{s\alpha3} & u_{s\beta3} \end{bmatrix}^T;$$

$R_3{}' = \frac{L_{r3}^2 R_s + L_{m3}^2 R_{r3}}{L_{s3} L_{r3}^2}; L_3{}' = \frac{L_{m3}}{\sigma_3 L_{s3} L_{r3}}; \sigma_3 = 1 - \frac{L_{m3}^2}{L_{s3} L_{r3}}$ is the magnetic flux leakage coefficient of the third-harmonic space, and $T_{r3} = \frac{L_{r3}}{R_{r3}}$ is the torque constant of the third-harmonic space. Define the system noise matrix $w(k)'$ as a 4×4 order matrix and the covariance matrix $Q' = \text{cov}(w') = E\{w', w'^T\}$ as a 4×1 order matrix. The measurement noise $v(k)'$ is a matrix of order 2×1 and its covariance matrix $R' = \text{cov}(v') = E\{v', v'^T\}$ is a matrix of order 2×2.

The initial parameter settings of the EKF observer are critical, especially the initial values of the system noise matrix Q, the measurement noise covariance matrix R, and the prediction covariance matrix P. The selection of these initial parameters directly determines the overall performance of the algorithm, and the improper selection of initial values can lead to scattering of the whole system.

For the system noise covariance matrix Q, it mainly includes the system external disturbances, motor parameter variation effects, and errors in the linearization discretization process. If Q becomes larger, it means that the system noise becomes stronger, indicating that the weighting effect of the measurement feedback is enhanced, and the EKF transient response becomes faster [21].

For the measurement noise covariance matrix R, it mainly includes the actual sensor measurement error, microcontroller sampling error, and other factors. If R is increased, it corresponds to a larger deviation in the current measurement, weakening the weight of the algorithm's predicted value, which will lead to a slower transient response [21].

For the error covariance matrix P, its initial state is generally chosen to be a diagonal array with all elements equal, which has a large effect on the convergence rate of the EKF and the amplitude of the transient state, with little effect on the steady state [22].

The selection of these matrices is related to the parameters of the motor body and is generally obtained using a trial-and-error approach. After a large number of trial-and-error values, the initial values were taken as follows:

$$Q = diag[0.5\,,\,0.5\,,\,5 \times 10^{-5}\,,\,5 \times 10^{-5}\,,\,5 \times 10^{-3}]$$
$$R = diag[0.05\,,\,0.05]$$
$$P = diag[1\,,\,1\,,\,1\,,\,1\,,\,1]$$
$$Q' = diag[0.5\,,\,0.5\,,\,5 \times 10^{-5}\,,\,5 \times 10^{-5}]$$
$$R' = diag[0.05\,,\,0.05]$$
$$P' = diag[1\,,\,1\,,\,1\,,\,1]$$

In the EKF speed sensorless control system, SVPWM modulation is used [23].The control strategy is IRFOC (indirect field-oriented control) [24,25]. The five-phase inverter is an H-bridge structure, and the overall flow chart is shown in Figure 4.

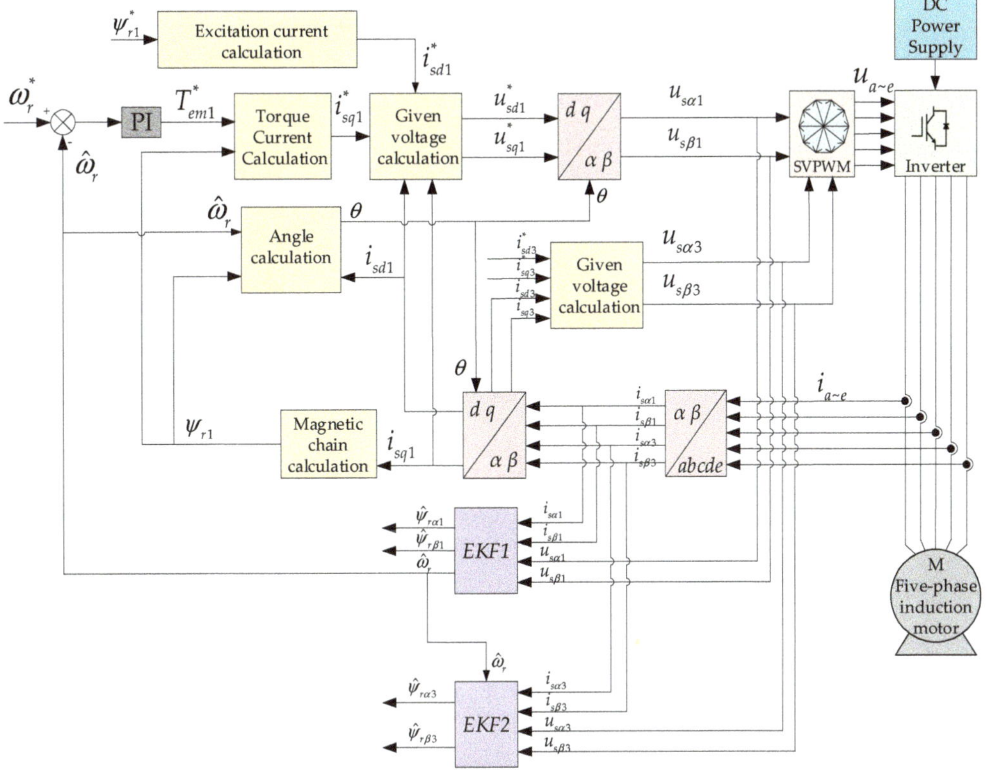

Figure 4. EKF speed sensorless control flow chart.

The current voltage conversion module includes two PI controllers, which use the difference between the measured current and the reference current. The PI controller estimates the voltage necessary to minimize this current difference. For the fundamental space, the current given signals i^*_{sd1} and i^*_{sq1} are calculated from the given torque signal T^*_{em1} and the given flux signal ψ^*_{r1}, respectively. The given torque signal is derived from the predicted speed and the error signal of the given speed through a PI controller. For the third-harmonic space, the given current signals i^*_{sd3} and i^*_{sq3} are set to 0 to achieve the effect of restraining the third-harmonic current.

4. Results and Discussion

In order to verify the effectiveness of the EKF algorithm in the control of a five-phase motor without a speed sensor, a simulation model of it was built in Matlab/Simulink in this study, and the EKF speed sensorless vector control system shown in Figure 4 was established. Among them, the motor parameters used were as seen in Table 1, and its modeling in Simulink is shown in Figure 5.

Table 1. Motor parameters.

Parameter	Symbol	Value
Stator resistance	R_s	0.95 Ω
Fundamental space rotor resistance	R_{r1}	0.78 Ω
Third-harmonic space rotor resistance	R_{r3}	0.52 Ω
Fundamental space spatial mutual inductance	L_{m1}	99.35 mH
Third-harmonic space mutual inductance	L_{m3}	11.04 mH
Fundamental space-stator leakage inductance	L_{sloss1}	6.87 mH
Fundamental space rotor leakage inductance	L_{rloss1}	4.04 mH
Third-harmonic space stator leakage inductance	L_{sloss3}	3.86 mH
Third-harmonic space rotor leakage inductance	L_{rloss3}	3.76 mH
Magnetic pole pairs	n_p	2
Rotational inertia	J	0.056 kg·m^2
Rated speed	n_N	1410 rpm
Rated voltage	U_N	380 V
Rated frequency	f_N	50 Hz
Rated field current	I_{sm}	3.86 A

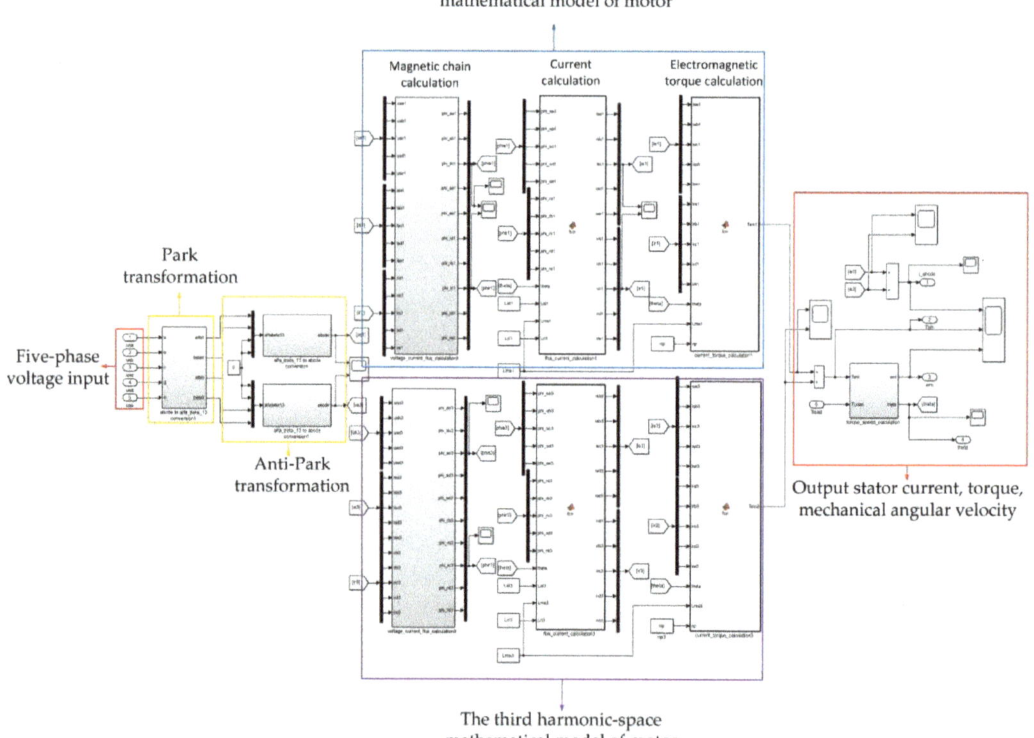

Figure 5. The simulation model of the five-phase induction motor.

The model is mainly divided into five modules: coordinate transformation modules, voltage–magnetic chain modules, magnetic chain–current modules, current–electromagnetic torque modules, and electromagnetic torque-speed modules. The coordinate transformation module is based on the five-phase Clark transformation matrix and its inverse transform matrix. The five-phase voltage is decoupled to the fundamental space and three-harmonic space of the αβ coordinate system, and then the transformation is converted them into the ABCDE five-phase voltage. In the voltage–magnetic chain module, the magnetic chain is calculated using the input voltage, current, and motor resistance. The magnetic chain–current module uses the inductive matrix and each phase magnetic chain to obtain the current. The torque–speed module mainly calculates motor rotor rotation speed.

4.1. Speed Prediction under Different Working Conditions

In order to verify whether the EKF algorithm can replace the traditional speed sensor, first, the feedback value of the speed loop was changed to the real speed of the motor. Here, EKF was used as an observer. The whole system adopted indirect field-oriented control structure. At that time, the comparison between the predicted speed of the EKF algorithm and the speed of indirect field vector control with a speed sensor is shown in Figure 6.

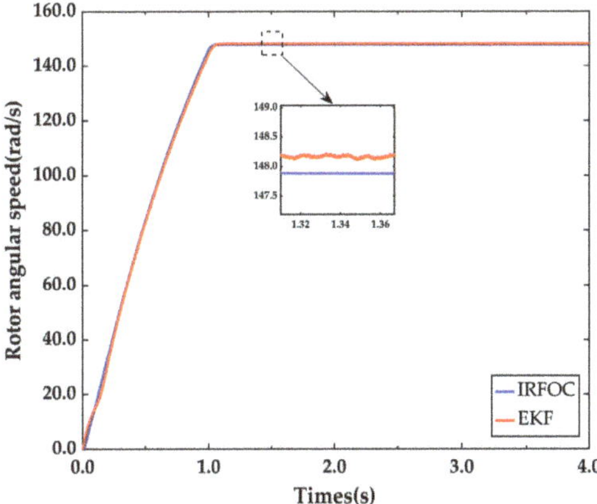

Figure 6. The performance of the EKF observer.

As shown in Figure 6, different from the vector control, the EKF algorithm introduced an error matrix in the iterative operation process, so the pulsation of speed waveform given by the EKF algorithm was larger than the real speed of IRFOC. Under the rated working condition, the EKF algorithm was accurate in observing the change in the rotating speed. Therefore, the EKF algorithm can replace the traditional speed sensor. Based on this premise, the structure of Figure 4 was adopted in the subsequent simulation, and the speed predicted by EKF was directly substituted into the speed closed-loop.

To verify the speed tracking performance of the algorithm, four working conditions shown in Table 2 were selected for simulation:

Table 2. Different working conditions.

Working Condition	Rotor Angular Speed Setting	Load Torque Setting
1	0~4 s:100 rad/s	0~4 s:0 N·m
2	0~4 s:100 rad/s	0~4 s:6 N·m
3	0~1.5 s:60 rad/s1.5~3 s:90 rad/s3~4 s:30 rad/s	0~4 s:0 N·m
4	0~1.5 s:100 rad/s1.5~4 s:−100 rad/s	0~4 s:0 N·m

The four working conditions in Table 2 were used to test the speed prediction performance of the EKF algorithm under no-load, on load, acceleration and deceleration, and forward and reverse rotation. It can be seen from Figure 7 that the EKF algorithm had a good speed prediction performance under these four working conditions. It can be seen from Figure 7a that, when the speed reached the given speed, there was a steady-state error of about 0.5 rad/s between the EKF algorithm and the actual speed. It can be seen from Figure 7b–d that, when the given speed changed suddenly or the load was added, a small error occurred between the predicted speed and the real speed, and then the predicted speed quickly converged to the actual speed. This showed that the overall robustness of the system was good, and it can be applied to the conditions requiring a wide range of speed regulation.

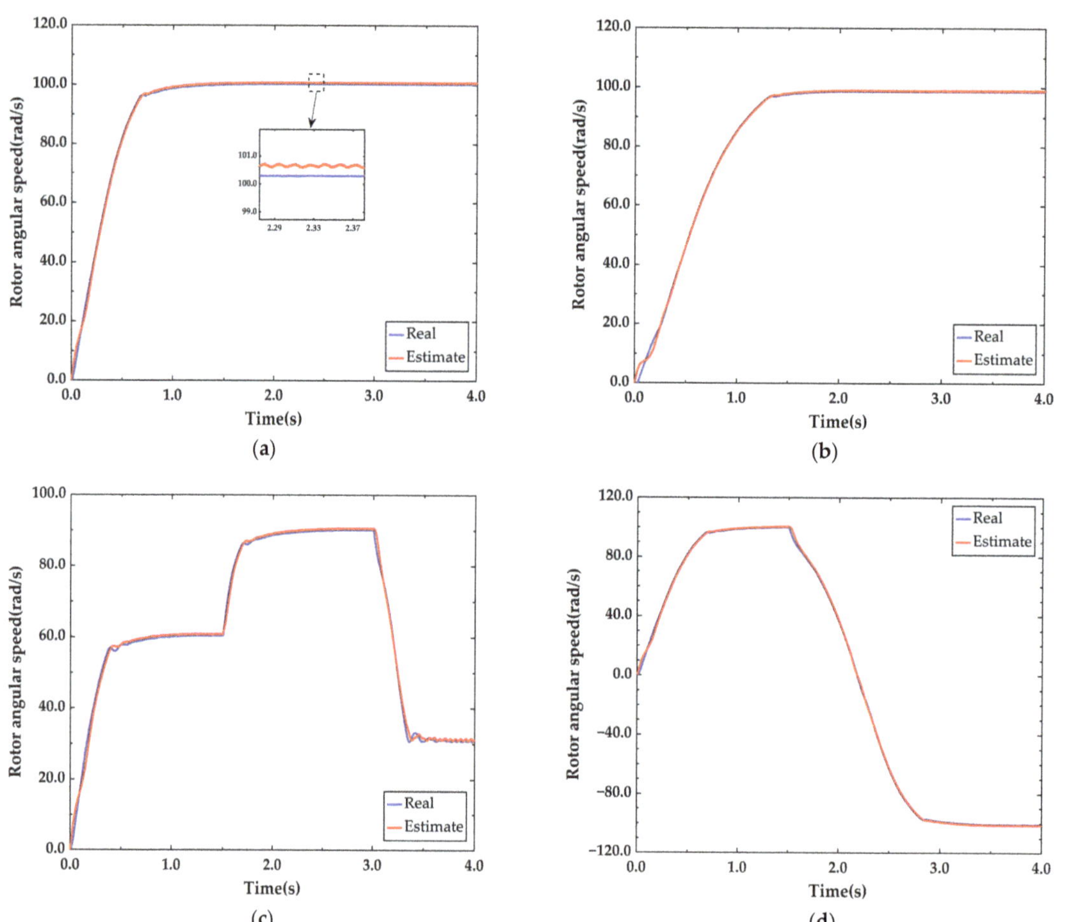

Figure 7. Speed prediction under different working conditions. (**a**) Condition 1; (**b**) Condition 2; (**c**) Condition 3; (**d**) Condition 4.

4.2. Rotor Flux Identifications

The rated field current of the motor used in this simulation was 3.86 A, so the rated fundamental space given flux linkage can be calculated as 0.96 Wb from Equation (15). In order to verify the flux linkage prediction ability of the EKF algorithm, the simulation was conducted by changing the field current to 50%, 64%, 84%, and 100% of the rated field current under condition 1 in Table 2.

$$\psi_{ref} = 2.5 I_{sm} L_{m1} \tag{15}$$

Similarly, according to formula (15), when the field current was 50%, 64%, and 84% of the rated value, the corresponding fundamental space given flux linkage was 0.48 Wb, 0.61 Wb, and 0.81 Wb. For the third-harmonic spatial flux linkage, its given value is always 0, so its predicted flux linkage was also about 0.

It can be seen from Figure 8 that the EKF algorithm could accurately predict the change in rotor flux linkage in the fundamental space in a steady state. Table 3 lists the predicted values and error values in four cases. It can be seen that the prediction error of flux linkage accounted for about 10% of the given value. In addition, the estimated value of the flux linkage fluctuated slightly in the steady state. This is because the rotor position angle in the IRFOC method estimated the position of the magnetic flux relative to the stator by integrating the slip frequency and the actual rotor angular speed. In the simulation experiment, the predicted rotor angular velocity was used instead of the actual rotor angular velocity. However, there was a small error between the predicted speed and the actual speed. Therefore, the calculated rotor position angular velocity also deviated from the reality, resulting in a small pulsation of the flux linkage in the steady state.

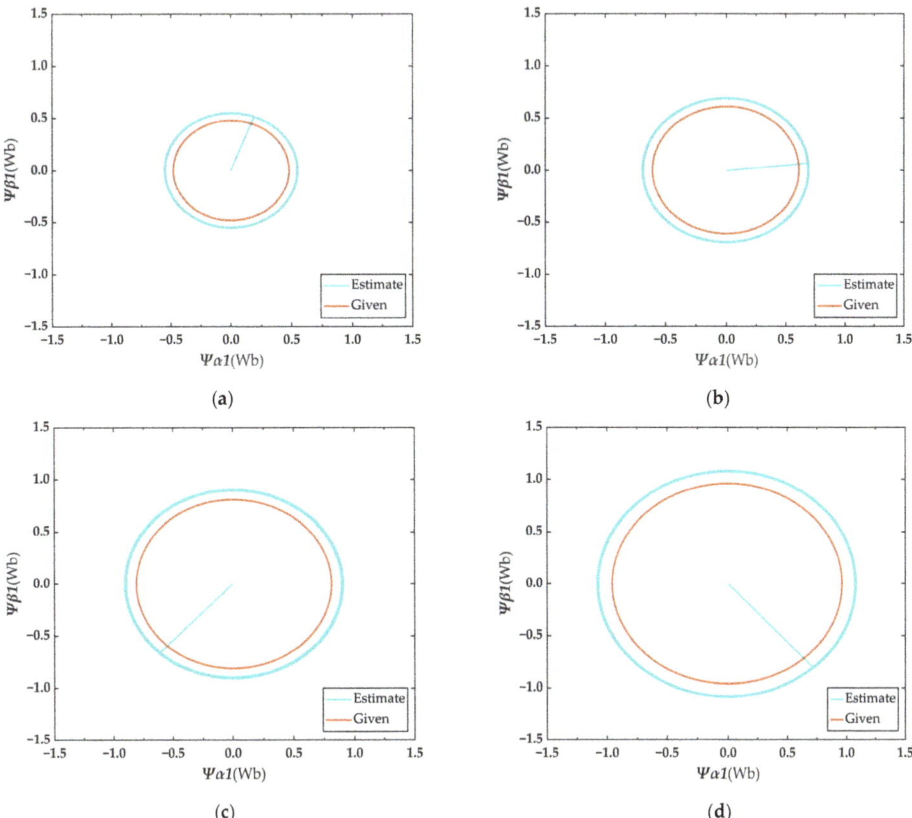

Figure 8. The prediction of fundamental space rotor flux linkage. (**a**) 50% of rated field current; (**b**) 64% of rated field current; (**c**) 84% of rated field current; (**d**) rated field current.

Table 3. Prediction error of fundamental space rotor flux linkage.

Working Condition (Flux Linkage Amplitude)	Predicted Flux Linkage Value	Error Value
50% of rated field current (0.48 Wb)	0.54 Wb	0.06 Wb
64% of rated field current (0.61 Wb)	0.68 Wb	0.07 Wb
84% of rated field current (0.81 Wb)	0.89 Wb	0.08 Wb
Rated field current (0.96 Wb)	1.06 Wb	0.10 Wb

It can be seen from Figure 9 that the predicted flux linkage of the third-harmonic in four cases was about 0, which also indicated that the third-harmonic current was effectively suppressed when the field current was changed.

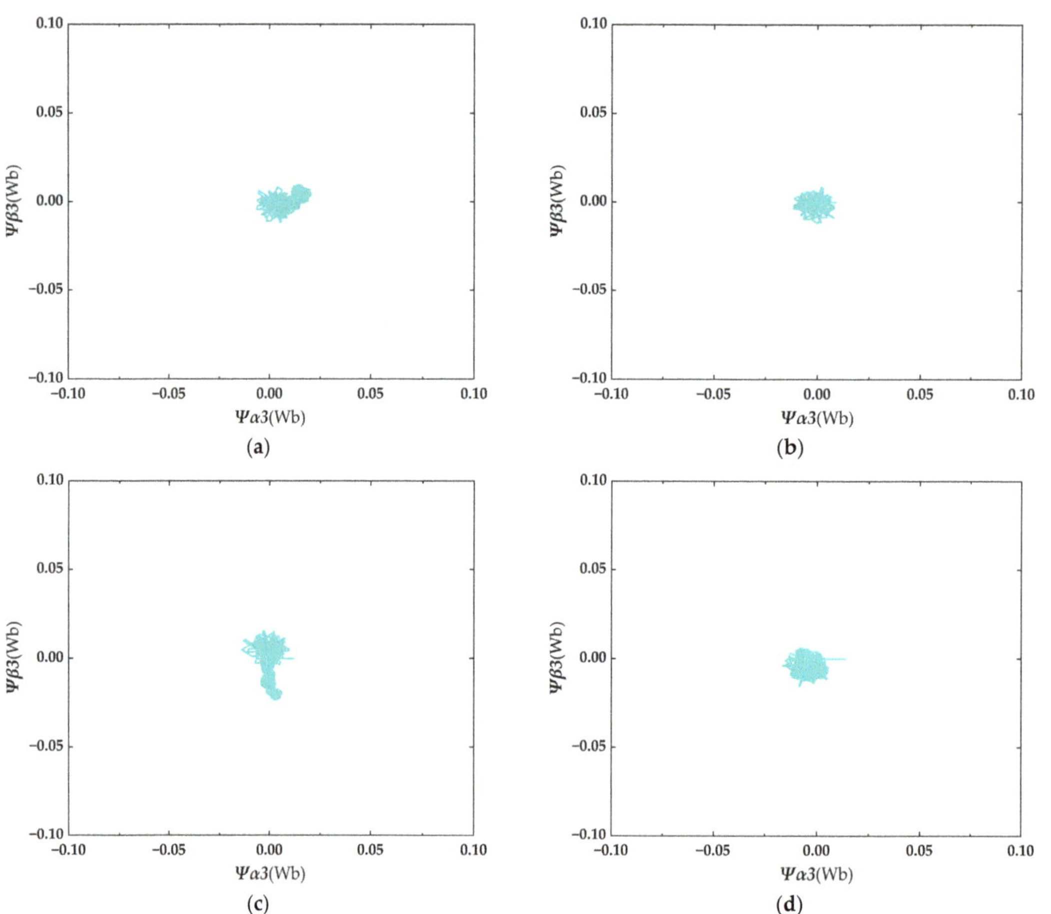

Figure 9. The prediction of third-harmonic space rotor flux linkage. (**a**) 50% of rated field current; (**b**) 64% of rated field current; (**c**) 84% of rated field current; (**d**) rated field current.

4.3. Third-Harmonic Current Suppression Effect

Figures 10 and 11 show the stator current waveform and the stator third-harmonic current waveform under condition 1 in Table 2. It can be seen from Figure 10 that the overall sinusoidal degree of the stator current was good, and its FFT analysis is shown in Figure 12. According to Figure 12, when the given speed was 100 rad/s, the corresponding stator current frequency was 31.83 Hz, and the THD of the stator current was 8.58%, indicating that the third-harmonic current had been effectively suppressed. It can be seen from Figure 10 that most of the third-harmonic currents were distributed below the amplitude of 0.6 A in the steady state.

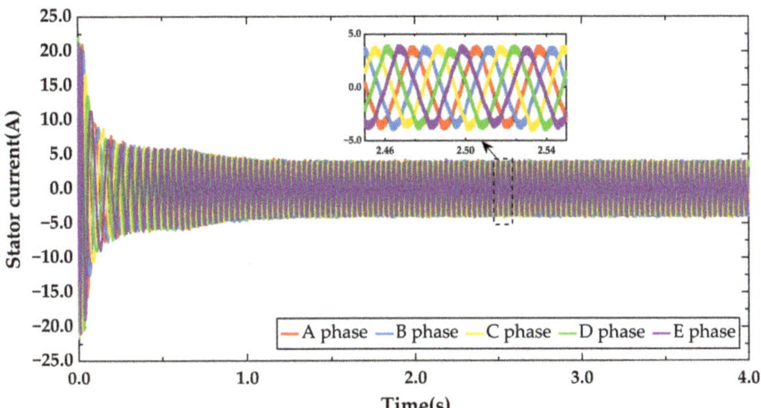

Figure 10. Stator current.

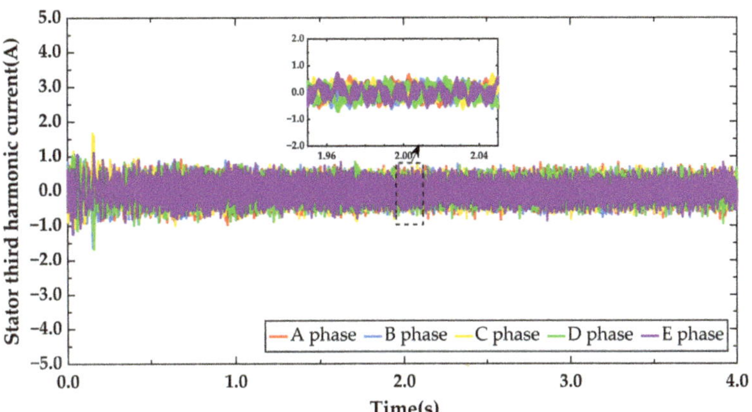

Figure 11. Stator current (third-harmonic space).

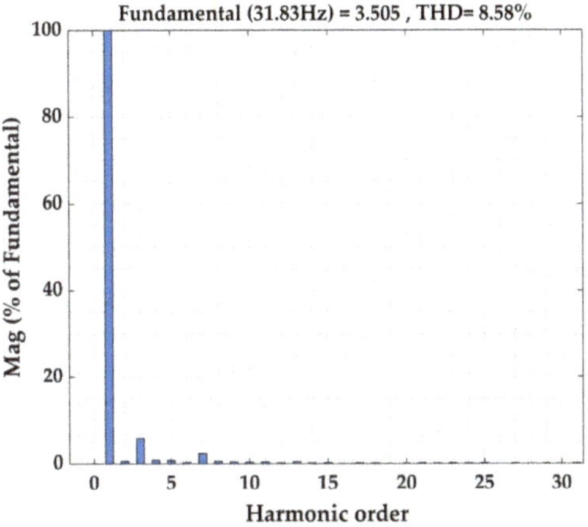

Figure 12. FFT analysis of stator current.

5. Conclusions

In this study, the EKF algorithm was extended to the parameter identification of the five-phase squirrel cage induction motor, and the theoretical part was derived in detail. On the premise of restraining the third-harmonic current of the motor, a new double EKF structure was proposed, which simplified the state equation of the system. The simulation results showed that the EKF algorithm was accurate in predicting the rotor angular speed, and the algorithm was suitable for the sensorless control of the five-phase squirrel cage induction motor, which needs a wide range of speed regulation. The EKF algorithm could observe the rotor flux linkage, but there was an error between the predicted value and the given value, which was about 10% of the given value. In view of the shortcomings of this paper, the follow-up research direction could be to reduce the order of the state equation of the whole system without the third-harmonic current suppression.

Author Contributions: Data curation, X.L. and J.Z.; Methodology, X.L.; Resources, H.C. and S.Z.; Software, H.X.; Writing—original draft, X.L.; Writing—review and editing, Y.X. All authors have read and agreed to the published version of the manuscript.

Funding: This research received no external funding.

Institutional Review Board Statement: Not applicable.

Informed Consent Statement: Not applicable.

Data Availability Statement: Not applicable.

Acknowledgments: Thanks to J.Z. for providing guidance on the data analysis of this article, Y.X., H.X., H.C. and S.Z. for revising and sorting this article.

Conflicts of Interest: The authors declare no conflict of interest.

References

1. Benbouhenni, H.; Bizon, N. A Synergetic Sliding Mode Controller Applied to Direct Field-Oriented Control of Induction Generator-Based Variable Speed Dual-Rotor Wind Turbines. *Energies* **2021**, *14*, 4437. [CrossRef]
2. De Pancorbo, S.M.; Ugalde, G.; Poza, J.; Egea, A. Comparative Study between Induction Motor and Synchronous Reluctance Motor for Electrical Railway Traction Applications. In Proceedings of the 5th International Electric Drives Production Conference (EDPC), Nuremberg, Germany, 15–16 September 2015.
3. Gnacinski, P.; Tarasiuk, T.; Mindykowski, J.; Peplinski, M.; Gorniak, M.; Hallmann, D.; Pillat, A. Power Quality and Energy-Efficient Operation of Marine Induction Motors. *IEEE Access* **2020**, *8*, 152193–152203. [CrossRef]
4. Yang, Z.; Shang, F.; Brown, I.P.; Krishnamurthy, M. Comparative Study of Interior Permanent Magnet, Induction, and Switched Reluctance Motor Drives for EV and HEV Applications. *IEEE Trans. Transp. Electrif.* **2015**, *1*, 245–254. [CrossRef]
5. Pandey, R.; Panda, A.K.; Patnaik, N. Comparative Performance Analysis of DTC fed Three-Phase and Five-Phase Induction Motor. In Proceedings of the 5th International Conference on Devices, Circuits and Systems (ICDCS), Coimbatore, India, 5–6 March 2020; pp. 162–166.
6. Sarwer, Z.; Sartaj, M.; Khan, M.R.; Zaid, M.; Shahajhani, U. Comparative Performance Study of Five-Phase Induction Motor. In Proceedings of the IEEE International Conference on Innovations in Power and Advanced Computing Technologies, Vellore, India, 22–23 March 2019.
7. Holtz, J. Sensorless control of induction motor drives. *Proc. IEEE* **2002**, *90*, 1359–1394. [CrossRef]
8. Piippo, A.; Salomaki, J.; Luomi, J. Signal injection in sensorless PMSM drives equipped with inverter output filter. *IEEE Trans. Ind. Appl.* **2008**, *44*, 1614–1620. [CrossRef]
9. Shivaramakrishna, K.V.; Chauhan, A.K.; Raghuram, M.; Singh, S.K. Sensorless Control of Induction Motor using EKF:Analysis of Parameter Variation on EKF Performance. In Proceedings of the IEEE International Conference on Power Electronics, Drives, and Energy Systems (PEDES), Trivandrum, India, 14–17 December 2016.
10. Accetta, A.; Cirrincione, M.; Pucci, M.; Vitale, G. Neural Sensorless Control of Linear Induction Motors by a Full-Order Luenberger Observer Considering the End Effects. *IEEE Trans. Ind. Appl.* **2014**, *50*, 1891–1904. [CrossRef]
11. Comanescu, M. Design and Implementation of a Highly Robust Sensorless Sliding Mode Observer for the Flux Magnitude of the Induction Motor. *IEEE Trans. Energy Convers.* **2016**, *31*, 656–664. [CrossRef]
12. Korzonek, M.; Tarchala, G.; Orlowska-Kowalska, T. A review on MRAS-type speed estimators for reliable and efficient induction motor drives. *ISA Trans.* **2019**, *93*, 1–13. [CrossRef] [PubMed]
13. Salim, R.; Mansouri, A.; Bendiabdellah, A.; Chekroun, S.; Touam, M. Sensorless passivity based control for induction motor via an adaptive observer. *ISA Trans.* **2019**, *84*, 118–127. [CrossRef] [PubMed]

14. Barut, M.; Bogosyan, S.; Gokasan, M. Speed-sensorless estimation for induction motors using extended Kalman filters. *IEEE Trans. Ind. Electron.* **2007**, *54*, 272–280. [CrossRef]
15. Shi, K.L.; Chan, T.F.; Wong, Y.K.; Ho, S.L. Speed estimation of an induction motor drive using an optimized extended Kalman filter. *IEEE Trans. Ind. Electron.* **2002**, *49*, 124–133. [CrossRef]
16. Feng, Y.X.; Liao, Y.; Zhang, X.K. A Third Harmonic Current Elimination Strategy for Symmetrical Six-Phase Permanent Magnet Synchronous Motor. *IEEE Access* **2021**, *9*, 167570–167579. [CrossRef]
17. Bounasla, N.; Barkat, S.; Benyoussef, E.; Tounsi, K. Sensorless Sliding Mode Control of a Five-Phase PMSM using Extended Kalman Filter. In Proceedings of the 8th International Conference on Modelling, Identification and Control (ICMIC), Algiers, Algeria, 15–17 November 2016; pp. 97–102.
18. Parsa, L.; Toliyat, H.A. Five-phase permanent-magnet motor drives. *IEEE Trans. Ind. Appl.* **2005**, *41*, 30–37. [CrossRef]
19. González, O.; Rodas, J.; Gregor, R.; Ayala, M.; Rivera, M. Speed sensorless predictive current control of a five-phase induction machine. In Proceedings of the 2017 12th IEEE Conference on Industrial Electronics and Applications (ICIEA), Siem Reap, Cambodia, 18–20 June 2017; pp. 343–348.
20. Wen-Jieh, W.; Chun-Chieh, W. A rotor-flux-observer-based composite adaptive speed controller for an induction motor. *IEEE Trans. Energy Convers.* **1997**, *12*, 323–329. [CrossRef]
21. Bolognani, S.; Tubiana, L.; Zigliotto, M. Extended Kalman filter tuning in sensorless PMSM drives. *IEEE Trans. Ind. Appl.* **2003**, *39*, 1741–1747. [CrossRef]
22. Janiszewski, D. Extended kalman filter based speed sensorless PMSM control with load reconstruction. In Proceedings of the 32nd Annual Conference of the IEEE-Industrial-Electronics-Society, Paris, France, 7–10 November 2006; pp. 4186–4189.
23. Zheng, L.; Fletcher, J.E.; Williams, B.W.; He, X. Dual-Plane Vector Control of a Five-Phase Induction Machine for an Improved Flux Pattern. *IEEE Trans. Ind. Electron.* **2008**, *55*, 1996–2005. [CrossRef]
24. Abbasi, H.; Ghanbari, M.; Ebrahimi, R.; Jannati, M. IRFOC of Induction Motor Drives Under Open-Phase Fault Using Balanced and Unbalanced Transformation Matrices. *IEEE Trans. Ind. Electron.* **2021**, *68*, 9160–9173. [CrossRef]
25. Consoli, A.; Scarcella, G.; Testa, A. Slip frequency detection for indirect field oriented control drives. *IEEE Trans. Ind. Appl.* **2004**, *40*, 194–201. [CrossRef]

Article

An Improved Model for Five-Phase Induction Motor Based on Magnetic Noise Reduction Part I: Slot Opening Width

Hansi Chen, Jinghong Zhao, Yiyong Xiong *, Xiangyu Luo and Qingfei Zhang

School of Electrical Engineering, Naval University of Engineering, Jiefang Road 717, Wuhan 430033, China; chs8000@163.com (H.C.); zhaojinghong@163.com (J.Z.); luoxiangyu0125@163.com (X.L.); zhangqf_1997@163.com (Q.Z.)
* Correspondence: xiongyiyong1989@163.com

Abstract: Based on the winding function considering the slot width and the air-gap permeability considering the slot opening width, the main radial electromagnetic force wave expressions of the induction motor are determined. The electromagnetic force-vibration prediction model of the induction motor is established. The natural frequency and acoustic radiation model of a finite-length cylindrical shell with two ends clamped is deduced. On this basis, an improved magnetic noise prediction model of cage induction motor is improved, which can calculate the combined effects of electromagnetic force on the axis and circumferential modes of the stator system. Aiming at two different noise reduction targets, an optimization method is proposed to reduce the overall electromagnetic noise of the motor without sacrificing efficiency and output torque. The feasibility of the model for electromagnetic noise prediction is verified by finite element simulation and experiments. For a 30/26 slots five-phase induction motor, low-noise analysis and optimization schemes of the opening width for two different targets are given. The results show that the larger slot opening scheme can also result in less magnetic noise for the right selection, which is contrary to the common design rule that recommends minimizing slot opening to reduce magnetic noise.

Keywords: five-phase induction motor; magnetic noise; slot optimization; electromagnetic vibration

Citation: Chen, H.; Zhao, J.; Xiong, Y.; Luo, X.; Zhang, Q. An Improved Model for Five-Phase Induction Motor Based on Magnetic Noise Reduction Part I: Slot Opening Width. *Processes* **2022**, *10*, 1496. https://doi.org/10.3390/pr10081496

Academic Editors: Haoming Liu, Jingrui Zhang and Jian Wang

Received: 20 June 2022
Accepted: 26 July 2022
Published: 29 July 2022

Publisher's Note: MDPI stays neutral with regard to jurisdictional claims in published maps and institutional affiliations.

1. Introduction

In the environment of prominent energy problems, new energy vehicles ushered in unprecedented broad prospects. In addition to power density and fault-tolerant performance, the +electromagnetic vibration and noise of electric vehicles, which are closely related to NVH characteristics, are also attracting more and more attention. The same as three-phase motors, the active noise reduction technology [1–4] of harmonic current injection into multiphase motors can only eliminate the electromagnetic force space harmonics of order 0 and $2p$. However, the stator elliptic mode is usually the mode with the most vibration noise radiation, so this technology cannot be applied to four-pole or six-pole motors. Other active noise suppression techniques often require more complex closed-loop control and electromechanical equipment [5], so it is necessary to find some low-noise rules that can be applied at the motor design stage. Rotor closing groove, increasing air gap length and rotor chute are commonly used in low noise design methods, but the side effect is to significantly reduce the performance of the motor.

Literature [6] mainly aimed at the tooth harmonic content of the three-phase induction motor electromagnetic wave [7,8], i.e., the interaction of air-gap permeance harmonics and fundamental magnetomotive force (MMF). The slot opening width of the optimization design principles are given, but due to the different phase number, the air-gap flux density characteristics and the electromagnetic harmonic content of the five-phase squirrel cage induction motor (FSCIM) are also different from that of the three-phase motor. Therefore, the electromagnetic vibration characteristics of the five-phase motor needs to be studied separately.

Aimed at reducing the magnetic noise of the FSCIM, an improved optimization method of low electromagnetic noise is proposed, which does not weaken the electromagnetic torque at the cost of changing the stator and rotor slot openings. Firstly, the five-phase winding function is derived from the superposition of a single conductor or a single coil, which includes the slot width. Then, combined with the stator and rotor current calculated by the single-phase harmonic expansion circuit [9–11] and the air-gap permeance function considering the slot openings, the radial air-gap flux density distribution is obtained. Furthermore, an improved analytical model of electromagnetic force, vibration and noise radiation was established to further consider the joint effect of electromagnetic force on the axial and radial modes of the stator system, and the feasibility of the model was verified by finite element simulation and noise measurement experiments. Next, on the basis of [6], the qualitative analysis of the rotor slot width and noise suppression basis was conducted. Finally, the slot opening width was optimized for the two kinds of integrated noise optimization objectives. Taking the five-phase cage induction motor with stator and rotor 30/26 slots as an example, the recommended schemes of low-noise slot opening design are given.

2. Analytical Model of Electromagnetic Vibration

2.1. Electromagnetic Force Basis

The electromagnetic force of the motor mainly includes magnetostriction and Maxwell force. Magnetostriction generally does not cause high-frequency magnetic noise [12,13], so that the impact on noise radiation is negligible. Maxwell force is composed of radial and tangential components, but since the radial component is usually an order of magnitude larger than the tangential component, and the tangential component mainly causes resonance in asymmetric windings and single-phase motors. In a five-phase induction motor with normal windings, electromagnetic noise due to radial components of Maxwell stress can be considered only [9,14]. The distribution of the exciting force in the air gap is mainly expressed in the form of electromagnetic force waves per unit area [15–17].

$$P_{eR}(t,\theta_s) = B_r^2(t,\theta_s)/2\mu_0 \tag{1}$$

where B_r is the radial component of the air-gap flux density; μ_0 is the air permeability; θ_s is the mechanical angular position on the circumference. The two-dimensional Fourier decomposition of the electromagnetic force density can be expressed as a superposition of a series of harmonics of the frequencies f_n and the m-th order in space, where m also represents the number of pole pairs of the harmonic:

$$P_{eR}(t,\theta_s) = \sum_n \sum_m P_{n,m} \cos(f_n t - m\theta_s + \varphi_{nm}) \tag{2}$$

where P_{nm} is the electromagnetic force wave amplitude.

If the rotor skew and end effect are ignored, the Maxwell stress will only cause the modal resonance in the circumferential direction of the stator system (composed of the stator yoke, stator teeth and stator windings). At this time, the stator system can be equivalently regarded as a ring for modal analysis. When the frequency of the m-th space harmonic of the electromagnetic force is equal to or similar to the natural frequency of the m-th order circumferential mode of the stator system, the vibration will resonate and the electromagnetic noise radiation will have a maximum value. Especially in the operating speed range of the induction motor, this situation should be avoided. As is known, the electromagnetic force amplitude is inversely proportional to the fourth power of the force wave order [7,18], so for small and medium-sized motors, only force waves with m less than 4 need to be considered.

2.2. Current Computation

The stator and rotor phase currents, including all spatial and time harmonics, can be calculated using the single-phase harmonic expansion circuit [19]. As shown in Figure 1, each voltage harmonic $U_{S,n}$ with frequency f_n generated by the five-phase PWM inverter circuit corresponds to an equivalent circuit containing the influence of v stator winding space harmonics.

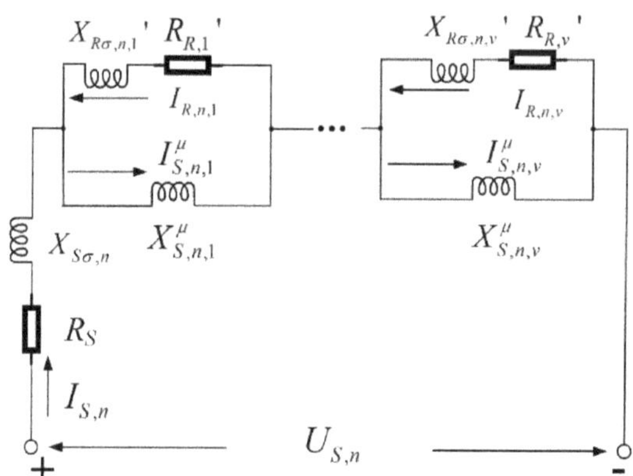

Figure 1. Harmonics expansion circuit.

According to Figure 1, the n main loop equations can be obtained as:

$$U_{S,n} = I_{S,n}(R_S + jX_{S\sigma,n}) + \sum_v jX_{S,n,v}^{\mu} I_{S,n,v}^{\mu}$$
$$= Z_{S,n}I_{S,n} + \sum_v Z_{S,n,v}^{\mu} I_{S,n,v}^{\mu} \tag{3}$$

the $n \times v$ node equations are:

$$I_{S,n} + I_{R,n,v} = I_{S,n,v}^{\mu} \tag{4}$$

the $n \times v$ small loop equations are:

$$0 = I_{R,n,v}\left(R_{R,v}' + jX_{R\sigma,n,v}'\right) + jX_{S,n,v}^{\mu} I_{S,n,v}^{\mu}$$
$$= I_{R,n,v}Z_{R,n,v} + Z_{S,n,v}^{\mu} I_{S,n,v}^{\mu} \tag{5}$$

where R_S and $R_{R,v}'$ are stator and rotor resistance, $X_{S\sigma,n}$ is the leakage reactance of stator to n-th time harmonic, $X_{S,n,v}^{\mu}$ is the harmonic magnetizing reactance of the stator with respect to the n-th time harmonic and v-th space harmonic [18,20] and $X_{R\sigma,n,v}'$ is the leakage reactance of the rotor with respect to n-th time harmonic and v-th space harmonic. The detailed expressions of the above circuit parameters can be calculated from [18].

$$X_{S,n,v}^{\mu} = 2\pi f_n L_{S,v}^{\mu} \tag{6}$$

$$L_{S,v}^{\mu} = L_{S,p}^{\mu}\left(\frac{pK_{wv}}{vK_{wp}}\right)^2 \tag{7}$$

where $L_{S,v}^{\mu}$ is v-th space harmonic leakage inductance, p is pole pair and K_{wv} is v-th winding factor.

For a certain time harmonic, the $2v + 1$ Equations (3)–(5) can be combined to obtain the following matrix:

$$U_n = Z_n \cdot I_n \tag{8}$$

$$U_n = \begin{pmatrix} U_{S,n} \\ 0 \\ 0 \end{pmatrix}, \ I_n = \begin{pmatrix} I_{S,n} \\ I^{\mu}_{S,n} \\ I_{R,n} \end{pmatrix}, \ Z_n = \begin{pmatrix} Z_{S,n} & Z^{\mu}_{S,n} & 0 \\ 1 & -I & I \\ 0 & D^{\mu}_{S,n} & D_{R,n} \end{pmatrix} \tag{9}$$

where I is an identity matrix, 0 is a zero matrix, $Z^{\mu}_{S,n}$ is the $Z^{\mu}_{S,n,v}$ column vector and $D^{\mu}_{S,n}$ and $D_{R,n}$ are $Z^{\mu}_{S,n,v}$ and $Z_{R,n,v}$ diagonal matrixes, respectively. The time and space harmonic currents of the stator and rotor can be obtained by substituting the parameters of each motor into the matrix (9).

2.3. Analytical Calculation of Electromagnetic Force

Air-gap flux density is the basis of motor energy exchange, and the radial component of flux density B_r can be expressed as [7,9,11,21]:

$$B_r(t, \theta_s) = \left(f^s_{mmf}(t, \theta_s) + f^r_{mmf}(t, \theta_s) \right) \Lambda(t, \theta_s) \tag{10}$$

where f^s_{mmf} and f^r_{mmf} are stator and rotor MMF, Λ is the air-gap permeability taking the slotted effects into account.

$$\Lambda(t, \theta_s) = \Lambda_0 + \sum_{k_1} \lambda_{k_1} + \sum_{k_2} \lambda_{k_2} + \sum_{k_1} \sum_{k_2} \lambda_{k_1} \lambda_{k_2} \tag{11}$$

The specific expressions of each part of air-gap permeability have been given in detail in [7] and [10]. The influence of stator and rotor slots is proportional to the width of the slot opening. λ_{k_1} and λ_{k_2} are inversely proportional to the number of magnetic conductivity harmonics k_1 and k_2. Based on the known distribution of windings, the winding function of single conductor or single coil $WFs(\theta_s)$ [11,22] can be used to obtain the expression of stator winding function of each phase by using the superposition principle, as shown in Figure 2. Then the stator/rotor MMF can be obtained by multiplying the winding function (12) by the current obtained from Section 2.2.

$$WF_s(\theta_s) = \begin{cases} \displaystyle\sum_{v=1}^{\infty} \frac{k_{sv}}{\pi v} \sin v\theta_s, \text{a conductor} \\ \displaystyle\sum_{v=1}^{\infty} \frac{2N_c}{\pi v} k_{sv} k_{yv} \cos v\theta_s, \text{a coil} \end{cases} \tag{12}$$

where k_{sv} and k_{yv} are the slotting coefficient and short distance coefficient of v pair poles harmonics, respectively.

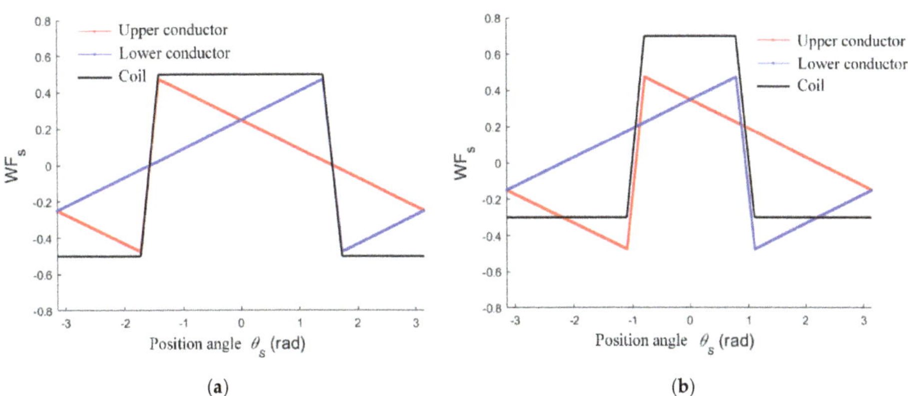

Figure 2. Winding function relation between coil and conductor at different pitch. (**a**) Whole pitch winding; (**b**) short pitch winding.

By substituting (9)–(11) into (1), the Maxwell force wave generated by all magnetic density harmonics can be determined, if only the force wave that plays a major role in noise is considered, that is, the low order force wave P_{belt} generated by the interaction between harmonics of phase belt of stator and rotor MMF and tooth harmonics of air-gap permeability. The characteristics of force waves are given in Table 1.

Table 1. Characteristics of phase belt force wave P_{belt}.

Symbol	Frequency f_P	Force Wave Mode m	Remark
P_{belt}	$f_n[k_2 Z_2(1-s)/p + l]$	$k_2 Z_2 - 2m_1 p k_1 + lp$	$l = 0, \pm 2$

By analyzing the expression of air-gap permeability, it can be seen that, regardless of the value of l, the amplitude of the phase belt force wave of slot effect and the width of slot opening always satisfy [6,23]:

$$P_{fp,m} \propto P_S P_R \propto \frac{\sin(\pi k_1 \beta_1)}{k_1} \frac{\sin(\pi k_2 \beta_2)}{k_2} \tag{13}$$

$$\beta_1 = 1 - \frac{b_{01}}{\tau_1}, \ \beta_2 = 1 - \frac{b_{02}}{\tau_2} \tag{14}$$

where P_S and P_R are the amplitude of harmonics of stator and rotor air gap permeability, β_1 and β_2 are the slot opening ratio. The slot parameters are shown in Figure 3.

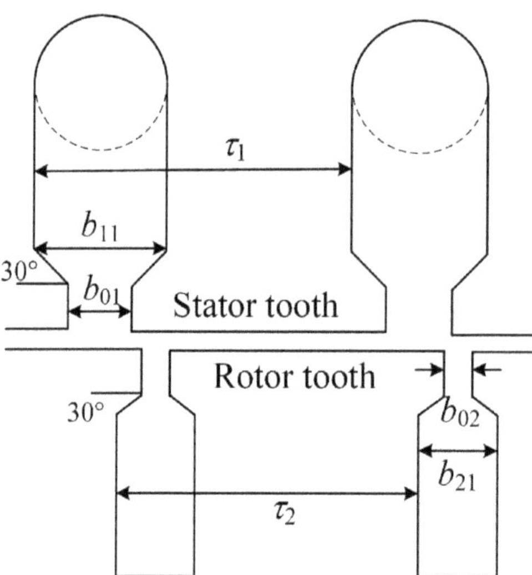

Figure 3. Schematic diagram of stator and rotor slots.

The slot size of the rotor and the rotor directly influence the amplitude of wave, while the number of pole slots influences the order of wave. The frequency of force wave is mainly influenced by the number of rotor slots Z_2, pole pair p, slip rate s and power supply frequency f_n. Generally speaking, in an induction motor from starting to stable operation, the slip ranges from 1 to less than 0.05. As the five-phase PWM power supply contains many rich time harmonics, the frequency range of the Maxwell force wave is too wide to cover the natural frequency of the stator system.

2.4. Natural Frequency of Stator System

The stator system is mainly composed of an iron core and windings—tooth and housing—which can be regarded as equivalent rings, respectively. The natural frequencies of the core and windings (tooth) can be obtained by solving the second-order characteristic equation of the cylindrical shell motion according to the Donnel–Mushtari theory [7,10].

It is worth noting that the vibration characteristics of the housing can be equivalent to a closed cylindrical shell with both ends constrained [24]. As to common motors, both ends are clamped (C). For the magnetic bearing motor, boundary conditions can be regarded as a simply supported (SS). The axial vibration mode a will be introduced in both cases, and the natural frequency of each (a, m) mode should be calculated.

The frequency and stiffness of the frame with end bells can be obtained, such as with the iron core; however $\Omega_{a,m}^2$ is redefined as the solution of the third-order characteristic motion equation of the cylindrical shell [15]. According to the Donnel–Mushtari theory, the characteristic motion equation is:

$$\Omega_{a,m}^6 - C_2\Omega_{a,m}^4 + C_1\Omega_{a,m}^2 - C_0 = 0 \tag{15}$$

with the constants as shown below:

$$C_2 = 1 + \frac{1}{2}(3 - \zeta_3)\left(m^2 + \lambda^2\right) + \kappa^2\left(m^2 + \lambda^2\right)^2 \tag{16}$$

$$C_1 = \frac{1}{2}(1 - \zeta_3)\left[\begin{array}{l}(3 + 2\zeta_3)\lambda^2 + m^2 + \left(m^2 + \lambda^2\right)^2 \\ +\dfrac{3 - \zeta_3}{1 - \zeta_3}\kappa^2\left(m^2 + \lambda^2\right)^3\end{array}\right] \tag{17}$$

$$C_0 = \frac{1}{2}(1 - \zeta_3)\left[\left(1 - \zeta_3^2\right)\lambda^4 + \kappa^2\left(m^2 + \lambda^2\right)^4\right] \tag{18}$$

for both boundary constraints:

$$\lambda = \begin{cases} a\pi\dfrac{R_3}{l_3} & SS \\[2ex] \dfrac{(a + 0.3)R_3\pi}{l_3} & C \end{cases} \tag{19}$$

Only the smallest real root from (15) is related to the natural frequency of vibration, and the stiffness and natural frequency of the frame (a, m) mode are:

$$K_{a,m,3} = \frac{\Omega_{a,m}^2 E_3 V_3}{R_3\left(1 - \varsigma_3^2\right)} \tag{20}$$

For small and medium induction motors, the manufacturing tolerance between the outer diameter of the stator core and the inner diameter of the frame is ±0.02 to ±0.1 mm, with the ends connected by an interference fit. There is no relative displacement between the core, winding, teeth and housing, which can be regarded as a whole. At this time, the stiffness and mass of the stator system can be approximately regarded as the algebraic sum of the mass and stiffness of each part [15], so the stator system (a, m) mode natural frequency can be expressed as:

$$f_{a,m} = \frac{1}{2\pi}\sqrt{\frac{K_{a,m}}{M_{dz}}} \approx \frac{1}{2\pi}\sqrt{\frac{K_{m,1} + K_{m,2} + K_{a,m,3}}{M_{dz,1} + M_{dz,2} + M_{dz,3}}} \tag{21}$$

$$\omega_{a,m} = 2\pi f_{a,m} \tag{22}$$

The subscripts 1, 2, and 3 represent the iron core, windings and tooth and housing, respectively.

2.5. Vibration Characteristics

The radial electromagnetic force wave of the motor acts on the inner surface of the stator core. When the m-order force wave with frequency f_P is close to the natural frequency of the stator system (a, m) mode, the resonance condition is satisfied, and the surface of the motor casing will generate a severe vibration, and its dynamic displacement amplitude can be expressed as:

$$Y_{P,a,m}^{D} = \frac{\pi D_{i1} l_1 P_{f_P,m}}{M_{dz}(2\pi f_{a,m})^2} A_{P,a,m} \tag{23}$$

$$A_{P,a,m} = \frac{1}{\sqrt{\left[1-(f_P/f_{a,m})^2\right]^2 + \left[2\zeta_{a,m}(f_P/f_{a,m})\right]^2}} \tag{24}$$

where $A_{n,a,m}$ is the dynamic amplification factor, $\zeta_{a,m}$ is the damping coefficient, l_1 is the length of the stator core and M_{dz} is the mass of the stator system.

2.6. Model Validation

Taking a five-phase motor as the prototype, the analytical model is verified by the finite element method. The basic parameters of the motor are shown in Table 2, and the slot shape is shown in Figure 3.

Table 2. Prototype parameters.

Symbol	Value	Symbol	Value
Z_1	30	b_{02}	1 mm
Z_2	26	b_{11}	6.1 mm
p	1	b_{21}	8.1 mm
P_N	4 kw	τ_1	$\pi D_{i1}/Z_1$
f_1	50 Hz	τ_2	$\pi D_2/Z_2$
D_1	175 mm	l_1	106 mm
D_{i1}	98 mm	l_2	146 mm
D_2	97.4 mm	l_3	219 mm
b_{01}	3.2 mm	-	-

The main low-order modal natural frequencies of the motor stator system are calculated using the analytical model and Ansys finite element software, respectively. As shown in Table 3 (the finite element results in parentheses), the root mean square error (RMSE) is 330.1 Hz, and the mean absolute percentage error (MAPE) was 4.38%. The results show that the analytical model in this paper can calculate the modal characteristics well and provide the basis for noise analysis. In order to consider the main low-order force wave characteristics, according to the force wave characteristics of Table 1, the main low-order force waves in the range from 2 to 10 for k_1 and k_2 are given in Table 4.

Table 3. Results of the natural frequency of stator system.

$f_{a,m}$	Mode a			RMSE	MAPE
Mode m	1	2	3		
0	8671 (8338)	8961 (8875)	9661 (9806)		
1	2481 (—)	3763 (—)	4385 (4554)		
2	2805 (2403)	3667 (3410)	4880 (4939)	330.1	4.38%
3	5884 (5432)	6197 (6252)	6697 (6954)		
4	10,582 (10,151)	10,736 (11,302)	11,031 (—)		

Table 4. Prototype parameters.

k_1	k_2	f_P	m
5	2	$f_n(2Z_2(1-s)/p+2) = 2570$ Hz	$2Z_2 - 10m_1 + 2p = 4$
8	3	$f_n(8Z_2(1-s)/p+2) = 3805$ Hz	$3Z_2 - 16m_1 + 2p = 0$
5	2	$f_n(2Z_2(1-s)/p-2) = 2325.8$ Hz	$2Z_2 - 10m_1 - 2p = 0$
8	3	$f_n(6Z_2(1-s)/p-2) = 3538.7$ Hz	$3Z_2 - 16m_1 - 2p = -4$
10	4	$f_n(7Z_2(1-s)/p-2) = 4751.6$ Hz	$4Z_2 - 20m_1 - 2p = 2$
5	2	$f_n(2Z_2(1-s)/p) = 2425.8$ Hz	$2Z_2 - 10m_1 = 2$
8	3	$f_n(3Z_2(1-s)/p) = 3638.7$ Hz	$3Z_2 - 16m_1 = -2$
10	4	$f_n(4Z_2(1-s)/p) = 4851.6$ Hz	$4Z_2 - 20m_1 = 4$

3. Magnetic Noise Radiation Calculation

3.1. Noise Radiation Model

For the reason that the motor noise is generally measured in the near field, the acoustic radiator cannot be treated as an infinite cylindrical model. Therefore, the sound power resulted from (f_n, m) order force wave acting on (a, m) mode can be expressed as:

$$W_m(n,a,m) = 0.5\rho_0 c_0 v_{a,m}^2 S_c \sigma_I(n,a,m) \tag{25}$$

where S_c is the surface area of the housing and σ_I is the relative acoustic emissivity depending on the mode (a, m) and the ratio of the length to the width of the housing.

For an infinitely long cylindrical shell, σ_I can be simplified as [15]:

$$\sigma_I(f_n, m) = \frac{2}{\pi k_0 R_3 \left| dH_m^{(2)}(k_0 R_3)/d(k_0 R_3) \right|^2} \tag{26}$$

where $H_{v_m}^{(2)}$ is the second kind of Hankel equation of m-th order and k_0 is the acoustic wave number.

For different boundary conditions at both ends, the relative acoustic emissivity can be obtained by substituting the axial wave number k_{z0} into the sound-wave radiation equation:

$$\sigma_I(n,a,m) = \int_{-k_0}^{k_0} \frac{2k_0 |\Gamma_a(k_z)|^2 dk_z}{\pi^2 R_3 l_3 \left| dH_m^{(2)}(k_r R_3)/d(k_r R_3) \right|^2} \tag{27}$$

$$|\Gamma_{v_n}(k_z)|^2 = \frac{k_{z0}^2 \left[1 + \dfrac{k_z^2}{k_{z0}^2} - \dfrac{2k_z}{k_{z0}} \sin(k_z l_3 - v_n \pi) \right]}{\left(k_z^2 - k_{z0}^2 \right)^2} \tag{28}$$

$$k_{z0} = \frac{(2v_n+1)\pi}{2l_3} \tag{29}$$

where z is the axial coordinate; k_r and k_z are the radial and axial components of the acoustic wavenumber; and $k_0^2 = k_r^2 + k_z^2$ [15].

Then the numerical solution of the analytical model can be obtained via Matlab, so as to further predict the sound power level of magnetic noise:

$$L_W(n,a,m) = 10\log_{10}(W_m(n,a,m)/W_0) \tag{30}$$

$$L_{WA}(n,a,m) = 10\log_{10}\left(\sum_n 10^{0.1(L_w(n,a,m)+\Delta L_A(n))}\right) \tag{31}$$

where $W_0 = 10^{-12}$ W is the reference sound power and $\triangle L_A(n)$ is a frequency (f_n)-dependent A-weighting factor [7,15,25].

3.2. Noise Prediction Verification Experiment

For the motors shown in Table 2, the five-phase SPWM control circuit is used to supply power to the motors to conduct actual noise measurement experiments. There are three vibration acceleration sensors at the top, side and bottom on the surface of the casing, and three sound pressure sensors are set at the top, axial and side 1.5 times the axial length of the motor. The overall layout of the specific test platform is shown in the Figure 4 is shown.

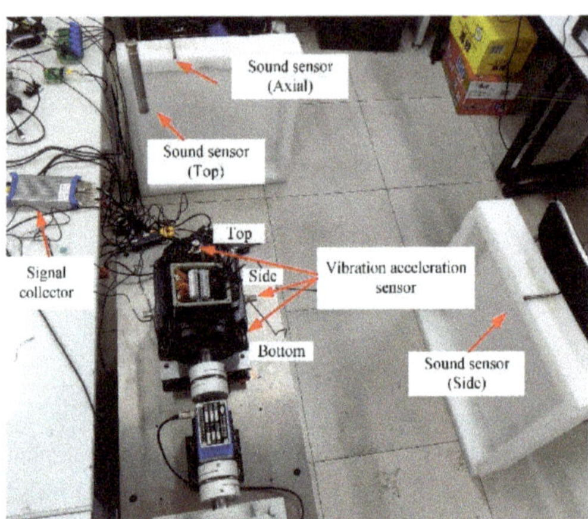

Figure 4. Five-phase motor vibration and noise measurement platform.

In terms of Figure 5, it can be found that the change trend of the vibration acceleration at the top and bottom of the casing is consistent, while the change trend at the side is just the opposite, which reveals that the electromagnetic force mainly excites the resonance of the 2-nd mode of the stator system. The Fourier spectrum analysis of the vibration acceleration and noise results under steady state conditions are given in Figure 6. It is found that the vibration frequency is mainly concentrated around 5000 Hz, which just corresponds to ($m = 2$, $a = 3$) the natural frequency of the modal. It is obviously consistent with the judgment of the macroscopic result in Figure 6. In addition, the high frequency belt in the noise spectrum is consistent with the vibration spectrum, but many components that do not exist in the vibration spectrum appear in the low frequency band, which is probably caused by the mechanical noise of the rotating shaft and the power switching.

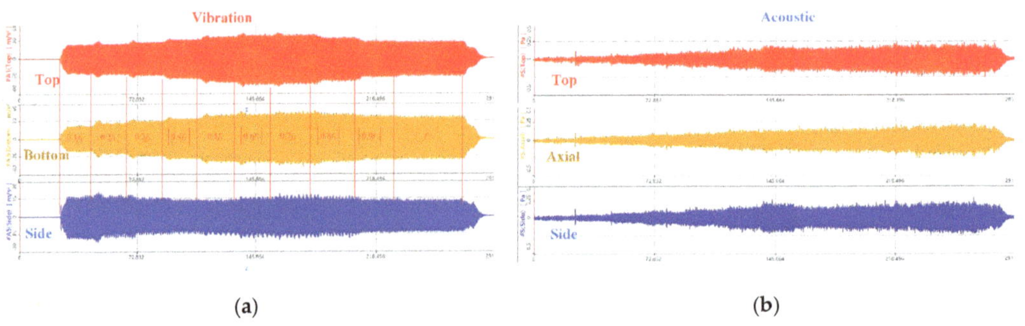

Figure 5. Vibration and noise transient measurement results. (**a**) Vibration acceleration in three directions at the top, bottom and sides of the enclosure; (**b**) Sound pressure in three directions: top, axial and side.

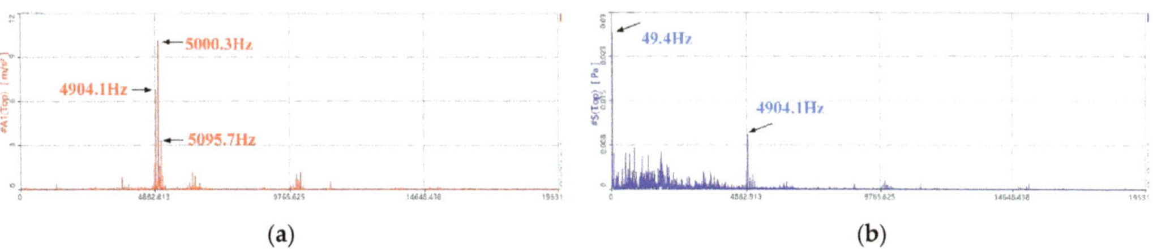

(a)   (b)

Figure 6. Comparison of vibration and noise spectrum in steady state. (**a**) Vibration acceleration spectrum; (**b**) sound pressure spectrum.

4. Low Noise Slot Opening Design

4.1. Low Noise Fundamental Analysis

According to Table 1, there are two main methods to reduce electromagnetic vibration noise in the design stage. One is to choose the pole-slot combination to avoid resonance between the electromagnetic force of (f_p, m) and motor structure of $(f_{a,m}, m)$; the second is to reduce the amplitude of the electromagnetic force wave by correctly selecting the slotting strategy of the stator and the rotor, thereby reducing the electromagnetic vibration noise. In this paper, based on the given pole-slot match, the electromagnetic vibration noise can be reduced by changing the width of the slot opening without the reduction of output performance.

According to (13), when $k_1\beta_1$ or $k_2\beta_2$ is an integer, the electromagnetic force of $m = k_2Z_2 - 2m_1pk_1 + lp$ will decrease to 0. Therefore, the amplitude of electromagnetic force of a certain order can be further expressed as [6]:

$$k_1\beta_1 = Z_{01}, \; k_2\beta_2 = Z_{02} \tag{32}$$

$$b_{01}{}^* = \tau_1\left(1 - \frac{Z_{01}}{k_1}\right), \quad Z_{01} \in [1, k_1 - 1] \tag{33}$$

$$b_{02}{}^* = \tau_2\left(1 - \frac{Z_{02}}{k_2}\right), \quad Z_{02} \in [1, k_2 - 1] \tag{34}$$

where Z_{01} and Z_{02} are integers and the superscript * represents the slot width value of exactly weakening force wave.

If k_1 or k_2 is equal to 1, then the width of the rotor slot opening cannot weaken any force wave. Therefore, to weaken the electromagnetic force amplitude through the width of the slot opening, k_1 and k_2 should be integers greater than 2. Substituting the value range of Z_{01} and Z_{02} into (33) and (34), the slot width optimization range for the electromagnetic force of order $m = k_2Z_2 - 2m_1pk_1 + lp$ can be obtained:

$$b_{01}{}^* \in \left[\frac{\tau_1}{k_1}, \tau_1\left(1 - \frac{1}{k_1}\right)\right] \tag{35}$$

$$b_{02}{}^* \in \left[\frac{\tau_2}{k_2}, \tau_2\left(1 - \frac{1}{k_2}\right)\right] \tag{36}$$

The optimal design of motor is a nonlinear, multi-physical field coupling and anisotropic problem, so focusing on the weakening of one order of force wave is likely to lead to the increase of the amplitude of other order of force wave. In addition, the five-phase motor is mainly controlled by PWM frequency conversion. In the process from start-up to stable operation, the power supply voltage contains abundant time harmonics, so the radial electromagnetic force wave and the stator mode are likely to resonance. To sum up, the optimal slot opening width is not necessarily the value that makes a certain order of force wave 0 in (33) and (34) but may be a value within the continuous range of (35) and (36). Therefore, two noise optimization objectives are set in this paper:

(1) The minimum total sound power level of each frequency on rated condition:

$$\min\left(\sum_m L_{WA}(f_m)\right) \tag{37}$$

(2) The sound power amplitude of the maximum noise generated during the start-up is the smallest:

$$\min(\max(L_{WA}(f_m))) \tag{38}$$

4.2. Slot Opening Width Schemes

For the optimization of the slot opening width, the optimal slot opening scheme with overall electromagnetic noise can be obtained by combining the above two optimization objectives in the range of fixed and rotor slot opening width determined by (35) and (36). In this paper, the test motor shown in Table 2 is taken as the object. In order to consider the influence of as many low-order force waves as possible, the vibration and noise prediction is made for the width of slot opening under all conditions from k_1 and k_2 from 2 to 10, and the following assumptions are made:

(1) The slip ratio changes linearly, ranging from 1 to 0.05.
(2) In each case, five-phase sequential currents of the same amplitude are passed through the stator windings.
(3) The pole-slot numbers are given, and the range of b_{01} and b_{02} are shown in Table 5.
(4) Without the consideration of saturation, the air-gap permeability does not contain harmonic content caused by saturation.

Table 5. Value range of stator and rotor slots opening width.

Symbol	Max	Min	Step
b_{01}	Max $(1 - \tau_1/\text{Max}(k_1), b_{11})$	$\tau_1/\text{Max}(k_1)$	0.01 mm
b_{02}	Max $(1 - \tau_1/\text{Max}(k_2), b_{21})$	$\tau_2/\text{Max}(k_2)$	0.01 mm

For the test motor shown in Table 2, the control variable method was used to calculate the curve of electromagnetic noise by taking the width of the stator and rotor slot opening as independent variables. From Figure 7, we can see that for a certain mode, changing the width of the slot opening can indeed weaken the vibration noise. Furthermore, slot opening width is significantly different for noise suppression sensitivity during start-up and steady-state. That is because, in addition to the amplitude of the electromagnetic force, the vibration noise is also affected by the natural frequency of the stator. In terms of the magnetic noise, the influence of the slot opening width shows a discrete and non-monotonic trend. Actually, it is closer to sinusoidal and there are periodic changes in the sensitivity with a large span, especially in the target (38). It is worth noting that the jitter in the waveform is caused by the same m from different k_{01}, k_{02}, because the curve reflects the total sensitivity of b_{01} to all k_{01}, k_{02} combinations, which result in not strictly sine.

It can be seen from the above that the amplitude, phase and period of the sensitivity curve of the slot opening width vary significantly with modes. So, it is reasonable to think that the slot width with the smallest overall magnetic noise is not necessarily the smallest. The top four quietest slot opening schemes of targets (37) and (38) for the test motor are given in Figure 8, which are different from conventional knowledge. Considering the engineering practice, b_{01} and b_{02} are set smaller than b_{11} and b_{21}.

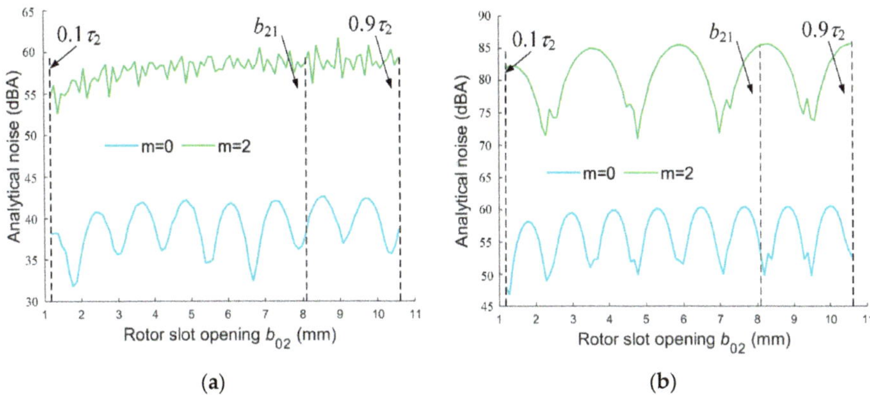

Figure 7. Curve of sound power level with b_{02} varying from $0.1\tau_2$ to $0.9\tau_2$. (**a**) Sound power level radiated by b_{02} associated with main mode for target (37); (**b**) Sound power level radiated by b_{02} associated with main mode for target (38).

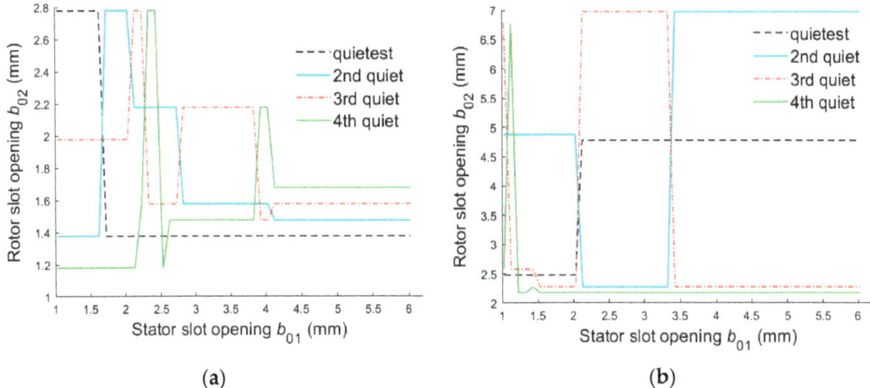

Figure 8. Top four quietest slot opening schemes. (**a**) For target (37); (**b**) for target (38).

By analyzing the noise prediction results of all schemes, the quietest scheme of the rotor slot opening is screened for the two different optimization objectives in Section 4.1. As expected, the main quiet slot opening schemes are not the smallest, and b_{02} shows a non-monotonic and discrete trend with the change of b_{01}. Furthermore, with the increase of b_{01}, the optimal solution of b_{02} gradually tends to be stable. Due to space limitations, only some quiet slot opening schemes are shown in Table 6.

Table 6. Main quiet slots opening width schemes (unit: mm).

b_{01}	b_{02} of (37)	b_{02} of (38)
$0.1\tau_1$	$0.236\tau_2, 0.117\tau_2, 0.168\tau_2, 0.1\tau_2$	$0.211\tau_2, 0.414\tau_2, 0.576\tau_2, 0.219\tau_2$
$0.139\tau_1$	$0.236\tau_2, 0.117\tau_2, 0.168\tau_2, 0.1\tau_2$	$0.211\tau_2, 0.414\tau_2, 0.219\tau_2, 0.194\tau_2$
$0.178\tau_1$	$0.117\tau_2, 0.236\tau_2, 0.168\tau_2, 0.1\tau_2$	$0.211\tau_2, 0.414\tau_2, 0.194\tau_2, 0.185\tau_2$
$0.217\tau_1$	$0.117\tau_2, 0.185\tau_2, 0.236\tau_2, 0.134\tau_2$	$0.406\tau_2, 0.194\tau_2, 0.593\tau_2, 0.185\tau_2$
$0.256\tau_1$	$0.117\tau_2, 0.185\tau_2, 0.134\tau_2, 0.126\tau_2$	$0.406\tau_2, 0.194\tau_2, 0.593\tau_2, 0.185\tau_2$
$0.295\tau_1$	$0.117\tau_2, 0.134\tau_2, 0.185\tau_2, 0.126\tau_2$	$0.406\tau_2, 0.194\tau_2, 0.593\tau_2, 0.185\tau_2$
$0.334\tau_1$	$0.117\tau_2, 0.134\tau_2, 0.185\tau_2, 0.126\tau_2$	$0.406\tau_2, 0.593\tau_2, 0.194\tau_2, 0.185\tau_2$
$0.373\tau_1$	$0.117\tau_2, 0.134\tau_2, 0.185\tau_2, 0.126\tau_2$	$0.406\tau_2, 0.593\tau_2, 0.194\tau_2, 0.185\tau_2$
$0.412\tau_1$	$0.117\tau_2, 0.126\tau_2, 0.134\tau_2, 0.143\tau_2$	$0.406\tau_2, 0.593\tau_2, 0.194\tau_2, 0.185\tau_2$
$0.451\tau_1$	$0.117\tau_2, 0.126\tau_2, 0.134\tau_2, 0.143\tau_2$	$0.406\tau_2, 0.593\tau_2, 0.194\tau_2, 0.185\tau_2$

Table 6. *Cont.*

b_{01}	b_{02} of (37)	b_{02} of (38)
$0.49\tau_1$	$0.117\tau_2, 0.126\tau_2, 0.134\tau_2, 0.143\tau_2$	$0.406\tau_2, 0.593\tau_2, 0.194\tau_2, 0.185\tau_2$
$0.529\tau_1$	$0.117\tau_2, 0.126\tau_2, 0.134\tau_2, 0.143\tau_2$	$0.406\tau_2, 0.593\tau_2, 0.194\tau_2, 0.185\tau_2$
$0.568\tau_1$	$0.117\tau_2, 0.126\tau_2, 0.134\tau_2, 0.143\tau_2$	$0.406\tau_2, 0.593\tau_2, 0.194\tau_2, 0.185\tau_2$

5. Conclusions

The analytical relationship between the electromagnetic–vibration–noise characteristics of the five-phase cage induction motor and stator/rotor slot opening widths is demonstrated and validated. Based on this, a method for magnetic noise reduction is improved, including the overall harmonics, to optimize the stator/rotor slot opening widths. This method can be applied to the design stage of a low-noise induction machine without the expense of increasing the slot leakage inductance or sacrificing the output performance.

In this paper, the optimization database of slot opening widths is given for a 30/26 slot five-phase induction motor prototype. The results show that the quietest scheme is not always the smallest slot opening, regardless of aims towards noise reduction during start-up or the steady state operation, which is different from common design rules. Because of the periodic sensitivity curve, a wider slot opening can inversely reduce the overall sound power level even more. In terms of the future work, the pole-slot scheme and slot opening width can be considered together to carry out multi-dimensional noise reduction research.

Author Contributions: Writing—original draft preparation, H.C.; validation, Y.X.; investigation, X.L.; supervision, J.Z.; data curation, Q.Z. All authors have read and agreed to the published version of the manuscript.

Funding: This research was funded by the National Natural Science Foundation of China under the grant 51507813.

Institutional Review Board Statement: Not applicable.

Informed Consent Statement: Not applicable.

Data Availability Statement: Not applicable.

Conflicts of Interest: The authors declare no conflict of interest.

References

1. Corton, R.; Sawezyn, H.; Belkhayat, D.; Brudny, J. Principe of Magnetic Noise Active Reduction Using Three Phase Systems Due to PWM Inverter Switching. In Proceedings of the 2nd International Seminar on Vibrations and Acoustic Noise of Electric Machinery (VANEM), Łódź, Poland, 1–3 June 2000; Volume 21.
2. Cassoret, B.; Corton, R.; Roger, D.; Brudny, J.F. Magnetic Noise Reduction of Induction Machines. *IEEE Trans. Power Electron.* **2003**, *18*, 570–579. [CrossRef]
3. Belkhayat, D.; Roger, D.; Brudny, J.F. Active Reduction of Magnetic Noise in Asynchronous Machine Controlled by Stator Current Harmonics. In Proceedings of the 1997 Eighth International Conference on (Conf. Publ. No. 444), Cambridge, UK, 1–3 September 1997.
4. Liu, H.; Wang, D.; Yi, X.; Zhang, Y. A Unified Fault-tolerant Control for 15-phase Induction Machine Under Various Phase Failure Conditions. *Proc. CSEE* **2019**, *39*, 327–336.
5. Cheng, Z.; Ruan, L.; Huang, S.; Yang, J. Research on Noise Reduction of 3.6 MW Evaporative Cooling Wind Motor Induced by Electromagnetic and Two-Phase Flow Resonance Based on Stator Optimization. *Processes* **2021**, *9*, 669. [CrossRef]
6. Le Besnerais, J.; Lanfranchi, V.; Hecquet, M.; Romary, R.; Brochet, P. Optimal Slot Opening Width for Magnetic Noise Reduction in Induction Motors. *IEEE Trans. Energy Convers.* **2009**, *24*, 869–874. [CrossRef]
7. Chen, Y. *Analysis and Control of Motor Noise*; Zhejiang University Press: Hangzhou, China, 1987; pp. 5–23.
8. Shin, K.-H.; Bang, T.-K.; Kim, K.-H.; Hong, K.; Choi, J.-Y. Electromagnetic Analysis and Experimental Study to Optimize Force Characteristics of Permanent Magnet Synchronous Generator for Wave Energy Converter Using Subdomain Method. *Processes* **2021**, *9*, 1825. [CrossRef]
9. Le Besnerais, J.; Lanfranchi, V.; Hecquet, M.; Brochet, P.; Friedrich, G. Acoustic Noise of Electromagnetic Origin in a Fractional-Slot Induction Machine. *COMPEL Int. J. Comput. Math. Electr. Electron. Eng.* **2008**, *27*, 1033–1052. [CrossRef]

10. Brudny, J.F. Modélisation de La Denture Des Machines Asynchrones. Phénomène de Résonance. *J. Phys. III* **1997**, *7*, 1009–1023. [CrossRef]
11. Bossio, G.; De Angelo, C.; Solsona, J.; Garcia, G.; Valla, M.I. A 2-D Model of the Induction Machine: An Extension of the Modified Winding Function Approach. *Energy Convers. IEEE Trans.* **2004**, *19*, 144–150. [CrossRef]
12. Belahcen, A. Vibrations of Rotating Electrical Machines Due to Magnetomechanical Coupling and Magnetostriction. *IEEE Trans. Magn.* **2006**, *42*, 971–974. [CrossRef]
13. Hilgert, T.G.D.; Vandevelde, L.; Melkebeek, J.A.A. Numerical Analysis of the Contribution of Magnetic Forces and Magnetostriction to the Vibrations in Induction Machines. *IET Sci. Meas. Technol.* **2007**, *1*, 21–24. [CrossRef]
14. Devillers, E.; Hecquet, M.; Le Besnerais, J.; Régniez, M. Tangential Effects on Magnetic Vibrations and Acoustic Noise of Induction Machines Using Subdomain Method and Electromagnetic Vibration Synthesis. In Proceedings of the 2017 IEEE International Electric Machines and Drives Conference (IEMDC), Miami, FL, USA, 21–24 May 2017; pp. 1–8.
15. Gieras, J.F.; Chong, W.; Lai, J.C. *Noise of Polyphase Electric Motors*; CRC Press: Boca Raton, FL, USA, 2005.
16. Alger, P.L. Induction Machines: Their Behavior and Uses. *J. Clin. Investig.* **1993**, *91*, 2314–2319.
17. Timár, P.L. *Noise and Vibration of Electrical Machines*; Elsevier: Amsterdam, The Netherlands, 1989.
18. Chen, S. *Motor Design*; Machinery Industry Press: Beijing, China, 2004; pp. 244–247.
19. Hubert, A. Contribution à l'Étude des Bruits Acoustiques Générés Lors de l'Association Machines Électriques-Convertisseurs Statiques de Puissance. Application à la Machine Asynchrone. Ph.D. Thesis, Université de Technologie de Compiègne, Compiègne, France, 2000.
20. Gojko, J.M.; Momir, D.D.; Aleksandar, O.B. Skew and Linear Rise of MMF across Slot Modelling-Winding Function Approach. *IEEE Trans. Energy Convers.* **1999**, *14*, 315–320. [CrossRef]
21. Le Besnerais, J.; Lanfranchi, V.; Hecquet, M.; Brochet, P. Optimal Slot Numbers for Magnetic Noise Reduction in Variable-Speed Induction Motors. *IEEE Trans. Magn.* **2009**, *45*, 3131–3136. [CrossRef]
22. Bao, X.; Cheng, Z.; Wang, H.; Di, C. Inductance Calculation of Eccentric Induction Motor Based on Modified Winding Function Approach. *Trans. China Electrotech. Soc.* **2016**, *31*, 8.
23. Le Besnerais, J. Fast Prediction of Variable-Speed Acoustic Noise Due to Magnetic Forces in Electrical Machines. In Proceedings of the 2016 XXII International Conference on Electrical Machines (ICEM), Lausanne, Switzerland, 4–7 September 2016.
24. Leissa, A.W.; Nordgren, R.P. Vibration of Shells. *J. Appl. Mech.* **1993**, *41*, 544. [CrossRef]
25. Timar, P.L.; Lai, C.S.J. Acoustic Noise of Electromagnetic Origin in an Ideal Frequency-Converter-Driven Induction Motor. *Electr. Power Appl. IEE Proc.* **1994**, *141*, 341–346. [CrossRef]

Article

Modeling and Analysis of New Power Devices Based on Linear Phase-Shifting Transformer

Jie Xue, Jinghong Zhao, Sinian Yan *, Hanming Wang, Changduo Zhou, Dongao Yan and Hansi Chen

School of Electrical Engineering, Naval University of Engineering, Jiefang Road 717, Wuhan 430000, China
* Correspondence: ysnian0504@126.com

Abstract: With the rapid development of new power systems, various new power devices have also been developed. It is very important to establish analytical models of new power devices to ensure or even improve the reliability and stability of the power system. A linear phase-shifting transformer (LPST) is a new type of power device that mainly relies on air gaps to transfer energy, so establishing an accurate air-gap magnetic field model is very important for improving the efficiency of this system. In this paper, an analytical model of an unequal-pitch linear phase-shifting transformer (UP-LPST) was established by combining the distributed magnetic circuit method (DMCM) and Schwartz–Christopher transformation (SCT). Taking the magnetic field strength as a variable, an accurate magnetic field analysis model for a UP-LPST considering saturation, cogging, and edge was established. Taking a 1 kw UP-LPST as a prototype, the accuracy of the model was verified by the finite element method and experiments. This modeling method could also be used to establish magnetic field models of other similar structures in new energy power systems, especially those with cogging structures.

Keywords: linear phase-shifting transformer (LPST); distributed magnetic circuit method (DMCM); cogging effect; Schwarz–Christoffel transformation (SCT); magnetic field

Citation: Xue, J.; Zhao, J.; Yan, S.; Wang, H.; Zhou, C.; Yan, D.; Chen, H. Modeling and Analysis of New Power Devices Based on Linear Phase-Shifting Transformer. *Processes* **2022**, *10*, 1596. https://doi.org/10.3390/pr10081596

Academic Editors: Haoming Liu, Jingrui Zhang and Jian Wang

Received: 13 July 2022
Accepted: 9 August 2022
Published: 12 August 2022

Publisher's Note: MDPI stays neutral with regard to jurisdictional claims in published maps and institutional affiliations.

1. Introduction

The operation of a new energy power system must have a high level of reliability. However, ensuring the stable operation of any line component of a power system is complex. Its stability and reliability are determined by different characteristic parameters of the system. The main influencing factors include different impedances of parallel lines in the system, the power factor, changes in the input power, and changes in the load [1–3]. A variety of electrical devices have been developed, with the phase-shifting transformer being one of the most important technologies. This device can control the current distribution between the branches of a parallel power system through the adjustment of the phase angle. At the same time, it can solve the overload problem caused by the unbalanced impedance of parallel transmission lines and improve the stability and efficiency of the power system, so it has received extensive attention from scholars at home and abroad [4–6].

As a new type of phase-shifting transformer, the linear phase-shifting transformer (LPST) has many advantages. Compared with the traditional phase-shifting transformer, its core structure is simple, the air gap adjustment is easy, the phase-shifting angle is wide, and the volume is small [7–9]. At the same time, it can effectively eliminate low-order harmonics, improve the quality of output waveforms, and reduce power grid harmonic pollution [10]. The LPST is a power device that mainly relies on air gaps to transfer energy, and there are several rectangular slots on the primary and secondary sides of the LPST to hold winding coils. This slotted structure produces a degree of cogging, which results in a significant increase in the tooth harmonic amplitude and distorts the air-gap magnetic field. More seriously, it reduces the quality of the output waveform and increases the energy loss [11], thereby reducing the stability and efficiency of the power system. Furthermore, because of the linear structure of an LPST, its core is not continuous. Therefore, the end effect is also an important factor affecting the output of the system, and it is particularly important to establish an accurate LPST magnetic field analysis model.

Due to the particularity of the LPST's structure, traditional modeling methods are not fully applicable. Therefore, we referred to the modeling method of linear motors to conduct our research. At present, the most common research methods include the direct method and the indirect method. In the direct method [12–14], the whole magnetic field is divided into the slot, the air gap, and the other subdomains using the method of partition modeling, and the expressions for each part of the magnetic field are obtained by solving the Laplace equation. The indirect method [15–17] involves multiplying the slotless air-gap magnetic field model by the air-gap's specific permeability to obtain the analytical formula of the cogging air-gap magnetic field. In [18], a combination of a linear and a sinusoidal air-gap ratio permeability function was used to represent the air-gap ratio permeability in a single slot to consider the influence of motor stator slotting. The authors of [19] determined the actual air-gap flux distribution on permanent magnets through the superimposed relative permeability algorithm. In [20], the air-gap permeability function on the surface of a smooth rotor was obtained using the mirror image method, which considered the interaction of slots. The authors of [21] designed a new air-gap relative permeability formula by applying an offset at the outer diameter of the rotor. In [22], the distribution of the air-gap magnetic field at the edge of a linear rotating permanent-magnet synchronous motor was calculated, and the relative permeability function of the air gap was solved using SCT. In [23], an analytical model of a fractional-slot linear phase-shifting transformer was established by the precise subdomain method. The model considered the influence of magnetic permeability, structural parameters, and the interaction between tooth slots on the magnetic field distribution. However, the analytical formula was too complicated and could not consider the influence of saturation.

In this paper, the slotless magnetic field and saturation were determined by the DMCM [24–27]. Taking a single slot on the primary side as an example, the single slot area was selected as the smallest unit. The irregular magnetic field was mapped into a regular magnetic field pattern by SCT. The air-gap relative permeance in the whole length range of the linear phase-shifting transformer was calculated by the above procedure. Then, the edge end was taken as an independent analytical model, and the distribution function of the air-gap relative permeance at the edge end was obtained. Based on the analytical model, the UP-LPST was taken as an example to analyze the effect of the slot on the air-gap magnetic field. At the same time, a UP-LPST with different slot spacing ratios was modeled and analyzed. The accuracy of the proposed model was verified by a comparison of the results obtained by the FEM and those obtained through experiments.

2. Analytical Model of Slotless Magnetic Field

The structure of a UP-LPST is basically the same as that of a linear motor. A diagram of its main structure is shown in Figure 1. The difference is that the length of the core on the primary and secondary sides of the UP-LPST is the same, and its core is symmetric about the air gap. Four groups of three-phase windings are distributed longitudinally along the core on the primary side, and one group of three-phase windings on the secondary side. In contrast to conventional phase-shifting transformers, the energy conversion of a UP-LPST is mainly realized by an air-gap traveling-wave magnetic field. When the primary winding of the UP-LPST is energized, a linear traveling-wave magnetic field can be generated in the core, and then a three-phase electromotive force is induced in the secondary side.

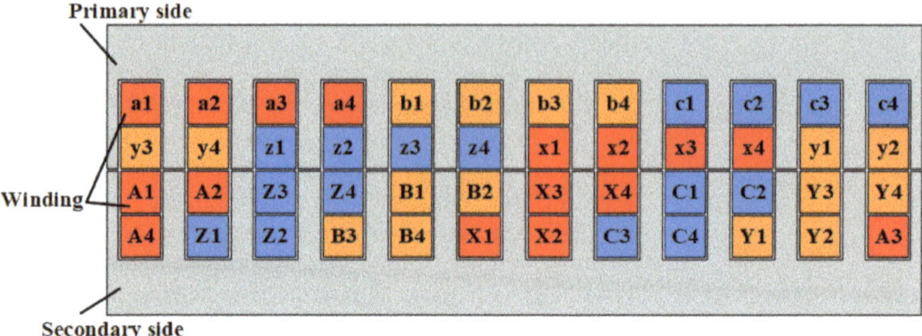

Figure 1. Schematic diagram of UP-LPST structure.

The UP-LPST studied in this paper comprised a pair of poles, as shown in Figure 1. The primary side was composed of four groups of three-phase bridge inverter circuits with a twelve-phase input (a1-x1, a2-x2, ... , c3-z3, c4-z4), and the windings were distributed with full pitch. The secondary side comprised a three-phase output (A1-X1-A2-X2-A3-X3-A4-X4, ...), and the windings were distributed in a combination of long-distance winding and short-distance winding. This winding distribution could effectively improve the three-phase asymmetry caused by the discontinuity of the core. The core structures of the primary and secondary sides were completely consistent and symmetric with respect to the air gap.

Firstly, in order to obtain the magnetic field without grooves, the following basic assumptions were made [26]:

1. The primary and secondary side end flux leakage is ignored.
2. The flux lines in the virtual teeth are all arranged in the longitudinal direction, and the flux lines in the yokes are all arranged in the normal direction.
3. The size of the primary- and secondary-side virtual teeth is the same.
4. The interaction of adjacent slots (virtual slots) is ignored.

The overall model was divided into five regions along the longitudinal direction for magnetic circuit calculation, as shown in Figure 2, where I is the air-gap region; II and III are the primary-side virtual teeth and the yoke region, respectively; and IV and V are the secondary-side virtual teeth and the yoke region, respectively.

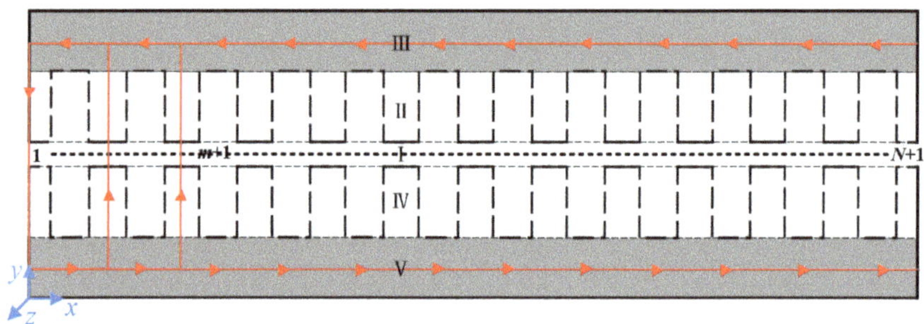

Figure 2. Schematic diagram of the block model of the UP-LPST.

For the loop shown in Figure 2:

$$F_{\Sigma}(m) = F_{\delta\_n}(m+1) + F_{j2\_l}(m) + F_{t1\_n}(m+1) + F_{t2\_n}(m+1) - F_{\delta\_n}(m) - F_{j1\_l}(m) - F_{t1\_n}(m) - F_{t2\_n}(m) \qquad (1)$$

where $F_{\delta\_n}(m)$ and $F_{\delta\_n}(m+1)$ are the air-gap normal magnetic pressure drops at node m and node $m + 1$, respectively; $F_{j1\_l}(m)$ and $F_{j2\_l}(m)$ are the longitudinal magnetic pressure drops of the yoke at the core node m on the primary and secondary sides, respectively; and

$F_{t1\_n}(m)$, $F_{t1\_n}(m+1)$, $F_{t2\_n}(m)$, and $F_{t2\_n}(m+1)$ are the normal magnetic pressure drops of the virtual teeth at the core node m and node $m+1$ on the primary and secondary sides, respectively. The positive direction of the coordinate axis is the positive direction of the magnetic pressure drop of each section.

Because the primary and secondary sides of the straight-line phase-shifting transformer were symmetric about the center line of the air gap, the model could be simplified to a one-sided model, as shown in Figure 3 [27].

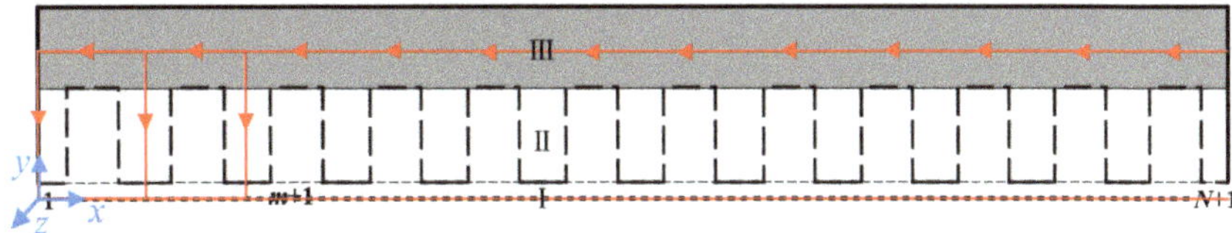

Figure 3. Diagram of the simplified model.

The total magnetic pressure drop of the circuit could be simplified as:

$$F_{\Sigma}(m) = F_{\delta\_n}(m+1) + F_{t1\_n}(m+1) - F_{\delta\_n}(m) - F_{j1\_l}(m) - F_{t1\_n}(m) \tag{2}$$

Each MMF could be expressed as follows:

$$\begin{cases} F_{\delta\_n}(m) = H_{\delta\_n}(m)\delta = \dfrac{B_{\delta0\_slotless\_n}(m)}{\mu_0}\delta \\[2ex] F_{t1\_n}(m) = H_{t1\_n}(m)h_{t1} = \dfrac{B_{t1\_n}(m)}{\mu_{t1}(m)}h_{t1} \\[2ex] F_{j1\_l}(m) = \dfrac{H_{j1\_l}(m)+H_{j1\_l}(m+1)}{2}dx = \dfrac{dx}{2}\left[\dfrac{B_{j1\_l}(m+1)}{\mu_{j1}(m+1)} + \dfrac{B_{j1\_l}(m)}{\mu_{j1}(m)}\right] \end{cases} \tag{3}$$

where H_{δ_n} is the magnetic field intensity in the normal direction of the air gap; B_{t1_n} and H_{t1_n} denote the flux density and magnetic field intensity in the normal direction of the primary-side virtual teeth, respectively; B_{j1_l} and H_{j1_l} denote the flux density and magnetic field intensity in the longitudinal direction of the primary yoke, respectively; δ is the length of the air gap; h_{t1} is the virtual tooth height of the primary core; μ_0 is the air permeability; and μ_{j1} denotes the permeability of each node of the primary yoke.

When the tooth is not saturated, the main flux within a pitch can be considered to pass entirely through the tooth. At this point, the tooth magnetic flux densities at node m are as follows:

$$B_{t0\_n}(m) = \frac{B_{\delta0\_slotless\_n}(m)Ht_1}{K_{Fe}Lb_t} \tag{4}$$

where H and L are the height and length of the core of the UP-LPST, respectively; b_t is the tooth width of the UP-LPST; t_1 is the tooth pitch; and K_{Fe} is the superposition coefficient of the core.

However, when the teeth are saturated, most of the main flux passes through the teeth, and the rest enters the yoke through the slot. At this time, the actual magnetic flux in the tooth becomes smaller, so it was necessary to revise (5) as follows:

$$B_{t\_n}(m) = B'_{t\_n}(m) - \mu_0 H_{t\_n}(m)k_\delta \tag{5}$$

where $B'_{t\_n}(m)$ is the normal apparent magnetic flux density of the tooth at node m, representing the magnetic flux density when all the flux enters the tooth; $B_{t\_n}(m)$ is the actual normal magnetic flux density of the tooth at node m; $H_{t\_n}(m)$ is the actual normal magnetic field intensity of the tooth at node m; and k_δ is the slot coefficient, $k_\delta = (H \cdot b_s)/(K_{Fe} \cdot L \cdot b_t)$.

According to the continuity of magnetic flux, the longitudinal magnetic flux density of the yoke at node m is equal to the total normal magnetic flux density of the air gap from node 1 to node m. The magnetic flux density of the yoke was calculated as follows:

$$B_{j1\_l}(x) = \frac{\phi_j(x)}{K_{Fe}h_{j1}D} = \frac{L}{K_{Fe}h_{j1}D}\int_x^0 B(x)dx \tag{6}$$

where D is the thickness of the UP-LPST.

Therefore, the magnetic flux density of the yoke of the primary core could be represented as follows:

$$B_{j1\_l}(m) = \begin{cases} 0 & m = 1 \\ \dfrac{Ldx\sum\limits_{k=1}^{m-1}\frac{B_{\delta0\_slotless\_n}(k)+B_{\delta0\_slotless\_n}(k+1)}{2}}{K_{Fe}h_{j1}D} & m = [2,N] \end{cases} \tag{7}$$

Meanwhile, it could be shown that the total air-gap flux in the whole length range of the UP-LPST is 0:

$$\phi_g(N) = Ldx\sum_{k=1}^{N}\frac{B_{\delta\_n}(k) + B_{\delta\_n}(k+1)}{2} = 0 \tag{8}$$

For the UP-LPST, the harmonic frequency of the fundamental wave current is $24k \pm 1$, so the lowest is the 23rd harmonic. Due to the high harmonic order, only the fundamental MMF was considered to simplify the calculation. Thus, the MMF of each air-gap node was calculated as:

$$F(m) = \frac{\tau}{\pi}J_1 \cos\left[\omega t - \frac{\tau}{\pi}x(m) + \frac{\pi}{2}\right] \tag{9}$$

where

$$\begin{cases} J_1 = \dfrac{12\sqrt{2}N_1 k_{w1}I_1}{p\tau} \\ x(m) = (m-1)dx \end{cases} \tag{10}$$

In this equation, N_1 is the number of turns on the primary side, I_1 is the primary measured current, k_{w1} is the winding coefficient, p is the polar logarithm, τ is the polar distance, and dx is the length of each block after segmentation. The normal magnetic flux density of the air gap at node m was calculated as follows:

$$B_{\delta0\_slotless\_n}(m) = \frac{F(m)\mu_0}{\delta K_s} \tag{11}$$

where K_s is the preset saturation coefficient and μ_0 is the vacuum permeability.

Finally, we judged whether the iteration precision value was satisfied:

$$\sum_{m=1}^{N}\left[\frac{F(m) - F_\Sigma(m)}{F(m)}\right]^2 < \varepsilon \tag{12}$$

When the actual error was greater than the accuracy requirement, the air-gap flux density was corrected as follows [26]:

$$B_{\delta0\_slotless\_n}(m) = B_{\delta0\_slotless\_n}(m)\left[1 + k_s\frac{F(m) - F_\Sigma(m)}{F(m)}\right] \tag{13}$$

where k_s is the iterative coefficient.

The iteration flow chart is shown in Figure 4. Firstly, select the DC bus voltage and calculate the effective value of the primary current. Then, calculate the initial air-gap flux density $B_{\delta0\_slotless\_n}(m)$ according to Equation (11). Calculate the magnetic density $B_{j1\_1}(m)$ and $B_{t1\_n}(m)$ of each node of the primary yoke and teeth according to Equations (6) and (8). Determine the magnetic permeability $\mu_{j1}(m)$ and $\mu_{t1}(m)$ of each node of the primary yoke and teeth using the *B-H* curve. Finally, carry out the iterative calculation until the judgment conditions are satisfied.

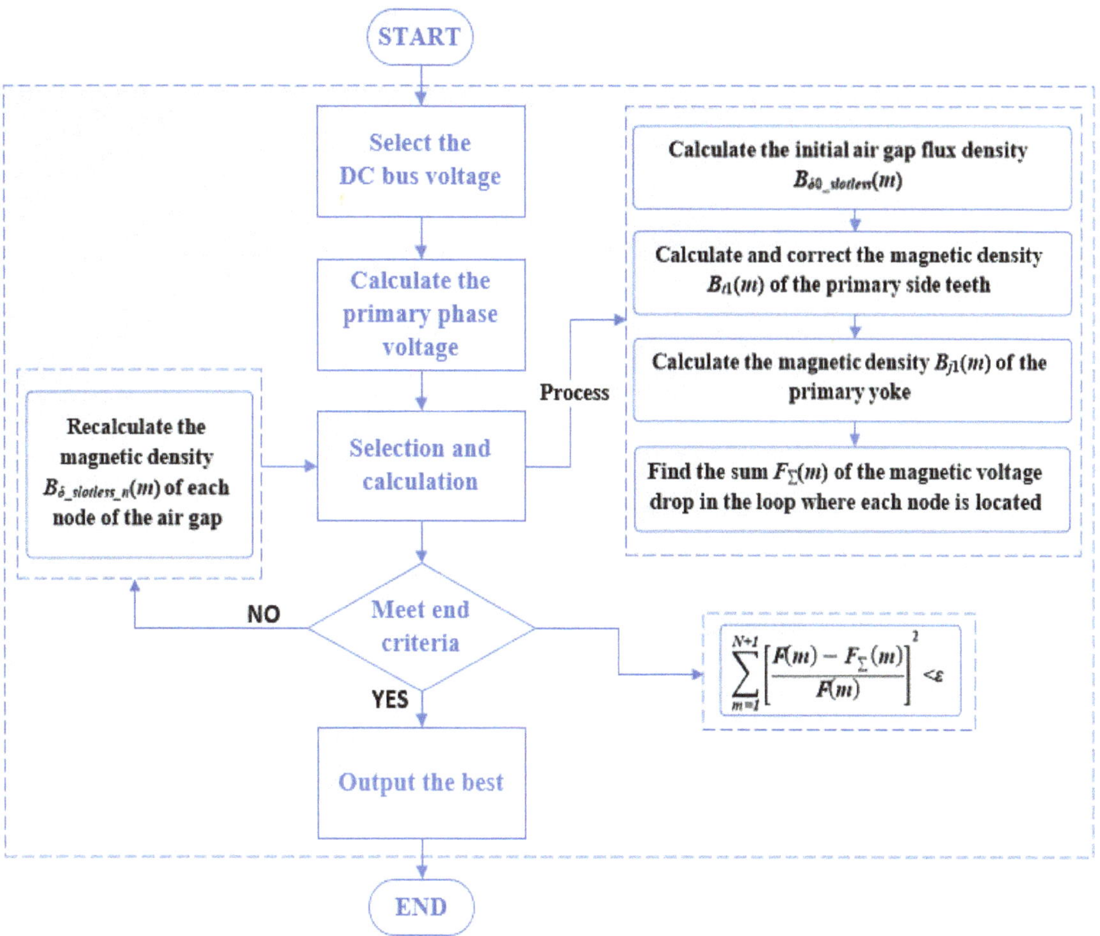

Figure 4. The iterative flowchart of the UP-LPST based on the DMCM.

By extending the DMCM to the LPST, the magnetic field distribution of the air gap and iron core yoke could be obtained. This paper only describes the air gap in detail. The final result is shown in Figure 5:

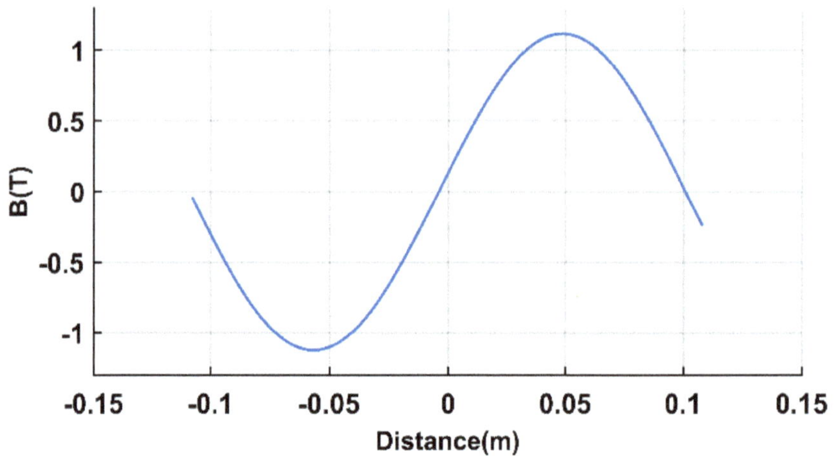

Figure 5. The slotless air-gap magnetic field within the length range of the UP-LPST.

3. Air-Gap Relative Permeance Based on SCT

3.1. Cogging Effect

According to [20], when the ratio of teeth width to air-gap length is greater than 2.44, the influence between adjacent slots can be ignored. However, the ratio of teeth width to air-gap length of the model used in this paper was much higher than 2.44. In order to facilitate the analysis, a single-slot model was adopted in this paper. Firstly, the following assumptions were made [20]:

1. The primary side is slotted, and the secondary side is a smooth plane.
2. The magnetic conductivity of the core on the primary and secondary sides is infinite.
3. Both the primary and secondary sides of the iron core are planes with equal magnetic potential, one of which is 0 and the other φ_0.

When the slot depth of the UP-LPST had been determined, the polygon of the z plane could be obtained, as shown in Figure 6a.

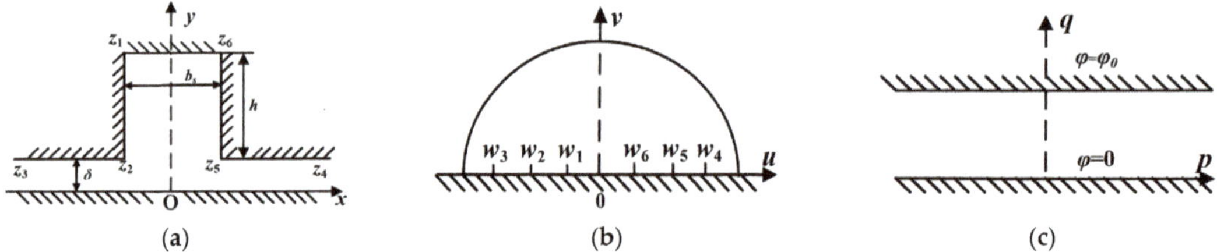

Figure 6. SCT analysis diagram of UP-LPST cogging effect. (**a**) z plane; (**b**) w plane; (**c**) t plane.

The relationship between the z plane and the w plane could be obtained from Table 1. below:

$$z = S \int \frac{\sqrt{w^2 - a^2}}{\sqrt{w^2 - 1}(w^2 - b^2)} dw + K = S' \int \frac{\sqrt{1 - \frac{w^2}{a^2}}}{\sqrt{1 - w^2}(1 - \frac{w^2}{b^2})} dw + K \qquad (14)$$

where

$$S' = -S\frac{a}{b^2}$$

Table 1. SCT table of z plane and w plane of finite slot depth model.

z Plane		θ *	w Plane	
Point	**Coordinate**		**Point**	**Coordinate**
z_1	$-b_s/2 + j(h + \delta)$	$\pi/2$	w_1	-1
z_2	$-b_s/2 + j\delta$	$3\pi/2$	w_2	$-a$
z_3	$-\infty + j0$ $-\infty + j\delta$	0	w_3	$-b$
z_4	$+\infty + j0$ $+\infty + j\delta$	0	w_4	b
z_5	$b_s/2 + j\delta$	$3\pi/2$	w_5	a
z_6	$b_s/2 + j(h + \delta)$	$\pi/2$	w_1	1

* The interior angle of a polygon.

Introducing the Jacobian elliptic function, let w be the inverse function of the intermediate variable k, as follows:

$$\begin{cases} w = snk \\ dw = cnk \cdot dnk \cdot dk = cnk\sqrt{1 - \frac{sn^2k}{a^2}}\,dk \end{cases} \tag{15}$$

Then (14) can be expressed as:

$$z = S'\int_0^k \frac{(1 - \frac{sn^2k}{a^2})}{(1 - \frac{sn^2k}{b^2})}\,dk = S'\int_0^k \left[1 + \left(\frac{1}{b^2} - \frac{1}{a^2}\right)\frac{sn^2k}{(1 - \frac{sn^2k}{b^2})}\right]dk \tag{16}$$

The transformation between the z–w plane is:

$$z = \frac{2\delta}{\pi}\left[\frac{sn\alpha dn\alpha}{cn\alpha}sn^{-1}w - \Pi(k, \alpha)\right] \tag{17}$$

where $sn\alpha$, $cn\alpha$, and $dn\alpha$ are Jacobian elliptic functions; $\Pi(k, \alpha)$ is the elliptic integral of the third kind; and δ is the air-gap length. Through the corresponding relation of the z–w plane, the relation between a, α, b_s, and δ could be obtained as follows:

$$\begin{cases} \dfrac{b_s}{\delta} - \dfrac{4K(\frac{1}{a^2})}{\pi}\left[\dfrac{sn\alpha \cdot dn\alpha}{cn\alpha} - Z(\alpha)\right] = 0 \\[2ex] \dfrac{h}{\delta} - \dfrac{2K'(\frac{1}{a^2})}{\pi}\left[\dfrac{sn\alpha \cdot dn\alpha}{cn\alpha} - Z(\alpha)\right] - \dfrac{\alpha}{K(\frac{1}{a^2})} = 0 \end{cases} \tag{18}$$

where $K(1/a2)$ is the elliptic integral of the first kind; $K'(1/a2)$ is the elliptic integral of the first kind of complementary modules; and $Z(\alpha)$ is the Jacobian zeta function. The abovementioned expressions are:

$$\begin{cases} \Pi(k, \alpha) = \int_0^{\sin\phi} \dfrac{dt}{(1-kt^2)\sqrt{(1-k^2t^2)(1-t^2)}} \\[2ex] K(\frac{1}{a^2}) = F(1, \frac{1}{a}) \\[2ex] E(x, \frac{1}{a}) = \int_0^x \dfrac{\sqrt{1 - \frac{t^2}{a^2}}}{(1-t^2)}\,dt \end{cases} \quad \begin{cases} Z(\alpha) = E(\alpha) - \dfrac{E(\frac{1}{a})}{K(\frac{1}{a})}K(\alpha) \\[2ex] F(x, \frac{1}{a}) = \int_0^x \dfrac{dt}{(1-t^2)(1-\frac{t^2}{a^2})} \end{cases} \tag{19}$$

where $F(x,1/a)$, $E(x,1/a)$, and $E(1/a_2)$ are the incomplete elliptic integral of the first kind, incomplete elliptic integral of the second kind, and complete elliptic integral of the second kind, respectively.

After logarithmic transformation, the corresponding relation of the *t-w* plane was written as:

$$t = \frac{\varphi_0}{\pi}[In(w-b) - In(w+b)] = \frac{\varphi_0}{\pi} In\left(\frac{w-b}{w+b}\right) \tag{20}$$

When the slot depth had been determined, the magnetic field density distribution of the air gap and slot in the UP-LPST was calculated as:

$$B_{\delta\_slot} = \mu_0 \left| \frac{dt}{dw} \cdot \frac{dw}{dz} \right| = \frac{\mu_0 \varphi_0}{\delta} \left| \frac{cn\alpha}{bsn\alpha \cdot dn\alpha} \sqrt{\frac{1-w^2}{1-\frac{w^2}{a^2}}} \right| \tag{21}$$

$$\lambda_{YB\_cogging\_n} = \frac{B_{\delta\_slot\_n}}{B_{\delta\_slotless\_n}} \tag{22}$$

3.2. End Effect

Because the core of the UP-LPST is not continuous, the end effect as its inherent property also needed to be analyzed. The specific SCT analysis model is shown in Figure 7.

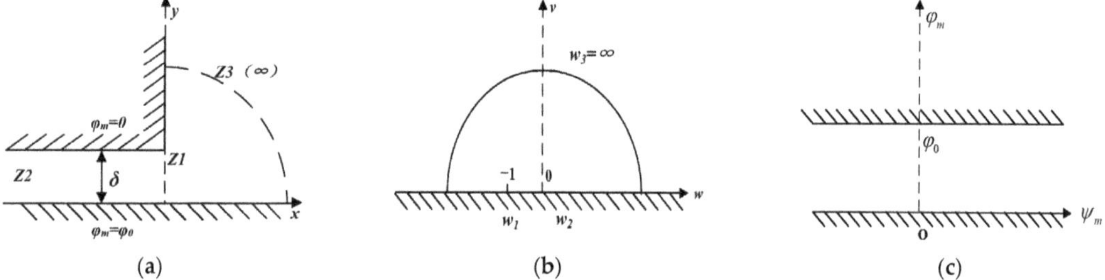

Figure 7. SCT analysis diagram of UP-LPST end effect. (**a**) z plane; (**b**) w plane; (**c**) t plane.

The relationship between the z plane and the w plane could be obtained from Table 2 below:

Table 2. SCT table of z plane and the w plane at the primary-side end.

z Plane		θ *	w Plane	
Point	**Coordinate**		**Point**	**Coordinate**
z_1	$j\delta$	$3\pi/2$	w_1	-1
z_2	$-\infty + j0$ $-\infty + j\delta$	0	w_2	0
z_3	$+\infty$ $+j\delta$	0	w_3	$\pm\infty$

* The interior angle of a polygon.

Similarly, the relationship of side *z–w–t* is:

$$z = \frac{\delta}{\pi}\left(2\sqrt{w+1} + In\frac{\sqrt{w+1}-1}{\sqrt{w+1}+1}\right) \tag{23}$$

$$t = \varphi_0 - \frac{\varphi_0}{\pi} In\, w \tag{24}$$

After considering the end effect, the magnetic flux density on the primary core was determined to be:

$$B_{end} = \mu_0 H = \mu_0 \left| \frac{dt}{dw} \cdot \frac{dw}{dz} \right| = \frac{\mu_0 \varphi_0}{\delta \sqrt{w+1}} \tag{25}$$

The magnetic flux density reached its maximum at point z_2 in the z plane:

$$B_{\max} = \frac{\mu_0 \varphi_0}{\delta} \tag{26}$$

With B_{max} as the base value, the magnetic field density at any point is as follows:

$$\frac{B_{end}}{B_{\max}} = \left| \frac{1}{\sqrt{w+1}} \right| \tag{27}$$

By inserting certain w values, the magnetic field density distribution curve could be drawn as shown in Figure 8.

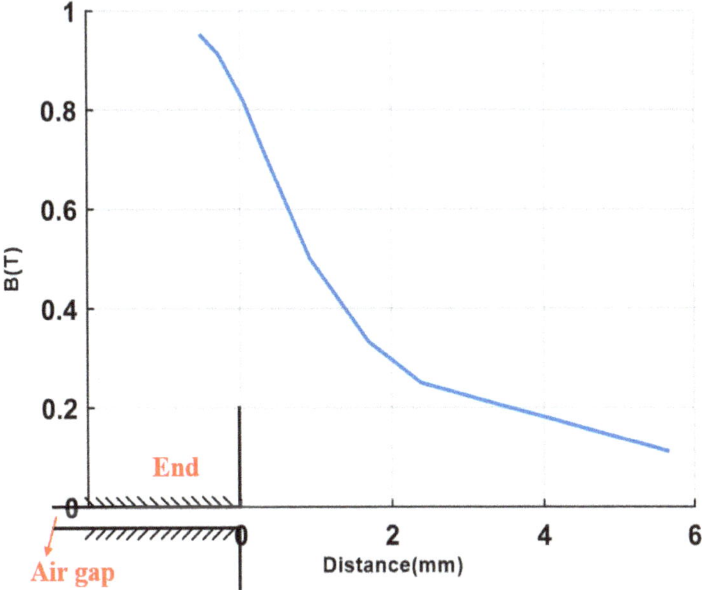

Figure 8. Distribution of air-gap magnetic field density at the UP-LPST end.

The relative permeability distribution function of the primary side considering the end effect could be obtained by fitting the curve in Figure 8.

$$\lambda'_{YB\_end}(x) = \begin{cases} e^{\frac{1}{2\delta}\left(x+\frac{L}{2}\right)} & x < -\frac{L}{2} \\ 1 & -\frac{L}{2} < x < \frac{L}{2} \\ e^{\frac{1}{2\delta}\left(x-\frac{L}{2}\right)} & \frac{L}{2} < x \end{cases} \tag{28}$$

The relative permeability of the air gap obtained when the secondary side was slotted separately was similar to that of the primary side, because the core structures of the primary and secondary sides of the UP-LPST are exactly the same. According to the above analysis, the air-gap magnetic field density $B(x)$ of the UP-LPST along the length direction could be calculated as follows:

$$B(x) = B_{\delta\_slotless\_n}(x) \cdot \lambda_{YB\_cogging\_n} \cdot \lambda_{YB\_end} \cdot \lambda_{FB\_cogging\_n} \cdot \lambda_{FB\_end} \tag{29}$$

4. Results

4.1. Analytical Results

In this paper, an unequal-pitch linear phase-shifting transformer was taken as an example. The specific parameters are shown in Table 3.

Table 3. Parameters of the UP-LPST.

Symbol	Value	Meaning
h	24 mm	Depth of the slot
b_s	12 mm	Width of the slot
t_1	18 mm	Tooth pitch
δ	0.3 mm	Air-gap length
L	216 mm	Longitudinal length of core
D	100 mm	Normal width of core
τ	105 mm	Pole pitch

Based on the equations presented in Section 2, the air-gap relative permeance at different normal positions within a range of tooth pitch could be obtained. In order to analyze the variation trend of the air-gap relative permeance λ in a single slot, four different positions were selected: the center line of the air gap ($y = 0$ mm), the outer surface of the primary side ($y = \delta/2$ mm), the center of the primary-side slot ($y = \delta/2 + h/3$ mm), and the bottom of the primary-side slot ($y = \delta/2 +$ hmm). Figure 9 shows that the air-gap permeability distribution was different for different air-gap radii. The closer to the opening surface of the slot, the greater the influence of the opening slot on the air-gap magnetic field, that is, the deeper the pit.

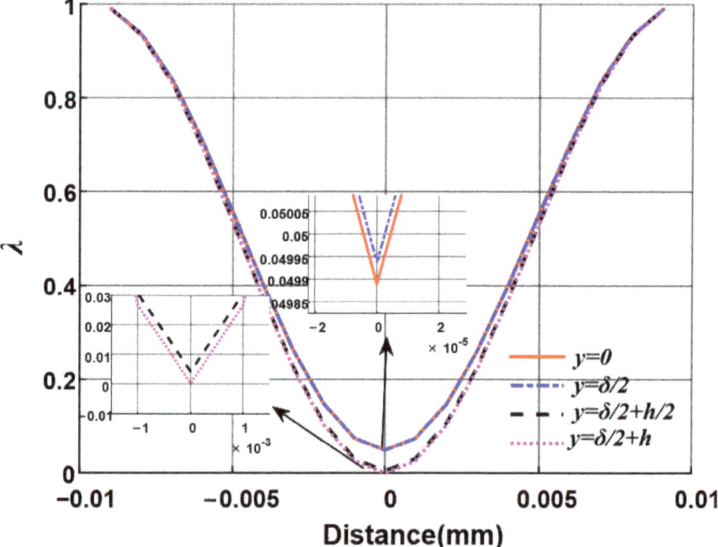

Figure 9. Air-gap relative permeance at different positions in a single slot.

The air-gap relative permeance in a single slot was decomposed by Fourier transformation with the tooth pitch as the period. In this way, the air-gap relative permeance across the whole length range of the UP-LPST could be obtained, as shown in Figure 10a. Figure 10b shows that the magnetic field density distribution at the center of the air gap considering the effects of cogging and end could be obtained by multiplying the calculated air gap relative permeability by the slotless magnetic field. The air-gap magnetic field was not highly sinusoidal when considering the effect of slot and end and was not affected by slot openings that were opposite teeth. However, when the slots were opposite each other, the air-gap magnetic field was affected by the interaction of the iron cores on both sides, presenting a concave shape. At the same time, due to the influence of "out and in", the magnetic flux density distribution at both ends was different.

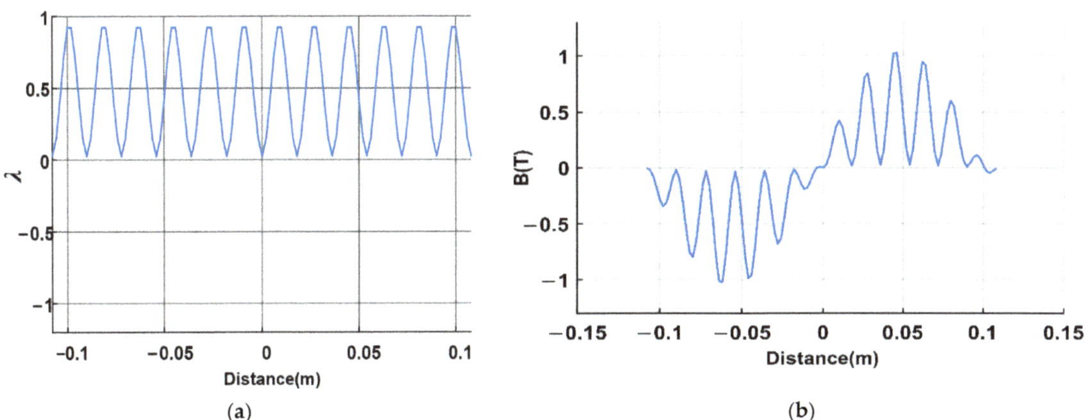

(a)   (b)

Figure 10. Distribution of air-gap relative permeance and flux density within the length range of the UP-LPST. (a) Air-gap relative permeance; (b) distribution of air-gap flux density under the influence of cogging and end.

Figure 11 presents the schematic diagram of harmonic amplitude and flux density distribution under different slot spacing ratios. The main harmonic amplitudes could be obtained by the Fourier decomposition of the air-gap magnetic fields with different slot spacing ratios. When the slot opening increased, the harmonic amplitude of the basic teeth increased significantly, which reduced the performance of the straight-line phase-shifting transformer. In Figure 11b, the lengths of the slot openings are 10 mm, 11 mm, 12 mm, and 13 mm. The minimum flux density was located in the middle of the slot opening, while the maximum flux density was located in the tooth, and the large slot opening had a greater impact on the flux density.

4.2. FEM Verification

To verify the correctness of the analytical results, an FEM model of the UP-LPST was constructed according to Table 1, as shown in Figure 12. A 100V DC voltage was applied to the FEM model, and the magnetic field distribution data at the center line of the air gap were extracted and compared with the analytical results. The comparison results are shown in Figure 13. The results showed that the overall distribution fitted well, but the details were slightly different. The errors were mainly focused on the top and edge of the sawtooth wave. This was because the magnetic field lines at the teeth and air gap were assumed to have only normal components. However, near the opening of the slot, the effect of magnetic focusing on the air-gap magnetic field was great, which was one of the reasons for the error. At the same time, this method did not consider the magnetic flux leakage effect of the UP-LPST, so there was a 2% error between the analytical results and the FEM results.

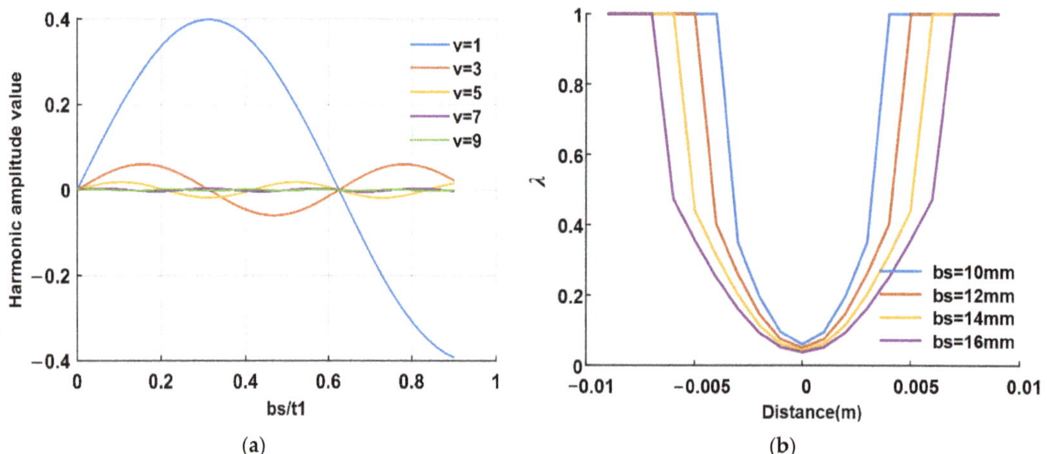

Figure 11. Distribution of harmonic amplitude and magnetic field density at different slot spacing ratios. (**a**) Harmonic amplitude value; (**b**) single-slot flux density distribution.

Figure 12. The 2D FEM simulation model.

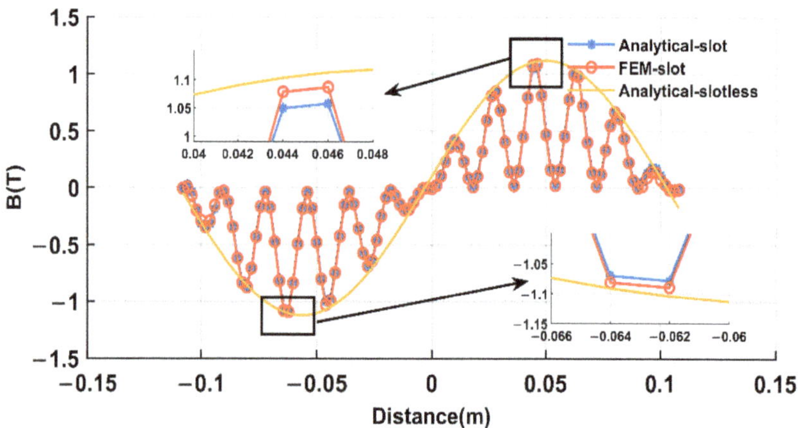

Figure 13. The air-gap flux density distribution of the UP-LPST.

Compared with the exact subdomain method (ESM) in [28], the error of the DMCM–SCT method was smaller due to the consideration of the influence of core saturation and edge, and the specific performance results are shown in Table 4.

Table 4. Comparison of the three methods.

Method	Core Saturation	End Effect	Calculation Dimension	Error
FEM	√	√	109,506	-
ESM	×	×	1136	<5%
DMCM–SCT	√	√	1300	<2%

According to Table 4, compared with ESM, the results obtained by DMCM–SCT were closer to those obtained by the finite element method. Although a certain calculation dimension was added, the overall effect was still better than ESM. However, the influence of end magnetic flux leakage was not considered in the calculation method, so there were still some errors in the results.

4.3. Test Verification

According to the data in Table 1, a prototype of the UP-LPST was manufactured, as shown in Figure 14. The test platform was mainly composed of a power supply driver module, a DSP, a signal amplification control circuit, a rectifier bridge, an inverter module, a terminal, and a UP-LPST. Since the air gap of the experimental prototype was only 0.3 mm, it was difficult to directly measure the actual air-gap flux density. Therefore, we carried out indirect verification by testing the no-load output voltage and current of the prototype.

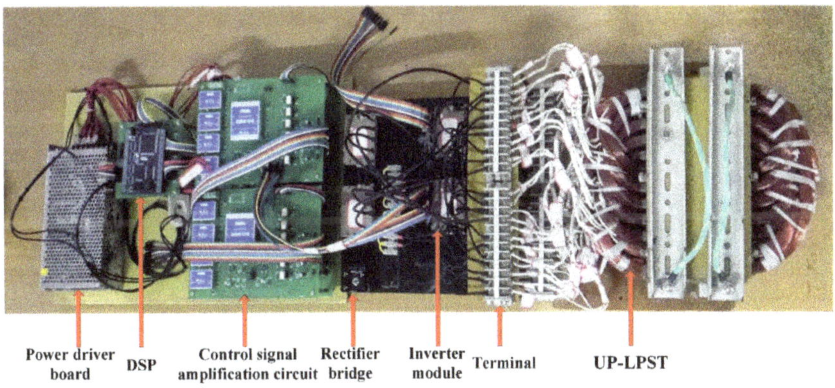

Figure 14. Experimental transformer.

Figure 15 shows the experimental waveform of the no-load output voltage of the prototype and the waveform of the FEM simulation calculation. By comparison, it could be found that the waveforms of the no-load output voltage calculated by FEM were basically the same as those measured by the experiment, but there were still slight differences. The main reason was that the manufacturing technology of the prototype was not ideal and the precision of core was not sufficient. At the same time, in the actual processing experiment, it was difficult to ensure that the length of the air gap was 0.3 mm, so the length of the air gap was not completely equal. However, the overall waveform direction was consistent, which indirectly verified the effectiveness of the analysis method in this paper.

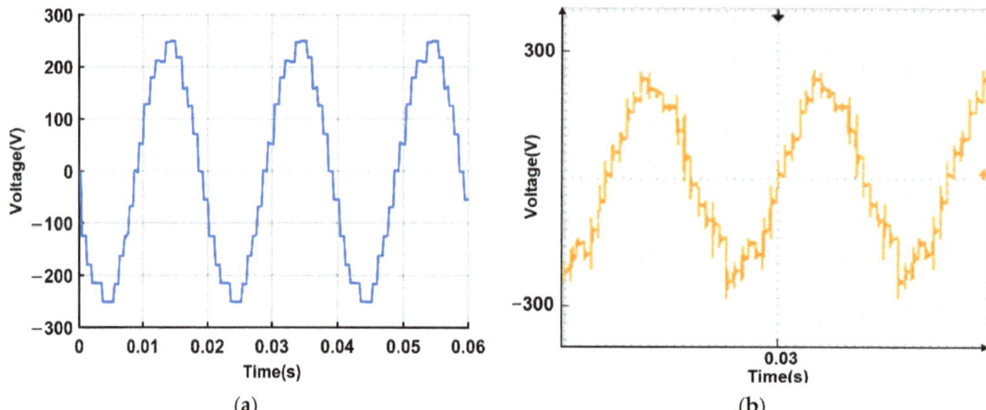

Figure 15. No-load phase-A output voltage diagram of finite element model and experiment. (**a**) FEM waveform; (**b**) experimental waveform.

5. Conclusions

The impact of connecting new energy sources to a power system on the power quality of the grid is mainly reflected in the voltage and current. The new power system will consume more reactive power during operation, which will cause a serious drop in the grid voltage. Therefore, a large number of rectifier and inverter devices need to be used in the grid connection. The existence of these devices will inevitably inject harmonic currents into the grid, resulting in the distortion of grid voltage and current waveforms. While affecting the power supply quality, it will also cause additional load losses to the power equipment flowing through the distorted current. The special structure and phase-shifting method of the linear phase-shifting transformer can effectively eliminate low-order harmonics and improve the output waveform quality. For the LPST, which mainly relies on air gaps for energy transfer, the establishment of an accurate magnetic field model has an important influence on the calculation and even the improvement of the system efficiency.

By focusing on the specifics of LPST energy transfer and analyzing its structure, the DMCM and SCT were extended to the LPST, and a magnetic field analysis model for the LPST was established. The model can simultaneously consider the effects of saturation, cogging, and edges on the magnetic field. Therefore, the model is more in line with the actual situation of the linear phase-shifting transformer. Under the conditions of a given input DC bus voltage, the accuracy of the model was proven by means of direct verification using a finite element model and indirect verification via an experiment. There was a degree of error in the results, because the magnetic leakage at the end of the linear phase-shifting transformer was not considered. However, the model can still be applied to phase-shifting transformers with similar structures, and even other power devices. Our model could play an important role in improving the efficiency of transformers and new power systems and reducing losses. However, this paper only analyzed the model under no-load conditions. An analysis under load conditions is the next step.

Author Contributions: Writing—original draft preparation, J.X.; investigation, H.W.; supervision, J.Z.; data curation, S.Y., C.Z., D.Y. and H.C. All authors have read and agreed to the published version of the manuscript.

Funding: This research was funded by the National Natural Science Foundation of Hubei under the grant 2018CFA008.

Institutional Review Board Statement: Not applicable.

Informed Consent Statement: Not applicable.

Data Availability Statement: Not applicable.

Conflicts of Interest: The authors declare no conflict of interest.

References

1. Mastny, P.; Moravek, J.; Drapela, J.; Vrana, M.; Vojtek, M. Problems of Verification of Operating Parameters of DC/AC Inverters and Their Integration into the Distribution System in the Czech Republic. In Proceedings of the CIRED Porto Workshop 2022: E-Mobility and Power Distribution Systems, Porto, Portugal, 2–3 June 2022; pp. 304–308. [CrossRef]
2. Iqbal, S.; Hossain, A.; Islam, J.; Surja, A.S.; Kabir, M. Enhancing Voltage Stability of Inter-Area Multi-Machine Power Systems using Reinforcement Learning-based STATCOM. In Proceedings of the 2022 International Conference on Advancement in Electrical and Electronic Engineering (ICAEEE), Gazipur, Bangladesh, 24–26 February 2022; pp. 1–4. [CrossRef]
3. Jalilian, A.; Muttaqi, K.M.; Sutanto, D.; Robinson, D. Distance Protection of Transmission Lines in Presence of Inverter-Based Resources: A New Earth Fault Detection Scheme During Asymmetrical Power Swings. *IEEE Trans. Ind. Appl.* **2022**, *58*, 1899–1909. [CrossRef]
4. Calar, M.; Durna, E.; Kayisli, K. 3-Phase Multi-Pulse Rectifiers with Different Phase Shifting Transformers and Comparison of Total Harmonic Distortion. In Proceedings of the 2022 9th International Conference on Electrical and Electronics Engineering (ICEEE), Alanya, Turkey, 29–31 March 2022; pp. 60–64. [CrossRef]
5. Yuan, D.; Yin, Z.; Wang, S.; Duan, N. Multi-level Transient Modeling of the Aeronautic Asymmetric 18-pulse Phase-shifting Auto-transformer Rectifier in Full-cycle Design. *IEEE Trans. Transp. Electrif.* **2022**, *8*, 3759–3770. [CrossRef]
6. Luo, M.; Dujic, D.; Allmeling, J. Leakage Flux Modeling of Medium-Voltage Phase-Shift Transformers for System-Level Simulations. *IEEE Trans. Power Electron.* **2018**, *34*, 2635–2654. [CrossRef]
7. Chen, G.; Ding, L.; Tang, F.; Teng, Y. The study of increasing Chongqing sectional transmission capacity using phase-shifting transformer. *Sichuan Electr. Power Technol.* **2014**, *37*, 49–54. (In Chinese)
8. Xiong, X.; Zhao, J.H.; Ding, H.; Sun, P.; Peng, W. Research on end effect of linear phase-shifting transformer. *J. Xi'an Jiaotong Univ.* **2017**, *51*, 110–115. (In Chinese)
9. Zhang, Z.M.; Zhao, J.H.; Ma, Y.Z.; Xu, H. Analytical modeling of inverter system of linear phase-shifting transformer. *Chin. J. Electr. Eng.* **2019**, *14*, 54–60. (In Chinese)
10. Zhao, J.H.; Xu, H.; Guo, G.Q. Design of 12/3-phase linear phase-shifting transformer. *J. Nav. Eng. Univ.* **2021**, *33*, 1–6. (In Chinese)
11. Zhao, J.H.; Ma, Y.Z.; Sun, P. Multiple superposition inverter system based on linear phase-shifting transformer. *Electr. Power Autom. Equip.* **2019**, *39*, 183–188. (In Chinese) [CrossRef]
12. Sterling, H. Harmonic Field Effects in Induction Machines. *IEEE Rev.* **1977**, *23*, 841. [CrossRef]
13. Wu, L.J.; Zhu, Z.Q.; Staton, D.A.; Popescu, M.; Hawkins, D. Subdomain Model for Predicting Armature Reaction Field of Surface-Mounted Permanent-Magnet Machines Accounting for Tooth-Tips. *IEEE Trans. Magn.* **2011**, *47*, 812–822. [CrossRef]
14. Wu, L.J.; Zhu, Z.Q.; Staton, D.; Popescu, M.; Hawkins, D. An Improved Subdomain Model for Predicting Magnetic Field of Surface-Mounted Permanent Magnet Machines Accounting for Tooth-Tips. *IEEE Trans. Magn.* **2011**, *47*, 1693–1704. [CrossRef]
15. Zhu, Z.Q.; Wu, L.J.; Xia, Z.P. An Accurate Subdomain Model for Magnetic Field Computation in Slotted Surface-Mounted Permanent-Magnet Machines. *IEEE Trans. Magn.* **2009**, *46*, 1100–1115. [CrossRef]
16. Jikun, Y.; Liyi, L.; Jiangpeng, Z. Analytical calculation of air-gap relative permeance in slotted permanent magnet synchronous motor. *Trans. China Electrotech. Soc.* **2016**, *31*, 45–52.
17. Fu, L.; Zuo, S.; Ma, C.; Tan, Q. Analytical calculation of armature reaction field including slotting effects in PMSM with concentrated fraction-al-slot winding. *Trans. China Electrotech. Soc.* **2014**, *29*, 29–35.
18. Zhu, Z.Q.; Howe, D. Instantaneous magnetic field distribution in brushless permanent magnet DC motors, part III: Effect of stator slotting. *IEEE Trans. Magn.* **1993**, *29*, 143–151. [CrossRef]
19. Lv, Y.; Cheng, S.; Wang, D.; Chen, J. A Fast Method for Calculating the Air-Gap Flux and Electromagnetic Force Distribution in Surface Permanent Magnet Motors. In Proceedings of the 2020 IEEE 9th International Power Electronics and Motion Control Conference (IPEMC2020-ECCE Asia), Nanjing, China, 29 November–2 December 2020; pp. 1973–1977. [CrossRef]
20. Tang, Y. *Electromagnetic Fields in Motors*, 2nd ed.; Science Press: Beijing, China, 1998.
21. Lee, S.-H.; Yang, I.-J.; Kim, W.-H.; Jang, I.-S. Electromagnetic Vibration-Prediction Process in Interior Permanent Magnet Synchronous Motors Using an Air Gap Relative Permeance Formula. *IEEE Access* **2021**, *9*, 29270–29278. [CrossRef]
22. Chang, J. New direct drive method for large telescope base on Arc PMLSM. *Univ. Chin. Acad. Sci.* **2013**.
23. Guo, G.; Zhao, J.; Wu, M.; Xiong, Y. Analytical Calculation of Magnetic Field in Fractional-Slot Windings Linear Phase-Shifting Transformer Based on Exact Subdomain Model. *IEEE Access* **2021**, *9*, 122351–122361. [CrossRef]
24. Jiang, H.; Su, Z.; Wang, D. Analytical Calculation of Active Magnetic Bearing Based on Distributed Magnetic Circuit Method. *IEEE Trans. Energy Convers.* **2020**, *36*, 1841–1851. [CrossRef]
25. Wang, N.; Wu, X.; Chen, J.; Guo, Y.; Cheng, S. A Distributed Magnetic Circuit Approach to Analysis of Multiphase Induction Machines with Nonsinusoidal Supply. *IEEE Trans. Energy Convers.* **2014**, *30*, 522–532. [CrossRef]
26. Guo, Y.; Wang, D.; Liu, D.; Wu, X.; Chen, J. Magnetic Circuit Calculation of Non-Salient Pole Synchronous Generator Based on Distributed Magnetic Circuit Method. In Proceedings of the 2011 International Conference on Electrical Machines and Systems, Beijing, China, 20–23 August 2011; pp. 1–6. [CrossRef]

27. Huang, C.; Sun, Z.; Xu, J.; Zeng, R.; Zhang, Y.; Han, Z.; Li, M. Calculation Method of Nonlinear Magnetic Circuit and Motor Performance Analysis for High-Power Linear Induction Motor. In Proceedings of the 2021 13th International Symposium on Linear Drives for Industry Applications (LDIA), Wuhan, China, 1–3 July 2021; pp. 1–6. [CrossRef]
28. Guo, G.; Zhao, J.; Xiong, Y.; Wu, M. Analytical Calculation of Open-Circuit Magnetic Field in Linear Phase-Shifting Transformer Based on Exact Subdomain Model. *IEEJ Trans. Electr. Electron. Eng.* **2021**, *17*, 72–81. [CrossRef]

 *processes*

Article

Layout Method of Met Mast Based on Macro Zoning and Micro Quantitative Siting in a Wind Farm

Wenjin Chen [1], Gang Qian [2], Weiwen Qi [2], Gang Luo [2], Lin Zhao [3] and Xiaoling Yuan [3,*]

1   State Grid Zhejiang Electric Power Company Limited, Hangzhou 310007, China
2   State Grid Shaoxing Power Supply Company, Shaoxing 312000, China
3   College of Energy and Electrical Engineering, Hohai University, Nanjing 211100, China
*   Correspondence: lingx@hhu.edu.cn

Abstract: In order to promote the wind monitoring accuracy and provide a quantitative planning method for met mast layout in practical projects, this paper proposes a two-stage layout method for met mast based on discrete particle swarm optimization (DPSO) zoning and micro quantitative siting. Firstly, according to the wind turbines layout, rotational empirical orthogonal function and hierarchical clustering methods are used to preliminarily determine zoning number. Considering the geographical proximity of wind turbines and the correlation of wind speed, an optimal macro zoning model of wind farm based on improved DPSO is established. Then, combined with the grid screening method and optimal layout evaluation index, a micro quantitative siting method of met mast is proposed. Finally, the rationality and efficiency of macro zoning method based on improved DPSO, as well as the objectivity and standardization of micro quantitative siting, are verified by an actual wind farm.

Keywords: met mast layout; REOF; DPSO macro zoning; micro quantitative siting

Citation: Chen, W.; Qian, G.; Qi, W.; Luo, G.; Zhao, L.; Yuan, X. Layout Method of Met Mast Based on Macro Zoning and Micro Quantitative Siting in a Wind Farm. *Processes* **2022**, *10*, 1708. https://doi.org/10.3390/pr10091708

Academic Editor: Zhiwei Gao

Received: 26 July 2022
Accepted: 24 August 2022
Published: 27 August 2022

Publisher's Note: MDPI stays neutral with regard to jurisdictional claims in published maps and institutional affiliations.

1. Introduction

1.1. Motivation of This Research

In order to achieve carbon neutrality and boost the construction of power system with a high proportion of renewable energy, wind power and other clean energy are developing rapidly, and the number and scale of wind farm constructions is increasing recently [1]. Met mast represents the basic equipment for wind resource monitoring and evaluation and plays an important role in the planning, construction and operation stage of wind farm [2,3]. The data of met mast not only represent an important basis for deciding whether to build a wind farm, but also the support for wind power prediction and closed-loop assessment of wind farms [4,5]. However, at present, some wind power enterprises are lack of emphasis on met mast. Meanwhile, the problem of setting up met mast arbitrarily is prominent, which greatly reduces the original value creation of met mast [6]. Therefore, it is necessary to arrange the met mast scientifically and rationally.

1.2. Literature Review

The layout of met mast in a wind farm is mainly concerned with two issues, namely the determination of the number of met mast and representative wind zone scope of each met mast and the micro siting of met mast within corresponding wind zones. For the first issue, it is mostly processed with methods based on macro zoning of wind farm. The number of met masts is consistent with zoning number of wind farm, and representative wind zone scope is presented through zoning result [7,8]. Currently, there are some studies on the macro zoning of wind farms. In [9,10], zoning of wind farms is conducted in the practical engineering field considering the empirical reference radius of a representative area range of met mast under different terrains. In [11], based on the spatial distribution density of wind turbines, density based spatial clustering of applications with noise (DBSCAN) is used to cluster wind turbines to realize zoning of wind farm. However, input parameters of DBSCAN algorithm are not easy to be selected, which greatly affects zoning results. In [12], considering wind speed correlation of wind turbines, rotational empirical orthogonal function (REOF) method is used to obtain spatial distribution characteristics of wind speed to achieve zoning of wind farm. However, this zoning method may lead to problem with zoning overlapping of wind turbines. For the second issue of micro siting of met mast, the related work at present is mostly based on qualitative analysis. In [13,14], alternative wind monitoring points are preliminarily selected out based on empirical layout principle, and computational fluid dynamics (CFD) tool is used to obtain important wind flow parameters of alternative wind monitoring points, then optimal met mast location is determined by correlation analysis. In [15], met mast location is screened out by landform similarity, wind climate similarity and other judging indexes. However, wake effect is rarely considered, which may select inappropriate met mast location considering real incoming wind speed cannot be obtained.

The existing problems in the current research are summarized as follows.

(1) The determination of the number of met mast is mostly dependent on engineering experience, and this method lacks reasonable quantitative calculation.
(2) The current zoning methods can not directly and automatically get zoning results, and human subjective judgment accounts for a certain proportion in the process.
(3) In the process of micro siting of met mast, the wake effect of wind turbines is ignored so the final selected met mast location cannot be guaranteed to be optimal. Meanwhile, quantitative layout indexes and the systematic siting method of met mast are absent in recent studies.

1.3. Contributions and Innovations

To fill research gaps, this paper proposes a relatively objective and efficient quantitative layout method of met mast. Firstly, the number of zoning is preliminarily determined based on REOF decomposition and hierarchical clustering (HC) method. The distance between wind turbines is redefined considering location proximity and wind speed correlation, successively a wind farm optimization zoning model based on inter-class dispersion degree and intra-class aggregation degree is established and solved by discrete particle swarm optimization (DPSO). Then, the micro quantitative siting strategy of alternative wind monitoring points based on grid screening method is proposed, and the optimal location of met mast is determined by the layout evaluation index. Finally, a real wind farm is used for simulation verification. The results demonstrate that the proposed method can reasonably determine the layout of met mast and has certain practicability.

The main contributions of this paper include the following:

(1) A quantitative calculation method of zoning number based on REOF and HC is proposed.
(2) Based on newly defined distance between wind turbines considering geographical location proximity and wind speed correlation, a DPSO zoning model is established, which helps to get zoning results directly.
(3) Considering various wind flow factors, including wake effect, a quantitative siting strategy for met mast is proposed and an evaluation index of micro siting is designed.

1.4. Organization of This Paper

The remainder of this paper is organized as follows. Optimal zoning of wind farms for the determination of the number of met mast and representative wind zone scope of each met mast is presented in Section 2. The micro quantitative siting method of met mast in each wind zone is proposed in Section 3. Simulation verification is given in Section 4, followed by the conclusion in Section 5.

2. Optimal Zoning of Wind Farm Based on Geographical Location Proximity and Wind Speed Correlation

2.1. Zoning Number Determination Based on REOF Decomposition and HC Method

At present, the zoning number of wind farm is mostly determined artificially by combining the site scope and topographic changes of wind turbine locations, which lacks objective basis [16]. Therefore, in this paper, REOF method considering wind speed distribution is combined with the agglomerative HC algorithm considering the placement of wind turbines to determine the zoning number of wind farms.

REOF decomposition is an effective method to analyze the regional structure of climate variable field [17]. REOF decomposition is achieved by varimax rotation, based on the calculation results of empirical orthogonal function (EOF) analysis. The spatial modes decomposed by REOF are rotation factor load vectors, and the high value of load vectors is concentrated in local area, so the spatial types are easier to identify. From the perspective of the variable field, after the varimax rotation, only a small area has high load in terms of decomposed typical spatial mode, and load value of the rest area is close to 0. The spatial structure of the climate variable field is simplified by REOF analysis. Based on the wind speed data of all wind turbine positions over the years, REOF is used to analyze the spatial distribution characteristics of wind speed. The steps include:

Step 1: The time-spatial matrix V containing the information of annual average wind speed at the locations of n wind turbines over t years, is anomaly processed, that is, all elements in original matrix minus the mean of elements of corresponding row, and acquired results are as new elements of processed matrix. V is shown in Equation (1). Then, EOF decomposition is performed;

$$V = \begin{bmatrix} v_{11} & v_{12} & \cdots & v_{1t} \\ v_{21} & v_{22} & \cdots & v_{2t} \\ \vdots & \vdots & \vdots & \vdots \\ v_{n1} & v_{n2} & \cdots & v_{nt} \end{bmatrix} \tag{1}$$

Step 2: By calculating error range of eigenvalue in Equation (2) and cumulative variance contribution rate, the double test of significance is carried out to judge whether the decomposed spatial mode is a valuable signal or noise.

$$e = \lambda \sqrt{\frac{2}{T^*}} \tag{2}$$

where: e represents the error range of eigenvalue λ; T^* represents effective degrees of freedom of data.

Step 3: The cumulative variance contribution rate is used to determine the number of high load vector, and the varimax rotation of selected high load vectors is made to obtain REOF decomposition result. According to the load value of the vector field obtained by REOF, the corresponding heat map is drawn to find several high load zones with significant characteristic differences.

The range of zoning number can be predicted according to REOF heat map, and then the zoning number can be further determined based on HC method.

Agglomerative HC is one of the typical unsupervised clustering algorithms, which adopts the bottom-up clustering strategy. In the process of initialization, each sample point is regarded as an independent cluster, and then clusters are continuously merged dependent on the principle of minimum distance until termination condition is reached [18]. Based on the actual space distance between wind turbines, the agglomerative HC algorithm is used to conduct coarse clustering for all wind turbine positions. The specific steps are as follows:

Step 1: n wind turbine positions are first divided into n clusters, and then the distance matrix between n clusters is calculated by adopting Euclidean distance based on three-dimensional data of the latitude, longitude and altitude of wind turbine positions.

Step 2: According to the distance matrix, the two clusters with the smallest distance are merged into one cluster, and the total number of clusters is reduced by 1.

Step 3: Based on cluster average algorithm in [18], the distance between any two clusters is calculated and a new distance matrix is obtained. If the number of clusters is 1 at the moment, clustering is already finished and go to the next step. Otherwise, Step 1 and 2 are repeated.

Step 4: Draw hierarchical pedigree diagram reflecting the kinship relationship between elements, according to the above clustering process.

According to the REOF heat map and hierarchical pedigree diagram, the optimal zoning number is determined considering some constraints. The constraints include that: The distances between different clusters should be relatively large. The number of wind turbines contained in a single cluster is generally between 10% and 80% of the total number of wind turbines, which can be adjusted slightly according to actual wind farm situation. The zoning number determined finally should conform to the range of zoning number estimated by REOF.

In addition, the coarse clustering result obtained by agglomerative HC can be used in the initialization of the DPSO algorithm in Section 2.3, which is beneficial for fast convergence of the algorithm.

2.2. The Distance Definition Considering Geographic Location Proximity and Wind Speed Correlation

Macro zoning of the wind farm mainly considers the correlation degree of wind flow distribution of different wind turbine positions, and wind turbines with strong correlation are divided into the same wind zone. The correlation degree can be judged from two aspects: one is based on the proximity of the geographical location of wind turbines; the other is based on the wind speed correlation of wind turbine positions. The zoning problem of wind farm can be regarded as the clustering problem of wind turbines. For clustering of data sets, the "distance" between samples is often used as an important classification standard. In principle, samples with large "distance" are divided into different clusters, and samples with small "distance" are divided into the same cluster. While the measurement of "distance" can actually be regarded as a measurement of similarity between samples. The higher the similarity between samples is, the smaller the distance is. In this paper, every wind turbine position is taken as a sample point. Moreover, in comprehensive consideration of geographical location proximity and wind speed correlation of wind turbines, a new distance is defined to measure the similarity of wind flow distribution between different wind turbine positions.

The coordinate matrix X of wind turbines is shown in Equation (3).

$$X = \begin{bmatrix} x_{11} & x_{12} & \cdots & x_{1n} \\ x_{21} & x_{22} & \cdots & x_{2n} \\ \vdots & \vdots & \vdots & \vdots \\ x_{m1} & x_{m2} & \cdots & x_{mn} \end{bmatrix} \tag{3}$$

where: m is the dimension number of coordinates; n is the number of wind turbines in the wind farm. m is usually equal to 3, representing three dimensions of longitude, latitude, and altitude

In order to eliminate dimensional differences, mean-variance normalization is carried out for each dimension. The normalized Euclidean distance between any two wind turbines is calculated by Equation (4).

$$d_{1,ij} = \sqrt{\sum_{k=1}^{m} \left(\frac{x_{ki} - x_{kj}}{S(x_k)} \right)^2} \tag{4}$$

where: $d_{1,ij}$ represents dominant distance between wind turbine i and wind turbine j. $S(x_k)$ is the standard deviation of all elements in row k of the matrix $\mathbf{X}$.

Next from the perspective of wind speed correlation, considering the wind speed correlation coefficient between two wind turbine positions, the correlation distance is calculated. Wind speed information is included in matrix $\mathbf{V}$, and Pearson similarity coefficient r_{ij} is calculated by Equation (5).

$$r_{ij} = \frac{\sum_{m=1}^{t} (v_{im} - E(v_i))(v_{jm} - E(v_j))}{\sqrt{\left(\sum_{m=1}^{t} (v_{im} - E(v_i))^2\right) \cdot \left(\sum_{m=1}^{t} (v_{jm} - E(v_j))^2\right)}} \tag{5}$$

where: $E(v_i)$ and $E(v_j)$ respectively represent the mean values of all elements in row i and row j of the matrix $\mathbf{V}$.

After the wind speed matrix is anomaly treated, all $E(v_i)$ are 0, $i \in [1, n]$, and Pearson similarity coefficient is degenerated into cosine similarity, as shown in Equation (6). The distance $d_{2,ij}$ that characterizes wind speed correlation is calculated by Equation (7).

$$\cos(\theta_{ij}) = \frac{v_i v_j^T}{\|v_i\| \cdot \|v_j\|} = \frac{\sum_{m=1}^{t} v_{im} v_{jm}}{\sqrt{\left(\sum_{m=1}^{t} v_{im}^2\right) \cdot \left(\sum_{m=1}^{t} v_{jm}^2\right)}} \tag{6}$$

$$d_{2,ij} = 1 - |\cos(\theta_{ij})| \tag{7}$$

where: $\cos(\theta_{ij})$ represents cosine similarity; $d_{2,ij}$ represents recessive distance between wind turbine i and wind turbine j

In order to make influence weight of dominant and recessive distance consistent, $d_{1,ij}$ and $d_{2,ij}$ are processed by maximum and minimum normalization method, as shown in Equation (8). A comprehensive distance between wind turbine positions is defined by Equation (9).

$$d'_{z,ij} = \frac{d_{z,ij} - \min\{d_{z,ij}\}}{\max\{d_{z,ij}\} - \min\{d_{z,ij}\}}, z = 1, 2 \tag{8}$$

$$d_{ij} = \max\{d'_{1,ij}, d'_{2,ij}\} \tag{9}$$

where: $\max\{d_{z,ij}\}$ represents the maximum of dominant distance (when $z = 1$) or recessive distance (when $z = 2$) between two wind turbines; $\min\{d_{z,ij}\}$ represents the minimum of dominant distance (when $z = 1$) or recessive distance (when $z = 2$) between two wind turbines; d_{ij} represents comprehensive distance between wind turbine i and wind turbine j.

2.3. Optimal Zoning of Wind Farm Based on Improved DPSO

Assuming the wind farm is divided into g clusters, $C = \{C_1, C_2, \ldots, C_g\}$, $|C_1|, |C_2|, \ldots, |C_g|$ are defined as the number of samples contained in the corresponding cluster. Considering the cohesion within clusters and dispersion between clusters, the evaluation indexes of zoning are established as shown in Equations (10) and (11).

1. Strong aggregation within zones

$$f_{1a} = \frac{\sum_{i \in C_a, j \in C_a} d_{ij}}{|C_a| \cdot (|C_a| - 1)}, i \neq j \tag{10}$$

where: f_{1a} represents convergence degree of wind turbines in zone a. The smaller the value of f_{1a} is, the higher the aggregation degree in this zone.

2. Strong dispersion between zones

$$f_{2a} = \frac{\sum_{i \in C_a} \left(\min_{C_b} \left\{ \frac{\sum_{j \in C_b} d_{ij}}{|C_b|} \right\} \right)}{|C_a|}, 1 \le b \le k, b \ne a \tag{11}$$

where: f_{2a} represents separation degree from zone a to other zones. The larger the value of f_{2a} is, the more discrete this zone is from other zones.

Silhouette coefficient is a parameter used to evaluate clustering method model and clustering result itself, which combines the degree of aggregation and the degree of dispersion beneficially [19]. Based on the modeling idea of silhouette coefficient, the paper establishes an optimization zoning model combined with evaluation indexes, and the objective function is shown in Equation (12).

$$\min F = \frac{\sum_{a=1}^{k} \left(\frac{f_{1a} - f_{2a}}{\max\{f_{1a}, f_{2a}\}} + 1 \right)}{k} \tag{12}$$

The value range of objective function F is $[0, 2]$. The closer F value is to 0, the better zoning result. Because the zoning result contains such information: stronger aggregation within zones and stronger dispersion between zones. Constraint conditions are shown in Equation (13):

$$\begin{cases} g > 1 \\ |C_a| > 1, & a \in [1, g] \text{ and } a \in Z \\ |C_a| \le 0.8n, & a \in [1, g] \text{ and } a \in Z \end{cases} \tag{13}$$

In order to solve optimal zoning model, an improved DPSO algorithm is adopted. The constraint conditions are processed by penalty function, that is, the penalty term is added to objective function, so that the particles which do not meet the constraint conditions cannot converge due to poor fitness. The solving process based on the improved DPSO algorithm is shown in Figure 1.

In order to improve the computational efficiency, on the basis of conventional DPSO algorithm, some improvements involving particle swarm initialization and particle position updating method are made as follows:

1. Particle swarm initialization considering reverse learning and HC result

The initial particle swarm based on conventional DPSO algorithm is generally generated randomly and it is difficult to ensure uniform distribution of initial particle swarm in the solution space. In order to overcome the above defects, the improved DPSO algorithm considers adopting the method of reverse learning to initialize particle swarm [20–22], and the specific steps are as follows:

- Generate χ (particle number) initial spatial solutions in the feasible search domains randomly;

Suppose g represents the number of clusters and n is the dimension number of solution, which is the same as the number of wind turbines, the feasible solution of the ith particle is expressed in Equation (14). All elements in P_i satisfy $p_{ij} \in [1, g], p_{ij} \in Z$ $(j = 1, 2, \ldots, n)$.

$$P_i = [p_{i1}, p_{i2}, \ldots, p_{in}] \tag{14}$$

- Calculate and generate the inverse solution of each initial solution;

The calculation of each dimensional component of the reverse solution is shown in Equation (15).

$$q_{ij} = 1 + g - p_{ij} \tag{15}$$

where: q_{ij} represents the reverse solution in the jth dimension of the ith particle.

- The coarse clustering solution of HC algorithm in Section 2.1 is incorporated into the initial solution set of particle swarm.
- Based on the union set generated by the above random solutions, reverse solutions, and coarse clustering solution, the objective function value is calculated and solutions with lower value are selected preferentially to form the initial population.

2. Particle position updating of DPSO algorithm

The updating of particle position is shown in Equations (16) and (17). The definition of relevant operators in the process of location updating is referred to [23].

$$U^{i+1} = wU^i + o_1(W_{\text{pbest}} - W^i) + o_2(W_{\text{gbest}} - W^i) \tag{16}$$

$$W^{i+1} = W^i + U^{i+1} \tag{17}$$

where: U^i is the updated particle velocity of the ith iteration; W^i is the updated particle position of the ith iteration; W_{pbest} is the current individual optimal particle position; W_{gbest} is the current global optimal particle position; w is inertial weight; o_1 and o_2 are cognitive learning factor and social learning factor respectively, whose value range is [0, 1].

After solving optimal zoning model of a wind farm, which wind turbines belong to the same cluster can be determined. For the convenience of calculation, each cluster of wind turbines is processed into rectangular zone. The maximum distance between east and west and the maximum distance between north and south of each cluster are extended by 5% as the length and width of the rectangular zone respectively, and finally the specific scope of each rectangular wind zone can be obtained.

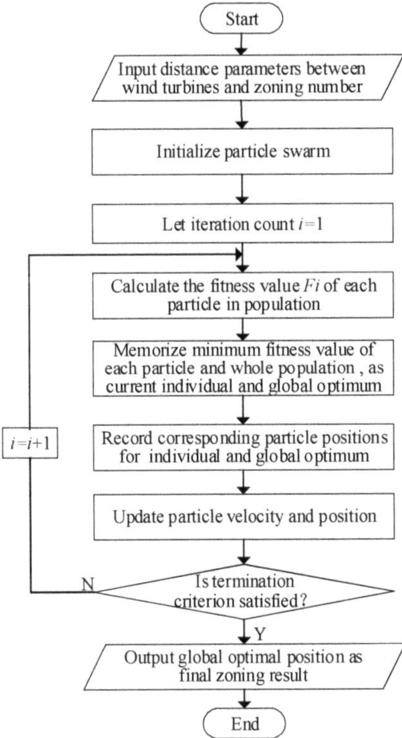

Figure 1. Improved DPSO algorithm.

3. Micro Quantitative Siting of Met Mast Based on Grid Screening Method

3.1. Select Alternative Met Mast Positions by Gridding

When macro zoning of wind farm is completed, micro siting of met mast is carried out in each zone. In order to simplify calculation, each rectangular zone is divided into many grids and the shape of grids is square. In order to ensure that every possible location suitable for building met mast can be obtained as much as possible, the side length L of grid should meet Equation (18), and the intersection points of grids are seen as the alternative wind monitoring points.

$$L = \min\{L_{ij}|i \in [1,n], j \in [1,n], i \neq j\}/\sqrt{2} \tag{18}$$

where: L_{ij} is the actual distance between wind turbine i and wind turbine j in wind farm.

Then the optimal wind monitoring point should be determined among all alternative grid points. The best wind monitoring point should have a good representation of the wind resources in the corresponding wind zone. The representativeness of wind monitoring points is mainly based on the following principles: spatial consistency principle, representativeness principle of wind condition parameters in prevailing wind direction, and screening principle considering wind speed reduction caused by wake effect. Based on these principles, this paper establishes six indicators, namely horizontal distance from wind turbines, altitude difference from wind turbines, wind acceleration factor of prevailing wind direction, turbulence intensity of prevailing wind direction, inflow angle of prevailing wind direction, and wind speed reduction rate caused by wake effect. The screening process of alternative wind monitoring points is shown in Figure 2.

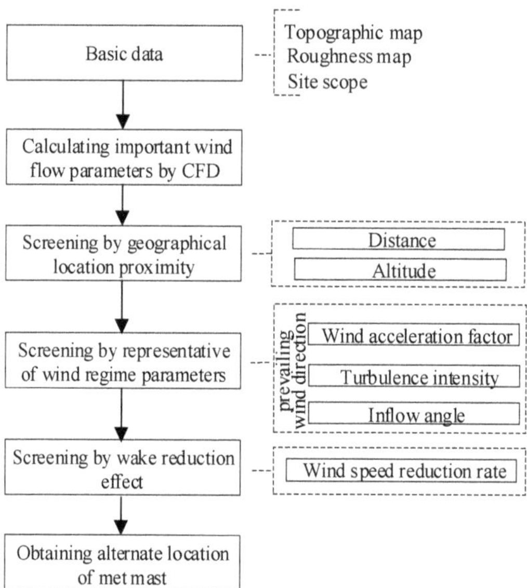

Figure 2. Location screening process of alternative met mast.

The wind condition parameters of prevailing wind direction and wind speed reduction rate caused by wake effect in each grid can be calculated by CFD software [24]. The steps of quantitative screening of alternative wind monitoring points are as follows.

Step 1: Exclude the alternative wind monitoring points within a distance away from wind turbines considering wake effect, as shown in Equation (19). The distance is equal to α times of rotor diameter.

$$R < \alpha D \tag{19}$$

where: D is the rotor diameter; R is the distance between the alternative wind monitoring points and wind turbines.

This screening index is mainly in response to standard [25]. Considering the wake effect of wind turbines, correction of free flow wind speed in front of wind turbines and so on, the value range of α is generally [2,4].

Step 2: Exclude alternative wind monitoring points whose altitude difference Δh from wind turbines is more than reference value H, as shown in Equation (20).

$$\Delta h > H \tag{20}$$

Step 3: Calculate the average wind acceleration factor $\bar{u}$ of the prevailing wind direction at all wind turbines in the wind farm, and keep alternative wind monitoring points whose wind acceleration factor is within the fluctuation range of plus or minus 5% of the average, as shown in Equation (21).

$$u_i \in [0.95\bar{u}, 1.05\bar{u})], i = 1, 2, \ldots, m \tag{21}$$

where: u_i is the wind acceleration factor in prevailing wind direction of ith alternative wind monitoring point; m is the number of alternative wind monitoring points reserved based on previous screening work.

Step 4: Calculate average turbulence intensity $\bar{I}$ in the prevailing wind direction of all alternative wind monitoring points reserved by above screening work, and keep the alternative wind monitoring points with turbulence intensity below $\bar{I}$, as shown in Equation (22).

$$I_i < \bar{I} \tag{22}$$

where: I_i is the turbulence intensity in prevailing wind direction of the reserved i^{th} alternative wind monitoring point.

Step 5: Calculate average $\bar{\varepsilon}$ of absolute value of inflow angle in prevailing wind direction of all alternative wind monitoring points reserved by above screening work, and keep the alternative wind monitoring points with absolute value of inflow angle below $\bar{\varepsilon}$, as shown in Equation (23).

$$\varepsilon_i < \bar{\varepsilon} \tag{23}$$

where: ε_i is the absolute value of inflow angle in prevailing wind direction of the reserved i^{th} alternative wind monitoring point.

Step 6: Calculate average wind speed reduction rate $\bar{\omega}$ of all alternative wind monitoring points reserved by above screening work, and keep the alternative wind monitoring points with wind speed reduction rate below $\bar{\omega}$, as shown in Equation (24).

$$\omega_i < \bar{\omega} \tag{24}$$

where: ω_i is the wind speed reduction rate caused by wake effect of the reserved i^{th} alternative wind monitoring point.

3.2. Micro Siting Evaluation of Met Mast in Wind Farm

The siting index is established to select the best position point of met mast from above finally reserved alternative wind monitoring points. Average wind speed distribution provides important information of wind resources. Meanwhile, the Weibull distribution, represented by shape parameters k and scale parameters c, is the most common wind speed distribution. The Weibull distribution is expressed by Equation (25).

$$P(v) = \frac{k}{c}\left(\frac{v}{c}\right)^{k-1} e^{-\left(\frac{v}{c}\right)^k} \tag{25}$$

where: $P(v)$ is the probability density of wind speed distribution.

As met mast should be representative of the wind resources in wind farm as much as possible, the average wind speed distribution of met mast should be as consistent as possible with the average wind speed distribution of wind turbines. That is, Weibull distribution parameters of met mast should be as consistent as possible with average Weibull distribution parameters of wind turbines in each wind zone.

In order to evaluate the representativeness of the reserved alternative wind monitoring points in the corresponding zone, the evaluation index Y of met mast siting is defined by Equation (26).

$$Y = 1 - \left(\frac{|v_{\text{ave}} - v_{\text{mo}}|}{v_{\text{ave}}} + \frac{|k_{\text{ave}} - k_{\text{mo}}|}{k_{\text{ave}}} + \frac{|c_{\text{ave}} - c_{\text{mo}}|}{c_{\text{ave}}} \right) \Big/ 3 \tag{26}$$

where: v_{mo}, k_{mo}, and c_{mo} are respectively the wind speed, the shape parameter, and the scale parameter of the alternative wind monitoring points; v_{ave}, k_{ave}, and c_{ave} are respectively the mean of wind speed, the mean of shape parameter, and the mean of scale parameter of all wind turbine positions.

The value of index Y is within the range of [0, 1]. The closer the Y value of the alternative wind monitoring point is to 1, the more suitable its location is for building a met mast.

In summary, the research framework of this paper is shown in Figure 3.

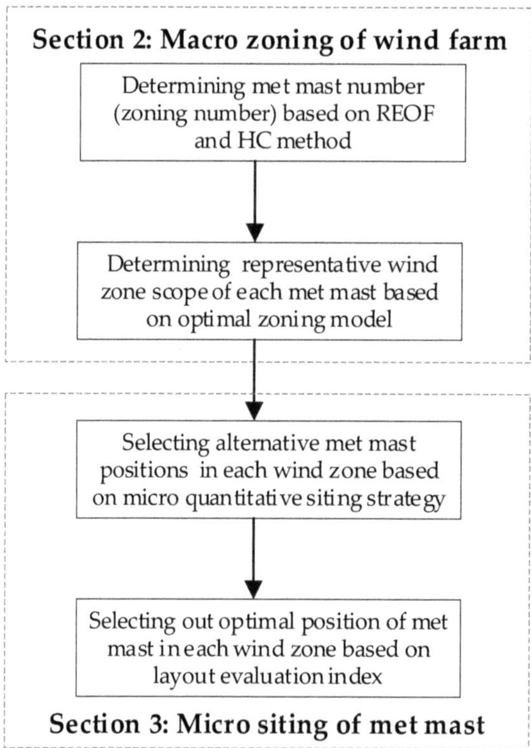

Figure 3. The research framework of this paper.

4. Simulation Verification

An island wind farm in Zhejiang province is selected. The wind farm is located in the range of east longitude $121°55'24'' \sim 121°57'44''$ and north latitude $29°47'17'' \sim 29°48'12''$, with altitude of 0~255 m. There are 17 wind turbines in the wind farm and the prevailing wind direction is about 300°. The topography of the wind farm is shown in Figure 4.

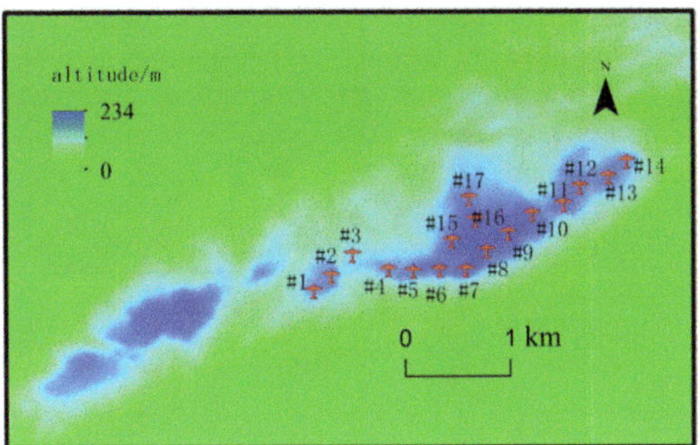

Figure 4. The topography of island wind farm.

Based on historical wind speed information of all wind turbine positions, the REOF method is used to analyze the spatial distribution characteristics of wind speed. Table 1 shows the variance contribution rate and cumulative variance contribution rate of the first two feature vectors of wind speed based on EOF and REOF. The cumulative variance contribution rate of the first two feature vectors is 99.95%, i.e., the first two feature vectors can effectively represent the overall characteristics of wind speed changes in wind farms. After rotation, the variance contribution of each load vector is more evenly distributed than before rotation. The total variance contribution does not change, and the rotation effect is significant. However, the variance contribution rate of the first feature vector is 74.93%, which still account for a large proportion of the total variance. It implies that the first vector takes majority responsibility for representing wind speed characteristics of wind farm. According to the spatial distribution information of rotating load vector field obtained by REOF, the corresponding heat map is made, as shown in Figure 5. It is obvious that there are two load centers with significantly different wind speed characteristics in wind farms. One is mainly concentrated at #7 wind turbine, and the other load center is mainly concentrated at #1 wind turbine.

Table 1. The Variance Contribution rate and Cumulative Variance Contribution Rate of the First Two Feature Vectors of Regional Wind Speed Based on EOF and REOF.

Serial Number	EOF Variance Contribution Rate	REOF Variance Contribution Rate	Cumulative Variance Contribution Rate
1	96.27%	74.93%	/
2	3.68%	25.02%	99.95%

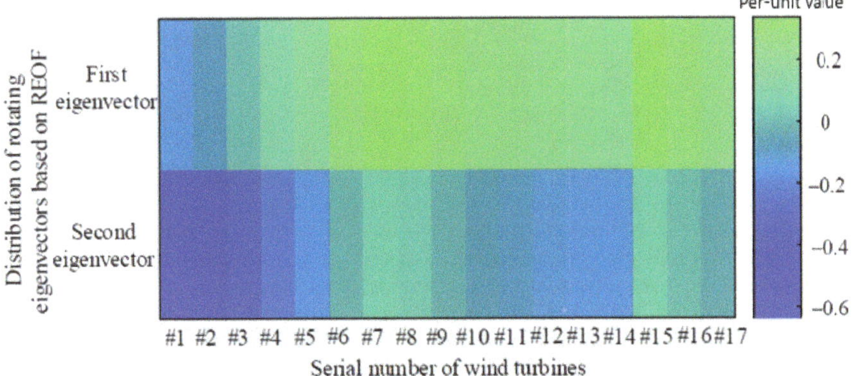

Figure 5. Heat map of load vector field spatial distribution based on REOF.

Based on location information (latitude, longitude and altitude) of wind turbines, HC algorithm is used to draw hierarchical pedigree diagram, as shown in Figure 6. Combined with the REOF information, it can be preliminarily judged the wind farm is suitable to be divided into two zones, and coarse clustering result of wind turbines is obtained based on the pedigree diagram.

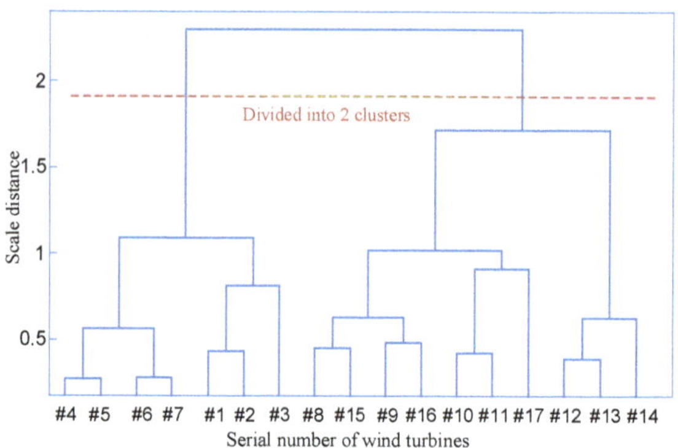

Figure 6. Hierarchical pedigree diagram of wind turbines clustering process.

Inertia weight coefficient $w = 0.7$ and the parameters $o_1 = 0.2$, $o_2 = 0.3$ in DPSO algorithm are taken to calculate the optimal zoning result. Meanwhile, DBSCAN zoning method is compared with DPSO method and respective result is shown in Figure 7 (Wind turbines with the same symbol are in the same wind zone in Figure 7b–d). Although DBSCAN as a classical clustering algorithm can automatically determine zoning number, the output zoning results are different when input parameters such as cluster density threshold d are set to different values. In this paper, two DBSCAN results when zoning number is 2 are selected and presented. As can be seen from Figure 7, under different d values, #8, #9, #10, #15, #16, and #17 wind turbines (the serial number of wind turbines is shown in Figure 4) are divided into completely different zone, indicating that the final zoning result of DBSCAN is very sensitive to parameter selection.

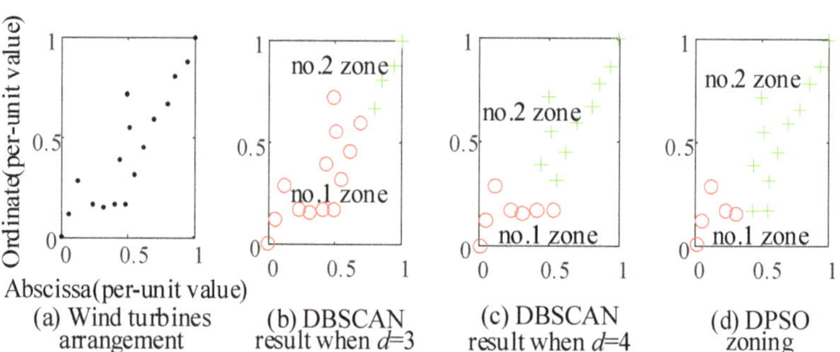

Figure 7. Comparison of zoning results based on DBSCAN and DPSO algorithm.

For clustering zoning without label, silhouette coefficient is generally considered to objectively evaluate zoning results [19]. The value range of silhouette coefficient is [−1, 1], the closer it is to 1, the better zoning effect is. Comparison of silhouette coefficient based on DBSCAN and DPSO algorithm is shown in Table 2. It can be seen that the silhouette coefficient of the DPSO zoning result is the largest, which proves that the zoning result is the optimal objectively. In addition, when $d = 4$, the DBSCAN zoning result is close to DPSO zoning, and the corresponding silhouette coefficient is suboptimal. The results of two zoning methods mirror each other to some extent, which further verifies the rationality of DPSO zoning.

Table 2. Comparison of silhouette coefficient based on DBSCAN and DPSO algorithm.

Serial Number	DBSCAN Algorithm $d = 3$	DBSCAN Algorithm $d = 4$	DPSO Zoning
Silhouette coefficient	0.5240	0.6163	0.6285

Compared with DBSCAN algorithm, DPSO zoning has the following advantages:

(1) In DBSCAN algorithm, parameters are very important, which is difficult to select and has a great influence on zoning results. However, the parameter selection of DPSO zoning has no influence on final optimization results, but only plays a role in the calculation efficiency. Moreover parameter selection is relatively simple.

(2) In the DBSCAN algorithm, different input parameters lead to different zoning results. Every result needs to be evaluated by a silhouette coefficient. The evaluation work is relatively heavy because of lots of repetitive work, and the optimality of the evaluated zoning result cannot be guaranteed because of the diversity of input parameters. However, the DPSO zoning model takes the evaluation index into account, which makes evaluation work easier. Moreover, the final optimal zoning result is presented directly by a clear and concise algorithm. The final zoning result of this wind farm is shown in Figure 8.

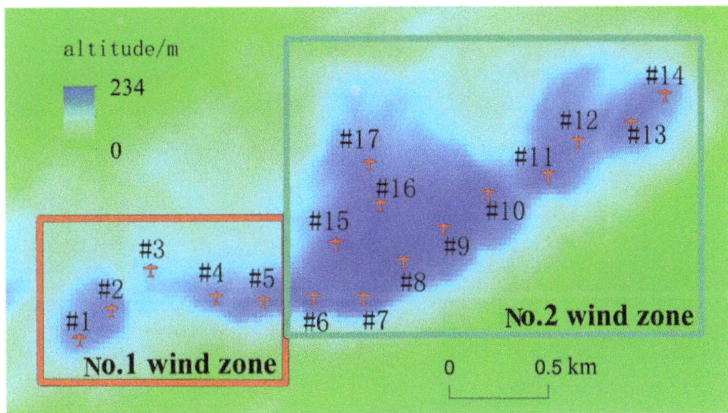

Figure 8. Zoning result of the island wind farm.

On the other hand, according to the REOF heat map, the first load center concentrated at #7 wind turbine corresponds to No. 2 wind zone in DPSO zoning result. The second load center concentrated at #1 wind turbine corresponds to No.1 wind zone. The reliability of DPSO zoning results is verified.

To further verify the superiority of the improved DPSO algorithm, Zhushan wind farm in Zhejiang province with 50 wind turbines is selected. The terrain and DPSO zoning result are shown in Figures 9 and 10 (Wind turbines with the same symbol are in the same wind zone in Figure 10b). It can be seen that Zhushan wind farm covers a large area and optimal zoning result is obviously related to the geographical location proximity between wind turbines, in line with practical experience. The convergence curves of the algorithm applied to island wind farm and Zhushan wind farm are shown in Figure 11. It can be found that:

(1) The convergence speed of the improved DPSO algorithm is faster than that of the conventional DPSO algorithm.

(2) In the island wind farm, when converging, there is a difference of about seven iterations between improved DPSO and conventional DPSO algorithm. Meanwhile, in the larger Zhushan wind farm, there is a difference of almost 70 iterations between improved DPSO and conventional DPSO algorithm. That is to say, the convergence speed of the improved DPSO algorithm is improved more obviously for wind farms with larger scale.

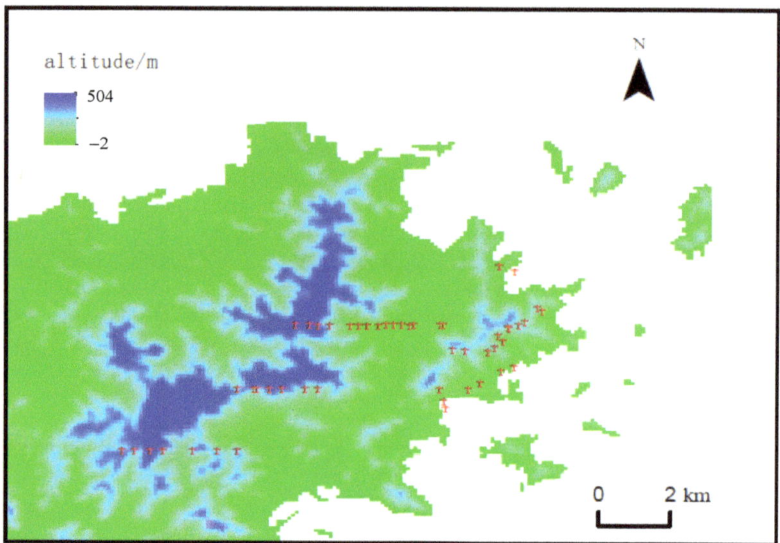

Figure 9. The topography of Zhushan wind farm.

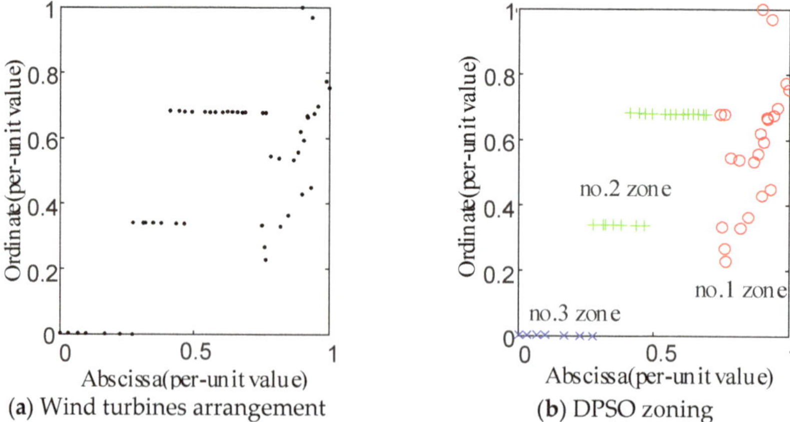

Figure 10. Optimal zoning result of Zhushan wind farm.

The results show that the improved DPSO algorithm can effectively improve the computational efficiency, and the larger the data scale is, the more obviously the efficiency of algorithm improves.

Micro siting of met mast is carried out in two zones of the island wind farm. The design of relevant parameters is referred in [26]. Firstly, Windsim software is used to simulate spatial wind flow distribution by the CFD numerical method. Wind condition parameters in the prevailing wind direction, including wind acceleration factor, turbulence intensity, inflow angle, and reduced wind speed considering wake effect, are obtained by gridding the calculation of the target area of the wind farm. Next based on obtained grid information, alternative wind monitoring points are screened out according to micro quantitative siting strategy of met mast. Finally, the optimal locations of met mast in each zone are obtained based on the calculation of siting evaluation index. The screening results of alternative wind monitoring points are shown in Table 3. The optimal location of wind monitoring point is shown in Figure 12. The two optimal locations of met masts in respective zones are relatively consistent with the above two load centers based on REOF method, which verifies the representativeness of met mast in final selected position.

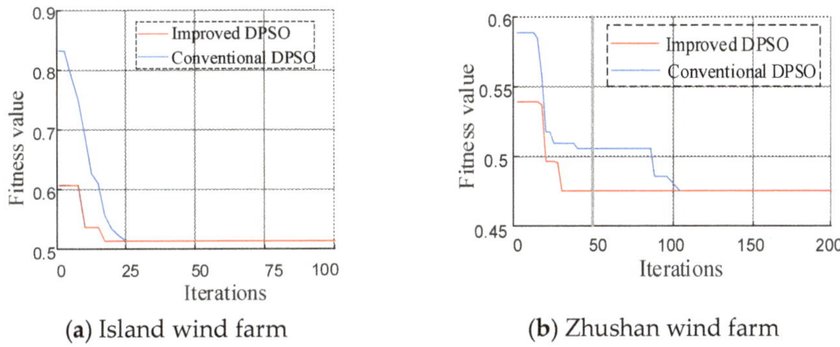

(**a**) Island wind farm   (**b**) Zhushan wind farm

Figure 11. Convergence curves of discrete particle swarm optimization algorithm.

Table 3. Screening results of alternative wind monitoring points.

	Alternative Wind Monitoring Points	Index	Selection of Wind Monitoring Points
No.1 wind zone	P_1	0.923761	P_1
	P_2	0.825439	
	P_3	0.905465	
No.2 wind zone	P_4	0.796873	P_3
	P_5	0.846572	

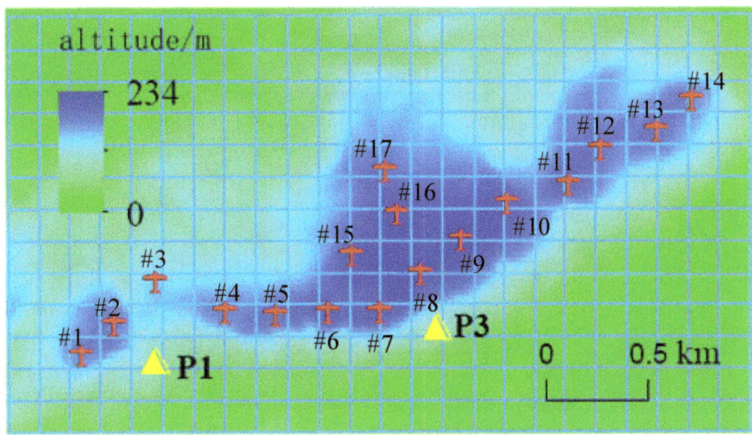

Figure 12. Optimal location of wind monitoring point.

To further prove the effectiveness of the proposed micro siting method, the met mast positions determined by proposed method (met mast at points P_1, P_3, corresponding longitude, latitude are respectively 121°55′49.2″ E, 29°47′18.2″ N/121°56′50.3″ E, 29°47′13.4″ N) and common method based on engineering experience in [14] (met mast at points P_6, P_7, corresponding longitude, latitude are respectively 121°56′10.1″ E, 29°47′80.3″ N/121°56′59.6″ E, 29°48′17.4″ N) are respectively used as CFD simulation inputs, and estimated results and errors regarding the annual power generation of wind farms are shown in Table 4. It can be seen that taking met mast positions (P_1, P_3) as input, the error of estimated power generation is the smallest. In addition, the error of estimated power generation with two met masts is smaller, compared with one met mast. It is indicated that two met masts are more representative for wind resources of the island wind farm. The reliability of the proposed method is verified to some extent.

Table 4. Annual energy production and error analysis of wind farm.

Input Data	Average Annual Power Generation/(10^4 kW·h)	Relative Error/%
	12,540.32 (actual measured power generation)	
(P_1, P_3)	12,314.59	−1.8
(P_6, P_7)	12,101.41	−3.5
P_1	11,449.31	−8.7
P_3	13,079.55	+4.3
P_6	11,474.39	−8.5
P_7	13,154.80	+4.9

5. Conclusions

In this paper, a method for the optimal layout of met mast in the wind farm is proposed. Firstly, a representative wind zone scope and the number of met mast are determined by macro zoning of wind farm. Then, a micro quantitative siting strategy is proposed and the optimal layout evaluation index is established to realize micro siting of met mast in each wind zone. The main conclusions are drawn as follows:

(1) The proposed optimal zoning method based on discrete particle swarm optimization provides a new zoning idea, which can provide a reliable zoning result for wind farms more directly and quickly compared with general zoning methods, such as the density based spatial clustering of applications with noise algorithm method.

(2) In the studied cases, the selected met mast position based on the proposed micro quantitative siting method is proven to be more accurate and representative by the test of wind farm power generation estimation, compared with the traditional qualitative analysis method.

The optimal layout method for met mast proposed in this paper has certain practical applicability, especially for wind farms with large scale or complex terrain. It can help to obtain more accurate data regarding wind resources, which means a lot for wind farm operation. In future work, the layout evaluation index will be further discussed and designed considering different functions of met mast, which will serve to improve micro siting work of met mast.

Author Contributions: Conceptualization, W.C. and X.Y.; methodology, G.Q. and L.Z.; software, L.Z.; validation, G.Q., W.Q., and L.Z.; formal analysis, L.Z. and G.L.; investigation, L.Z.; resources, W.Q.; data curation, W.Q. and L.Z.; writing—original draft preparation, L.Z.; writing—review and editing, X.Y.; visualization, G.L.; supervision, W.C. and X.Y.; project administration, W.C.; funding acquisition, W.C. All authors have read and agreed to the published version of the manuscript.

Funding: This work was supported by the Science and Technology Project of State Grid Zhejiang Electric Power Company Limited, China (Grant No. 5211SX2000ZM).

Institutional Review Board Statement: Not applicable.

Informed Consent Statement: Not applicable.

Data Availability Statement: Not applicable.

Conflicts of Interest: The authors declare no conflict of interest.

Nomenclature

$\lvert C_a \rvert$	Number of samples contained in zone a.
$\cos\left(\theta_{ij}\right)$	Cosine similarity about wind speed of wind turbine i and j.
$d_{1,ij}$	Dominant distance between wind turbine i and j.
$d_{2,ij}$	Recessive distance between wind turbine i and j.
d_{ij}	Comprehensive distance between wind turbine i and j.
$E(v_i)$	Mean value of all elements in row i of the matrix V
e	Error range of eigenvalue.
F	Objective function for optimization.
f_{1a}	Convergence degree of wind turbines in zone a.
f_{2a}	Separation degree from zone a to other zones.
g	Zoning number.
Δh	Altitude difference between the alternative wind monitoring points and wind turbines.
L	Side length of grid.
l_i	Turbulence intensity in prevailing wind direction of the reserved i^{th} alternative wind monitoring point.
m	dimension number of space coordinates
n	Number of met mast.
o_1/o_2	Cognitive/social learning factor and learning factor.
$P(v)$	Probability density of wind speed distribution.
P_i	Feasible solution of the ith particle.
q_{ij}	Reverse solution in the jth dimension of the ith particle.
R	Horizontal distance between the alternative wind monitoring points and wind turbines.
r_{ij}	Pearson similarity coefficient of wind speed of wind turbine i and j.
$S(x_k)$	Standard deviation of all elements in row k of the matrix X.
U^i	Updated particle velocity of the ith iteration.
u_i	Wind acceleration factor in prevailing wind direction of ith alternative wind monitoring point.
V	Time-spatial wind speed matrix at all wind turbine positions.
W^i	Updated particle position of the ith iteration.
W_{pbest}	Individual optimal particle position.
W_{gbest}	Global optimal particle position.
w	Inertial weigh.
X	Coordinate matrix of wind turbines.
Y	Siting evaluation index of met mast.
α	Multiples of rotor diameter
ε_i	Absolute value of inflow angle in prevailing wind direction of the reserved i^{th} alternative wind monitoring point.
ω_i	Wind speed reduction rate caused by wake effect of the reserved i^{th} alternative wind monitoring point.
χ	Particle number in DPSO algorithm.

References

1. Yao, J.; Yao, F. Status quo, development and utilization efficiencies of wind power in China. *Processes* **2021**, *9*, 2133. [CrossRef]
2. Liu, Y.; Yang, J.; Jiang, C.; Niu, S.; Li, H.; Chen, S. Review on met mast site selection methods in grid-connected wind farm. In Proceedings of the 2019 IEEE 3rd International Electrical and Energy Conference (CIEEC), Beijing, China, 7–9 September 2019; pp. 1134–1137.
3. Gao, Z.; Liu, X. An overview on fault diagnosis, prognosis and resilient control for wind turbine systems. *Processes* **2021**, *9*, 300. [CrossRef]
4. Fan, G.; Wang, Y.; Yang, B.; Zhang, C.; Fu, B.; Qi, Q. Characteristics of wind resources and post-project evaluation of wind farms in coastal areas of Zhejiang. *Energies* **2022**, *15*, 3351. [CrossRef]
5. Dong, Y.; Zhang, L.; Liu, Z.; Wang, J. Integrated forecasting method for wind energy management: A case study in China. *Processes* **2020**, *8*, 35. [CrossRef]
6. Jangamshetti, S.; Ran, V. Optimum siting of wind turbine generator. *IEEE Trans. Energy Convers.* **2001**, *16*, 8–13. [CrossRef]
7. Bebi, E.; Alcani, M.; Malka, L.; Leskoviku, A. An evaluation of wind energy potential in Topoja area, Albania. *Sci. Bus. Soc.* **2022**, *7*, 21–25.

8. Bailey, B.H. *Wind Resource Assessment Handbook: Fundamentals for Conducting a Successful Monitoring Program*; National Renewable Energy Laboratory (NREL): Golden, CO, USA, 1997; pp. 23–24.

9. Sun, S.; Liu, S.; Liu, J.; Schlaberg, H.I. Wind field reconstruction using inverse process with optimal sensor placement. *IEEE Trans. Sustain. Energy* **2019**, *10*, 1290–1299. [CrossRef]

10. Khan, K.S.; Tariq, M. Wind resource assessment using SODAR and meteorological mast—A case study of Pakistan. *Renew. Sust. Energ. Rev.* **2018**, *81*, 2443–2449. [CrossRef]

11. Han, P.; Xia, Y.; Zhang, Y.; Luo, K. Equivalent model of wind farm based on DBSCAN. In Proceedings of the 2017 IEEE Innovative Smart Grid Technologies-Asia (ISGT-Asia), Auckland, New Zealand, 4–7 December 2017; pp. 1–6.

12. Bechrakis, D.; Sparis, P. Correlation of wind speed between neighboring measuring stations. *IEEE Trans. Energy Convers.* **2004**, *19*, 400–406. [CrossRef]

13. Yang, J.Y.; Woo, Y.M.; Sheng, K.; Tang, Y.H. Research on the met mast siting used in post assessment of mountain wind farm. In Proceedings of the World Wind Energy Conference, Shanghai, China, 7–9 April 2014; pp. 443–447.

14. Zhang, H. *Wind Resources and Micro-Location*; China Machine Press: Beijing, China, 2013; pp. 42–43.

15. Sun, S.; Liu, S.; Chen, M.; Guo, H. An optimized sensing arrangement in wind field reconstruction using CFD and POD. *IEEE Trans. Sustain. Energy* **2020**, *11*, 2449–2456. [CrossRef]

16. Terciyanlı, E.; Demirci, T.; Küçük, D.; Saraç, M.; Çadırcı, I.; Ermiş, M. Enhanced nationwide wind-electric power monitoring and forecast system. *IEEE Trans. Industr. Inform.* **2014**, *10*, 1171–1184. [CrossRef]

17. Wu, J.; Chen, Y.; Guo, P.; Wang, X.; Hu, X.; Wu, M. Sea surface wind speed retrieval based on empirical orthogonal function analysis using 2019–2020 CYGNSS data. *IEEE Trans. Geosci. Remote Sens.* **2022**, *60*, 5803213. [CrossRef]

18. Tso, W.; Demirhan, C.; Heuberger, C.; Powell, J.; Pistikopoulos, E. A hierarchical clustering decomposition algorithm for optimizing renewable power systems with storage. *Appl. Energy* **2020**, *270*, 115190. [CrossRef]

19. Dinh, D.T.; Fujinami, T.; Huynh, V.N. Estimating the optimal number of clusters in categorical data clustering by silhouette coefficient. In *International Symposium on Knowledge and Systems Sciences*; Springer: Singapore, 2019; pp. 1–17.

20. Cao, Y.; Zhang, H.; Li, W.; Zhou, M.; Zhang, Y.; Chaovalitwongse, W.A. Comprehensive learning particle swarm optimization algorithm with local search for multimodal functions. *IEEE Trans. Evol. Comput.* **2019**, *23*, 718–731.

21. Bangyal, W.H.; Nisar, K.; Ibrahim, A.A.; Haque, M.R.; Rodrigues, J.J.; Rawat, D.B. Comparative analysis of low discrepancy sequence-based initialization approaches using population-based algorithms for solving the global optimization problems. *Appl. Sci.* **2021**, *11*, 7591. [CrossRef]

22. Ashraf, A.; Pervaiz, S.; Haider Bangyal, W.; Nisar, K.; Ibrahim, A.A.; Rodrigues, J.J.; Rawat, D.B. Studying the impact of initialization for population-based algorithms with low-discrepancy sequences. *Appl. Sci.* **2021**, *11*, 8190. [CrossRef]

23. Guo, Z.; Liang, Y.; Bian, X.; Wang, D. Multi objective optimization for arrangement of the asymmetric-paths winding based on improved discrete particle swarm approach. *IEEE Trans. Energy Convers.* **2018**, *33*, 1571–1578. [CrossRef]

24. Iranzo, A. CFD applications in energy engineering research and simulation: An introduction to published reviews. *Processes* **2019**, *7*, 883. [CrossRef]

25. National Power Dispatch and Communication Center. *Functional Specification of Wind Power Forecasting System: Q/GDW 10588-2015*; State Grid Corporation of China: Beijing, China, 2015.

26. Zhang, X. *The Research and Application on Optimal Site Selection of a Met Mast in a Large-scale Interconnected Wind Farm*; North China Electric Power University: Beijing, China, 2018.

Article

Community Integrated Energy System Multi-Energy Transaction Decision Considering User Interaction

Yuantian Li and Xiaojing Wang *

School of Electrical Engineering, Xinjiang University, Urumqi 830017, China
* Correspondence: wangxiaojing345@163.com; Tel.: +86-180-9961-3605

Abstract: With the gradual liberalization of China's energy market, the distributed characteristics of each entity in the community integrated energy system are more and more obvious, and the traditional centralized optimization is difficult to reveal the interaction between the entities. This paper aims to improve the profit of the community operator and the users' value-added benefit of energy use, and proposes a multi-energy transaction decision of a community integrated energy system considering user interaction. First, a refined model of user interaction, including energy conversion, is established, and then the optimization model of multi-energy transaction decision between the community operator and the users is constructed based on the master–slave game. The upper layer aims to maximize the profit of the community operator according to the energy use strategies' feedback from the users, decides the retail energy prices of the community operator to the users, and optimization variables include equipment output and energy purchased from the power grid and natural gas grid. The lower layer aims to maximize the value-added benefit of energy use for users. The users optimize their energy use strategies based on the retail energy prices published by the community operator. The model is solved by the differential evolution algorithm combined with the CPLEX solver. Finally, different scenarios are analyzed in a numerical example, and the results show that the strategy proposed in this paper to set community prices increases the community operator's profit and profit margin by 5.9% and 7.5%, respectively, compared to using market energy prices directly. At the same time, the value-added benefit to users also increases by 15.2%. The community operator and users can achieve a win–win situation.

Keywords: integrated energy system; multi-energy trading; consumer psychology; convertible load

Citation: Li, Y.; Wang, X. Community Integrated Energy System Multi-Energy Transaction Decision Considering User Interaction. *Processes* **2022**, *10*, 1794. https://doi.org/10.3390/pr10091794

Academic Editors: Haoming Liu, Jingrui Zhang and Jian Wang

Received: 16 July 2022
Accepted: 28 August 2022
Published: 6 September 2022

Publisher's Note: MDPI stays neutral with regard to jurisdictional claims in published maps and institutional affiliations.

1. Introduction

With the disadvantages of low economic benefits and high energy consumption of traditional energy systems becoming increasingly prominent, integrated energy systems (IES) that can realize flexible energy conversion and efficient utilization have become the focus of energy research and development [1]. The community integrated energy system (CIES) near the user side contains a variety of energy coupling equipment, which couples and complements electricity, heat, natural gas, and other energy sources, enabling local consumption of renewable energy and providing users with comprehensive energy services. It is an important direction for the future development of the intelligent community [2]. Therefore, it has become a hot research issue to study how to improve the economics of the community integrated energy system, and to formulate transaction strategies between the community operator and the users to guide the users to rationally use energy to achieve a win–win situation [3,4].

At present, domestic and foreign scholars have focused on improving the economics of the integrated energy system, mainly on the refined modeling of equipment on the power supply side and demand side management. In terms of refined equipment modeling, this technology proposes a general dynamic energy efficiency model of an integrated energy system, which lays the foundation for the optimization of integrated energy system operation and the formulation of trading strategies [5]. Chen et al. established an optimization model considering coupled dynamic energy efficiency, and the results showed that considering the dynamic energy efficiency of equipment can improve the energy utilization rate, which is more consistent with the actual optimization results [6]. In demand side management, Wang Yongli et al. considered the electric and heating demand response, and the established source–load interactive model can reduce the operating cost of the service provider and improve wind power consumption [7]. Guo Zihao et al. considered the multi-energy flow coupling characteristics and the users' flexible load and considered the source–load interaction to optimize the operator's benefit [8]. Liu et al. considered the controllable degree of flexible load in the scheduling process and the constraint of user satisfaction to give full play to the demand response ability of the users. The results show that the proposed scheduling scheme can reduce the cost of the community operator [9]. It can be seen that coordinated optimization on both sides of supply and demand can improve the system economy. However, the references [7–9] are mostly centralized optimization and do not consider the energy trading and pricing problems of the operator in the market environment. There is a lack of research on the impact of the operator's retail energy prices on users' energy-use strategies.

With the development of the electricity market, the community operator can be regarded as a distribution-side entity or retail entity with self-production and self-selling capabilities [10]. It can guide users to participate in interaction by formulating reasonable retail energy prices, adjusting energy use strategies, and achieving demand side management. Aiming at the transaction problem between the operator and the users, Li Yuan et al. used the master–slave game method to establish an operator energy pricing model, including electric vehicles and P2G, which can improve the system economy [11]. References [12,13] analyzed the interaction mechanism between the community operator and the users in the electricity market based on the master–slave game model, with the operator as the leader and the users as the follower. In reference [14], the master–slave game model of the transaction between the community operator and consumers was established, and transaction strategies considering the demand response ability of consumers were proposed. Fu et al. constructed a user model containing four types of loads: electricity, heat, cold, and gas. Combined with the operator revenue optimization model, a master–slave game pricing mechanism between operators and users is proposed [15]. Fleischhacker et al. proposed an energy value allocation and stabilization algorithm based on a cooperative game. By investing in distributed energy, community operators can share value among their members [16]. Based on the master–slave game, Anoh et al. constructed an energy trading strategy between operators and consumers in the microgrid to optimize the interests of producers and consumers [17]. Wei et al. proposed a multi-leader and multi-follower Stackelberg game approach to solve the multi-energy trading problem. Multiple energy operators act as leaders to determine real-time energy prices, while multiple consumers act as followers to optimize their energy usage strategies [18]. However, the above models do not consider the role of convertible load in the process of user interaction. Li Peng et al. included convertible loads in consideration of integrated demand response, which improved user interaction but did not consider the impact of energy prices on convertible load [19]. Under the incentive of multiple retail energy prices, users will preferentially use energy with lower prices to meet the same energy demand. The actual amount of interaction will be influenced by consumer psychology [20], so considering the convertible load can further tap into the potential of user interaction.

Based on the above research, this paper establishes a refined model of user interaction considering energy conversion and constructs an optimization model of multi-energy transaction decisions between the community operator and the users based on the master–slave game. Taking the community operator as the leader, the optimization goal is the maximum daily profit, and the optimization variables are retail energy prices, equipment output, etc. The users are the followers, the optimization goal is the maximum value-added benefit of energy use, and the optimization variables are the users' energy use strategies of electricity, heat, and natural gas. Finally, the validity of the proposed model is verified by an example.

The rest of this paper is organized as follows. Section 2 summarizes the models of community integrated energy system and of user interaction. Section 3 establishes the optimization model for the multi-energy transaction decision between the community operator and the users. Section 4 sets up different scenarios to analyze the trading strategy proposed in this paper. Finally, the conclusions are drawn in Section 5.

2. Models of Community Integrated Energy Systems and User Interactions

2.1. CIES Model Based on Energy Hub

The concept of an energy hub (EH) was first proposed by Geidl et al. [21], it simplifies the energy flow relationship. An EH with multiple input and output ports is modeled by a coupling matrix that can easily describe the transformation and coupling relationship between energy input and output [22]. Therefore, to analyze the energy coupling and input–output energy flow relationship in the system, the energy hub model is used to describe the CIES model abstractly, as shown in Figure 1. In this paper, the electricity, heat, and natural gas demanded by the users in winter are supplied by the community operator in CIES, who has certain renewable energy units according to natural and geographical conditions. In the actual operation process, the community operator purchases electric energy and natural gas energy from the energy market and uses the energy conversion equipment to convert the energy into the energy required by the users according to the multi-energy complementary characteristics. The renewable energy equipment of CIES includes wind turbines (WT) and photovoltaic (PV); energy conversion equipment includes combined heat and power (CHP) units, gas boilers (GB), and electric heat pumps (EHP); energy storage devices include electricity storage and heat storage.

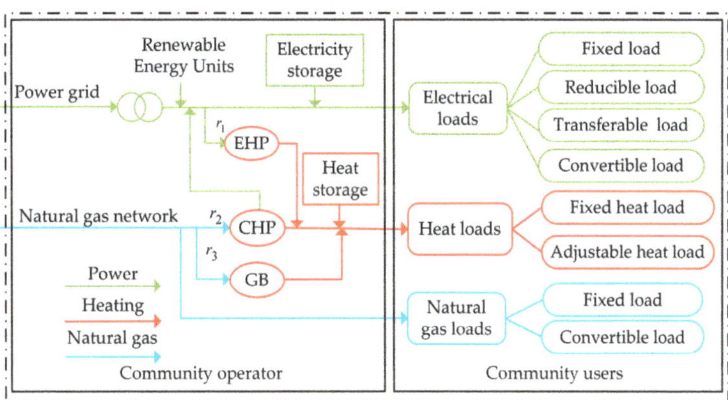

Figure 1. Community Integrated Energy System Model.

According to the energy flow, the community energy supply model can be represented by the following matrix:

$$
\begin{bmatrix} P^{e}_{out,t} \\ P^{h}_{out,t} \\ P^{g}_{out,t} \end{bmatrix} = \begin{bmatrix} 1-r_1 & 0 & r_2\eta_{CHP,e} \\ r_1\eta_{EHP} & 1 & r_2\eta_{CHP,h} + r_3\eta_{GB} \\ 0 & 0 & 1-r_2-r_3 \end{bmatrix} \begin{bmatrix} P^{e}_{in,t} \\ P^{h}_{in,t} \\ P^{g}_{in,t} \end{bmatrix} - \begin{bmatrix} P^{ES,n}_{c/d,e,t} \\ P^{ES,n}_{c/d,h,t} \\ 0 \end{bmatrix} \tag{1}
$$

In the above formula: $P^{e}_{out,t}$, $P^{h}_{out,t}$, and $P^{g}_{out,t}$ are the electricity, heat, and natural gas power supplied by the community to the users, respectively; $P^{e}_{in,t}$ is the sum of purchased power $P^{e}_{net,t}$, wind power $P^{e}_{w,t}$, and photovoltaic power $P^{e}_{v,t}$; $P^{h}_{in,t}$ is the heat power purchased by the community. This paper considers that the heat energy is only supplied by the community, so it is taken as 0; $P_{in,t}^g$ is the power of the community to purchase natural gas; $P_{c/d,e,t}^{ES,n}$ and $P_{c/d,h,t}^{ES,n}$ are the actual charging and discharging power of electric and thermal energy storage respectively; r_1, r_2, and r_3 are dispatch factors, which represent the proportion of heterogeneous energy flow into energy conversion equipment; η_{EHP} and η_{GB} are the efficiency of electric heat pump and gas boiler; $\eta_{CHP,e}$ and $\eta_{CHP,h}$ are the electrical and thermal efficiencies of the CHP unit.

2.2. The Models of User Interaction That Account for Energy Conversion

User interaction in energy communities makes sense and can improve the economy of the community operator [8,11]. In this paper, the community operator guides users to adjust energy use strategies by formulating reasonable retail energy prices. In general, user interaction strategies are often limited to responses in a single form of energy, such as load reduction and transfer, which have a greater impact on users' actual energy use. With the development of user terminal equipment, the user side can realize the conversion of energy forms. When the community operator publishes retail energy prices, the multi-energy complementary users consider the energy conversion efficiency of the terminal equipment to obtain the difference in equivalent energy prices. Users can use the corresponding terminal equipment to achieve secondary energy conversion and choose the appropriate way to meet their own load demand. For example, users can choose to use electric heating or natural gas heating to achieve the same hot water demand according to the equivalent electricity price and the equivalent natural gas price, etc. At the same time point, the users' actual energy use demands do not change, and the users' actual energy use has little impact, which can improve the flexibility and economy of the users' energy use. Therefore, it is of great significance to construct user-interactive models that consider the conversion of energy use.

According to the above analysis, multi-energy complementary users can choose to convert, reduce, transfer, and other ways to achieve interaction.

2.2.1. Convertible Load Model

This paper considers the users' electricity–gas convertible load and improves the convertible model of reference [23]. That is, the influence of consumer psychology is considered when optimizing the convertible load model. Based on the principle of consumer psychology, the difference between the equivalent electricity price and the equivalent natural gas price affects the response of the user's convertible load. The users' interactive response range is divided into saturation zone, linear zone, and dead zone [24,25]. When the difference between the equivalent electricity price and the equivalent natural gas price is lower than the dead zone threshold, users are unwilling to respond interactively. When it exceeds the threshold, users start to respond interactively. In the linear zone, the users' convertible load increases with the difference between the equivalent electricity price and the equivalent natural gas price, and it shows a linear upward trend. When the compensation limit is exceeded, the users' electricity–gas convertible amount tends to be saturated. In this paper, the energy use rate $\lambda_{con,t}^{g,e}$ of the users' gas load to electric load and the energy use rate $\lambda_{con,t}^{e,g}$ of the users' electric load to gas load are used to characterize the influence of the difference between the equivalent electricity price and the equivalent natural gas price $\pi_{con,t}$ on the users' mode of energy use.

$$\lambda_{con,t}^{g,e} = \begin{cases} 1 & \pi_{con,t} \leq -\pi_{con,max}^{g,e} \\ \dfrac{\pi_{con,t}+\pi_{con,min}^{g,e}}{\pi_{con,min}^{g,e}-\pi_{con,max}^{g,e}} & -\pi_{con,max}^{g,e} < \pi_{con,t} < -\pi_{con,min}^{g,e} \\ 0 & -\pi_{con,min}^{g,e} \leq \pi_{con,t} \leq 0 \end{cases} \tag{2}$$

$$\lambda^{e,g}_{con,t} = \begin{cases} 0 & 0 \leq \pi_{con,t} \leq \pi^{e,g}_{con,min} \\[2mm] \dfrac{\pi_{con,t} - \pi^{e,g}_{con,min}}{\pi^{e,g}_{con,max} - \pi^{e,g}_{con,min}} & \pi^{e,g}_{con,min} < \pi_{con,t} < \pi^{e,g}_{con,max} \\[2mm] 1 & \pi^{e,g}_{con,max} \leq \pi_{con,t} \end{cases} \tag{3}$$

In the above formula: $\pi^{g,e}_{con,min}$, $\pi^{e,g}_{con,min}$, $\pi^{g,e}_{con,max}$, and $\pi^{e,g}_{con,max}$ are the dead zone threshold and saturation zone limit of the difference between the equivalent electricity price and the equivalent natural gas price when the users respond to convertible load.

According to the calorific value equivalence theorem and the energy conservation theorem, the constraints that the convertible load needs to satisfy are as follows:

$$\begin{cases} L^e_{con,t} = L^{e,n}_{con,t} - \mu^{e,g}_{con,t} \Delta L^e_{con,t} + \mu^{g,e}_{con,t} \Delta L^{g,e}_{con,t} \\ L^g_{con,t} = L^{g,n}_{con,t} - \mu^{g,e}_{con,t} \Delta L^g_{con,t} + \mu^{e,g}_{con,t} \Delta L^{e,g}_{con,t} \\ \Delta L^e_{con,t} = \lambda^{e,g}_{con,t} L^{e,n}_{con,t} \\ \Delta L^g_{con,t} = \lambda^{g,e}_{con,t} L^{g,n}_{con,t} \\ \Delta L^{g,e}_{con,t} = \Delta L^g_{con,t}/I, \Delta L^{e,g}_{con,t} = I \Delta L^e_{con,t} \\ \mu^{e,g}_{con,t} + \mu^{g,e}_{con,t} \leq 1 \end{cases} \tag{4}$$

In the above formula: $L^{e,n}_{con,t}$ and $L^{g,n}_{con,t}$ are the power before the convertible electrical load and convertible natural gas load response; $L^e_{con,t}$ and $L^g_{con,t}$ are the power after the response of the convertible electrical load and convertible natural gas load; $\mu^{e,g}_{con,t}$ and $\mu^{g,e}_{con,t}$ are 0–1 auxiliary variables for interactive response; $\Delta L^e_{con,t}$ and $\Delta L^g_{con,t}$ are convertible electrical load and convertible natural gas load response quantities; $\Delta L^{g,e}_{con,t}$ and $\Delta L^{e,g}_{con,t}$ are the increased power of the convertible electrical load and convertible natural gas load after the response; I is the electricity–gas conversion coefficient, which is taken as 1.25 in this paper.

2.2.2. Reducible Electrical Load Model

The reducible electrical load is the load that users can partially reduce [7]. The model is as follows:

$$\begin{cases} L^e_{adj,t} = L^{e,n}_{adj,t} - \mu^e_{adj,t} \Delta L^e_{adj,t} \\ 0 \leq \Delta L^e_{adj,t} \leq L^{e,n}_{adj,t} \end{cases} \tag{5}$$

In the above formula: $L^{e,n}_{adj,t}$ is the load power that can be reduced before the users respond; $\mu^e_{adj,t}$ is a 0–1 variable of whether to reduce; $\Delta L^e_{adj,t}$ is the load power actually reduced by the users.

Considering that power load reduction has a great impact on user satisfaction, the maximum duration of power load reduction is constrained in this paper:

$$\sum_{\tau=t}^{t+t^e_{adj,max}} (1 - \mu^e_{adj,\tau}) \geq 1 \quad t = 1, 2, \cdots, T - t^e_{adj,max} \tag{6}$$

In the above formula: $t^e_{adj,max}$ is the maximum duration of electrical load reduction.

2.2.3. Transferable Electrical Load Model

After the community publishes the electricity price, users will transfer part of the electrical load from higher to lower hours to reduce the cost of energy use, such as washing machines, electric vehicles, and other loads. This part of the load is called a transferable power load [7]. The transferable electrical load model is as follows:

$$\begin{cases} L_{\text{tran},t}^{e} = L_{\text{tran},t}^{e,n} + L_{\text{tran},t}^{e,in} - L_{\text{tran},t}^{e,out} \\ \sum_{t=1}^{T} L_{\text{tran},t}^{e,in} = \sum_{t=1}^{T} L_{\text{tran},t}^{e,out} \\ 0 \leq L_{\text{tran},t}^{e,in} \leq L_{\text{tran,max}}^{e,in} \\ 0 \leq L_{\text{tran},t}^{e,out} \leq L_{\text{tran,max}}^{e,out} \end{cases} \tag{7}$$

In the above formula: $L_{\text{tran},t}^{e,n}$ and $L_{\text{tran},t}^{e}$ are the power before and after the response of the transferable electrical load at time t; $L_{\text{tran},t}^{e,in}$ and $L_{\text{tran},t}^{e,out}$ are the actual transfer-in and transfer-out power of the transferable electrical load at time t; $L_{\text{tran,max}}^{e,in}$ and $L_{\text{tran,max}}^{e,out}$ are the maximum load that can be transferred in and out at time t.

2.2.4. Heat Load Model

Some users in the community have high requirements for thermal comfort, the indoor temperature cannot be adjusted, and they are willing to bear the additional cost of thermal comfort. This part of the heat load is a fixed heat load. Another part of the user is willing to adjust the thermal comfort range, which is an adjustable heat load.

The adjustable heat load adopts the first-order building thermodynamic model [26]:

$$L_{\text{adj},t}^{h} = N_1 \frac{1}{R} \left(\frac{T_{\text{in},t+1} - e^{-\Delta t/\tau_1} T_{\text{in},t}}{1 - e^{-\Delta t/\tau_1}} - T_{\text{out},t} \right) \tag{8}$$

In the above formula: N_1 is the number of users with adjustable heating temperature; $\tau_1 = RC_{\text{air}}$, C_{air} is the heat capacity of indoor air, which can be taken as 1.2 kWh/°C, R is the equivalent thermal resistance of the house, which can be taken as 6.8 °C/kW; $T_{\text{in},t}$ is the indoor temperature of the heating that can be adjusted at time t; $T_{\text{out},t}$ is the outdoor temperature.

The interactive response potential of heat load is mainly related to the human body's heat-using psychology for temperature perception, which has a certain elasticity. To better describe the user's thermal response potential, this paper introduces predicted mean vote (PMV), and the relationship between room temperature and PMV index value $\lambda_{\text{PMV},t}$ is as follows [27]:

$$T_{\text{in},t} = T_{\text{com,s}} - \frac{M(2.43 - \lambda_{\text{PMV},t})(\lambda_{\text{clo}} + 0.1)}{3.76} \tag{9}$$

In the above formula: $T_{\text{com,s}}$ is the average temperature of human skin in a comfortable state, which can be taken as 33.5 °C; λ_{clo} and M are the thermal resistance of the clothes and the metabolic rate of the human body, take 0.11 (m^2·°C)/W and 80 W/m^2 respectively.

Considering the recommendations of the ISO-7730 standard and the daily routine of the users, this paper limits the time sharing of the PMV index value, which is expressed as:

$$\begin{cases} |\lambda_{\text{PMV},t}| \leq 1, t \in [1,7] \cup [21,24] \\ |\lambda_{\text{PMV},t}| \leq 0.5, t \in [8,20] \end{cases} \tag{10}$$

The fixed heat load model is:

$$L_{\text{fir},t}^{h} = N_2 \frac{1}{R} \left(\frac{T_{\text{set}} - e^{-\Delta t/\tau_1} T_{\text{set}}}{1 - e^{-\Delta t/\tau_1}} - T_{\text{out},t} \right) \tag{11}$$

In the above formula: N_2 is the number of users with non-adjustable heating temperature; T_{set} is the most comfortable indoor temperature of the users who cannot be adjusted.

To prevent the indoor temperature of adjustable users from being lower than the most comfortable temperature, the following constraints are imposed on the indoor temperature:

$$\sum_{t=1}^{T} \frac{T_{\text{in},t}}{T} = T_{\text{set}} \tag{12}$$

To sum up, the refined model of interaction considering user-side energy conversion is described as a matrix as follows:

$$\begin{bmatrix} L_t^e \\ L_t^h \\ L_t^g \end{bmatrix} = \begin{bmatrix} P_{\text{out},t}^e \\ P_{\text{out},t}^h \\ P_{\text{out},t}^g \end{bmatrix} = \begin{bmatrix} L_{\text{fir},t}^e \\ L_{\text{fir},t}^h \\ L_{\text{fir},t}^g \end{bmatrix} + \begin{bmatrix} L_{\text{adj},t}^e \\ L_{\text{adj},t}^h \\ 0 \end{bmatrix} + \begin{bmatrix} L_{\text{tran},t}^e \\ 0 \\ 0 \end{bmatrix} + \begin{bmatrix} L_{\text{con},t}^e \\ 0 \\ L_{\text{con},t}^g \end{bmatrix} \tag{13}$$

In the above formula: L_t^e, L_t^h and L_t^g are the electricity load, heat load, and natural gas load of the users; $L_{\text{fir},t}^e$ and $L_{\text{fir},t}^g$ are fixed electrical load and natural gas load; $L_{\text{fir},t}^h$ is the fixed heat load.

3. Optimization Model of Multi-Energy Transaction Decision between the Community Operator and the Users

3.1. Model Architecture of the Multi-Energy Transaction between the Community Operator and the Users Based on the Master–Slave Game

The master–slave game is an effective method to solve the problem of how to make decisions when there is an interest relation or conflict. The master–slave game is a dynamic non-cooperative game, and the unequal status of participants is the most fundamental difference between the master–slave game and the classical game. In the master–slave game, each subject has a different status and decision-making sequence. The leader has a leadership advantage and can occupy the first or advantageous position in the game, and the follower must follow the leader to make decisions. Not only does the retail price of energy set by the community operator affect consumer demand, but demand also affects price. Both parties have independent interest demands, and both make decisions with the goal of maximizing their own interests and influencing each other, and there is a master–slave game relationship.

The interaction relationship between the community operator and the users based on the master–slave game is shown in Figure 2. In the energy market environment, the community operator guides users to interactively use energy by setting reasonable retail energy prices. The energy purchase strategies of the energy market and equipment output are optimized under the maximum profit, and the users adjust the interactive energy use strategies according to the energy sales price of the community operator. In the master–slave game, the community operator, as the leader, guides users to interactively use energy by adjusting the retail prices of electricity and heat energy. As followers, users who receive retail prices of electricity and heat energy released by the community operator will change their energy usage habits to a certain extent and adjust their interactive energy usage strategies. At the same time, the users' changes in their own energy use needs will affect the retail energy price formulation strategies of the community operator. Additionally, the cycle goes back and forth when neither the community operator nor the users can improve their own interests by changing their own strategies; the equilibrium solution of the master–slave game model is reached. Finally, the community operator determines the final retail energy prices, the energy purchased in the energy market, and the output of each device. The users determine the interactive energy use strategies according to the retail energy prices released by the community operator.

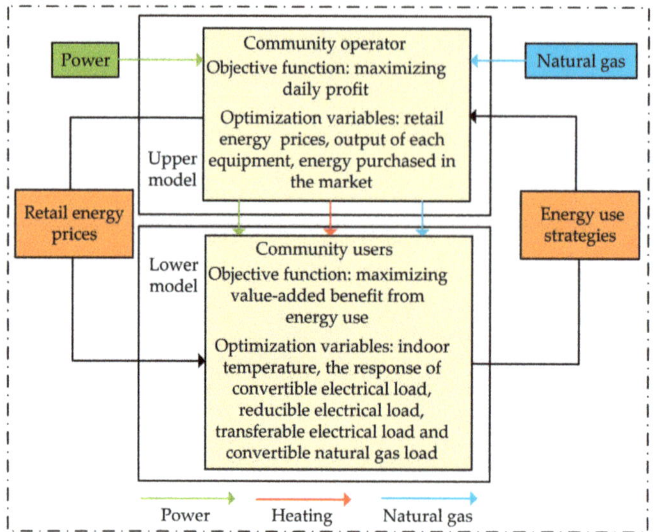

Figure 2. Multi-energy transaction decision-making model architecture between the community operator and the users.

3.2. Pricing Model of Community Operator's Retail Energy Prices

3.2.1. Objective Function

The community operator guides users to interact, optimize energy transactions, and maximize daily profit by formulating reasonable retail energy prices in the upper-level model. The objective function is:

$$C_1 = \max(C_{\text{sell}} + C_{\text{add}} - C_{\text{ob}} - C_{\text{mend}} - C_{\text{wv}}) \tag{14}$$

In the above formula: C_1 is the daily profit of the community operator; C_{sell} is the energy retail income; C_{add} is the additional income for thermal comfort; C_{ob} is the cost of purchasing energy; C_{mend} is the cost of equipment operation and maintenance; C_{wv} is the cost of renewable power abandonment.

(1) Energy retail income

$$C_{\text{sell}} = \sum_{t=1}^{T} \sum_{i \in E} F_{i,t} L_t^i \tag{15}$$

In the above formula: $F_{i,t}$ is retail energy prices of the community; E is the load type of the users, $E = \{e, h, g\}$.

(2) Additional income for thermal comfort

$$C_{\text{add}} = \sum_{t=1}^{T} F_{\text{add}} L_{\text{fir},t}^{\text{h}} \tag{16}$$

In the above formula: F_{add} is the additional unit service price at which the heating load is not adjustable for the users to maintain the most comfortable temperature.

(3) Cost of purchasing energy

$$C_{\text{ob}} = \sum_{t=1}^{T} [F_{\text{net},e,t} P_{\text{net},t}^{\text{e}} + F_{\text{net},g,t} P_{\text{in},t}^{\text{g}}] \tag{17}$$

In the above formula: $F_{net,e,t}$ and $F_{net,g,t}$ are the unit cost of purchasing electricity and natural gas at time t.

(4) Cost of equipment operation and maintenance

$$C_{mend} = \sum_{t=1}^{T} \sum_{j=1}^{J} F_{m,j} P_{j,t} \tag{18}$$

In the above formula: $F_{m,j}$ is the unit operation and maintenance costs of equipment j; $P_{j,t}$ is the output of equipment j at time t.

(5) Cost of renewable power abandonment

$$C_{wv} = \sum_{t=1}^{T} [\Delta P_{w,t} F_w + \Delta P_{v,t} F_v] \tag{19}$$

In the above formula: $\Delta P_{w,t}$ and $\Delta P_{v,t}$ are the amount of abandoned wind and light at time t; F_w and F_v are the unit cost of abandoned wind and light.

3.2.2. Multiple Energy Price Constraints

To prevent the community operator from maliciously raising prices and ensuring the benefit of users, multiple energy prices need to be constrained:

$$F_{i,t,min} \leq F_{i,t} \leq F_{i,t,max} \tag{20}$$

$$\sum_{t=1}^{T} \frac{F_{i,t}}{T} \leq F_{i,av} \tag{21}$$

In the above formula: $F_{i,t,max}$ and $F_{i,t,min}$ are the maximum and minimum prices for electricity and heat prices to protect the benefits of the users and the operator; $F_{i,av}$ is the average price of electricity and heat in the energy market.

3.2.3. Energy Conversion Equipment Constraints

The traditional energy conversion equipment model considers the output efficiency of the equipment to be a fixed value. To ensure the accuracy of the equipment output, this paper adopts the dynamic energy-efficiency model of energy conversion equipment. The output efficiency of energy conversion equipment is mainly related to the load rate [28,29]. It is expressed by means of polynomial fitting, as shown in Formula (22).

$$\eta_{x,n} = \sum_{k=0}^{n} \alpha_{x,k} \left(\frac{P_x}{P_{x,N}} \right)^k \tag{22}$$

In the above formula: $\eta_{x,n}$ is the dynamic energy efficiency of equipment x using n-order polynomial fitting; $\alpha_{x,k}$ is the fitting coefficient; P_x and $P_{x,N}$ are the actual output power and rated output power of the equipment.

The input–output relationship of energy conversion equipment considering dynamic energy efficiency is as follows:

(1) Combined heat and power unit

The output power efficiency of the CHP unit can be fitted by a fourth-order fitting [28]:

$$P_{CHP,t}^{e} = \eta_{CHP,4} P_{CHP,t}^{g} \tag{23}$$

In the above formula: $P_{CHP,t}^{g}$ is the natural gas power entering the CHP unit; $P_{CHP,t}^{e}$ is the output electric power of the CHP unit.

In this paper, the strategy of determining the heat by electricity is adopted, and the thermoelectric ratio ψ_{CHP} can be described by the second-order fitting of the electrical load rate of the CHP unit:

$$\psi_{\text{CHP}} = \sum_{k=0}^{2} \left(\alpha_{\psi,k} N_{\text{CHP},e}^{k} \right) \tag{24}$$

In the above formula: $N_{\text{CHP},e}$ is the electrical load rate of the CHP unit.

Therefore, the output thermal power $P_{\text{CHP},t}^{\text{h}}$ of the CHP unit is:

$$P_{\text{CHP},t}^{\text{h}} = \psi_{\text{CHP}} P_{\text{CHP},t}^{\text{e}} \tag{25}$$

(2) Gas boiler

The output thermal efficiency of the gas boiler can be fitted by the first-order:

$$P_{\text{GB},t}^{\text{h}} = \eta_{\text{GB},1} P_{\text{GB},t}^{\text{g}} \tag{26}$$

In the above formula: $P_{\text{GB},t}^{\text{g}}$ is the natural gas power entering the gas boiler; $P_{\text{GB},t}^{\text{h}}$ is the output thermal power of the gas boiler.

(3) Electric heat pump

The output thermal efficiency of the electric heat pump is related to the load rate and temperature. In this paper, only the influence of the load rate is considered, and second-order fitting can be used [29]:

$$P_{\text{EHP},t}^{\text{h}} = \eta_{\text{EHP},2} P_{\text{EHP},t}^{\text{e}} \tag{27}$$

In the above formula: $P_{\text{EHP},t}^{\text{e}}$ is the electric power entering the electric heat pump; $P_{\text{EHP},t}^{\text{h}}$ is the output heat power of the electric heat pump.

3.2.4. Device Operation Constraints

(1) Energy Conversion Equipment Constraints

$$\begin{cases} P_i^{\min} \leq P_{i,t} \leq P_i^{\max} \\ -P_i^{\text{down}} \leq P_{i,t} - P_{i,t-1} \leq P_i^{\text{up}} \end{cases} \tag{28}$$

In the above formula: $P_i^{\max}$ and $P_i^{\min}$ are the upper and lower limits of the output of the coupling device at time t; P_i^{up} and P_i^{down} are the upper and lower limits of the climbing power of the coupling device at time t.

(2) Energy Storage Device Constraints

$$\begin{cases} S_{i,t}^{\text{ES}} = (1 - \sigma_i^{\text{ES}}) S_{i,t-1}^{\text{ES}} + P_{\text{c/d},i,t}^{\text{ES,n}} \Delta t \\ P_{\text{c/d},i,t}^{\text{ES,n}} = \eta_{\text{c},i}^{\text{ES}} P_{\text{c},i,t}^{\text{ES}} - \frac{1}{\eta_{\text{d},i}^{\text{ES}}} P_{\text{d},i,t}^{\text{ES}} \\ C_{i,\min}^{\text{ES}} \leq S_{i,t}^{\text{ES}} \leq C_{i,\max}^{\text{ES}} \\ 0 \leq P_{\text{c},i,t}^{\text{ES}} \leq \varepsilon_{\text{c},i,t} P_{\text{c},i,t}^{\max} \\ 0 \leq P_{\text{d},i,t}^{\text{ES}} \leq \varepsilon_{\text{d},i,t} P_{\text{d},i,t}^{\max} \\ S_{i,0}^{\text{ES}} = S_{i,T}^{\text{ES}} \end{cases} \tag{29}$$

In the above formula: $S_{i,t}^{\text{ES}}$ is the energy storage value of the energy storage device at time t; σ_i^{ES} is the self-loss rate of the energy storage device; $\eta_{\text{c},i}^{\text{ES}}$ and $\eta_{\text{d},i}^{\text{ES}}$ are the charging and discharging efficiency of the energy storage device; $P_{\text{c},i,t}^{\text{ES}}$ and $P_{\text{d},i,t}^{\text{ES}}$ are the charging and discharging power of the energy storage device; $C_{i,\max}^{\text{ES}}$ and $C_{i,\min}^{\text{ES}}$ are the upper and lower limits of the capacity of the energy storage device; $\varepsilon_{\text{c},i,t}$ and $\varepsilon_{\text{d},i,t}$ are 0–1 auxiliary variables; $P_{\text{c},i,t}^{\max}$ and $P_{\text{d},i,t}^{\max}$ are the upper limit of energy storage charging and discharging power respectively.

3.2.5. Renewable Energy Output Constraints

In the below formula: $P_{\mathrm{w},t}^{\mathrm{e,pre}}$ and $P_{\mathrm{v},t}^{\mathrm{e,pre}}$ are the predicted power values of wind power and photovoltaic.

$$\begin{cases} 0 \le P_{\mathrm{w},t}^{\mathrm{e}} \le P_{\mathrm{w},t}^{\mathrm{e,pre}} \\ 0 \le P_{\mathrm{v},t}^{\mathrm{e}} \le P_{\mathrm{v},t}^{\mathrm{e,pre}} \end{cases} \tag{30}$$

In addition, the upper-layer model also needs to satisfy the power balance constraint of Equation (13). The above nonlinear model can be approximated by piecewise linearization.

3.3. Energy Use Strategies Model Considering User Interaction

According to the retail energy prices released by the community operator, users consider increasing or decreasing load, converting load, and transferring load. For details, see the refined model of user interaction considering energy conversion in Section 2. The users' goal is to optimize their own interactive energy-use strategies and maximize the value-added benefit of energy use. This paper defines the value-added benefit of a users' energy use, which is the users' total energy use utility C_{ben} minus the total cost. The total cost includes the cost of purchasing energy C_{ub} and the additional cost of thermal comfort C_{app}.

$$C_2 = \max[C_{\mathrm{ben}} - (C_{\mathrm{ub}} + C_{\mathrm{app}})] \tag{31}$$

$$C_{\mathrm{ben}} = \sum_{t=1}^{T} \sum_{i \in E} [f_1^i L_t^i - \frac{f_2^i}{2}(L_t^i)^2] \tag{32}$$

$$C_{\mathrm{ub}} = C_{\mathrm{sell}} = \sum_{t=1}^{T} \sum_{i \in E} F_{i,t} L_t^i \tag{33}$$

$$C_{\mathrm{app}} = C_{\mathrm{add}} = \sum_{t=1}^{T} F_{\mathrm{add}} L_{\mathrm{fir},t}^{\mathrm{h}} \tag{34}$$

In the above formula: f_1^i and f_2^i are the constant coefficients of the users' preference for type i energy, reflecting the users' preference for energy demand.

3.4. Model-Solving Process

According to the reference [30], it is easy to prove that there is a unique equilibrium solution for the optimization model of multi-energy transaction decisions between the community operator and the users based on the master–slave game in this paper.

This paper uses MATLAB software programming, and the method of combining the differential evolution algorithm and CPLEX solver is used to solve the proposed model. The solution process is as follows:

(1) Initialize the parameters of the community operator and the users, $k = 0$, set the maximum number of iterations $k_{\max} = 100$, use the differential evolution algorithm to randomly generate retail energy prices of 10 groups of the community operator, and transmit them to the energy use strategies model considering user interaction.

(2) $k = k + 1$.

(3) The users receive retail energy prices published by the community operator. Use the CPLEX solver to solve the energy use strategies model and the optimal value-added benefit C_2^k, and return the energy use strategies to the model of the community operator.

(4) The community operator optimizes the output of equipment and the amount of electricity and gas purchased in the market according to the energy use strategies of the users and calculates the optimal profit C_1^k of the community operator.

(5) Use the variation and crossover of the differential evolution algorithm to generate a group of new retail energy prices and repeat the processes in (3) and (4). Additionally, calculate the optimal value-added benefit C_2^{k*} of the users and the optimal profit C_1^{k*} of the community operator under the new retail energy prices.

(6) Perform selection operation: compare the optimal solutions of the community operator before and after mutation and crossover; if $C_1^{k*} \geq C_1^k$, then $C_1^{k+1} = C_1^{k*}$, $C_2^{k+1} = C_2^{k*}$; if $C_1^{k*} < C_1^k$, then $C_1^{k+1} = C_1^k$, $C_2^{k+1} = C_2^k$.

(6) If $k \geq k_{\max}$, end the program; otherwise, return to flow (2).

4. Case Analysis

4.1. Parameter Settings

This paper selects a community integrated energy system in the northern winter of China as the research object. The forecasting curves of the renewable energy output, initial load, and outdoor temperature curves are shown in Figure 3. Other required parameter data are shown in the table in the Appendix A. The equipment parameters of the community operator are shown in Table A1, which is used to deal with the community operator's device model. The data parameters for the users are shown in Table A2, which is the parameter data required by the user model. The equipment efficiency fitted from Table A2 is shown in Figure A1. The time-of-use electricity price in the energy market is shown in Table A3 [31], it is the price that the community operator trades with the energy market. To protect the benefit of the users, the upper limit of the electricity price set by the community operator is not higher than the time-of-use electricity price in the energy market; the lower limit is not less than 0.2 RMB/(kWh). The heat price of the energy market is 0.35 RMB/(kWh) [32], and the thermal price range set by the community ranges from 0.2 to 0.5 RMB/(kWh). The natural gas price within the community is the same as the market gas price, which is 0.34 RMB/(kWh) [33].

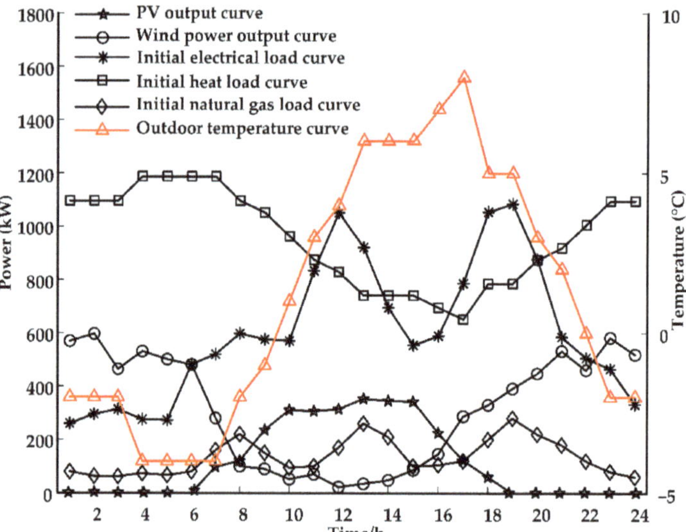

Figure 3. Forecasting curves of renewable energy output, initial load, and outdoor temperature.

4.2. Scene Settings

To illustrate that the multi-energy trading strategies proposed in this paper can improve the profit of the community operator and the value-added benefit of the users, and can improve renewable energy utilization, three scenarios are set for comparative analysis:

- Scenario 1: Regardless of user interaction [6], the energy price sold by the community operator to the users is the market price.
- Scenario 2: Considering user interaction [15] and ignoring the influence of the users' electricity–gas convertible load, the community operator and the users compete to determine the electricity price and heat price of the community.

- Scenario 3: Considering user interaction, using the refined model of user interaction considering energy conversion and considering the impact of electricity–gas convertible load, the community operator and the users play games to formulate the internal electricity price and heat price in the community.

4.3. Simulation Analysis

4.3.1. Analysis of the Multi-Energy Transaction Results of the Community Operator

Under the model proposed in this paper in Scenario 3, the electricity and heat prices traded between the community operator and the users are shown in Figure 4.

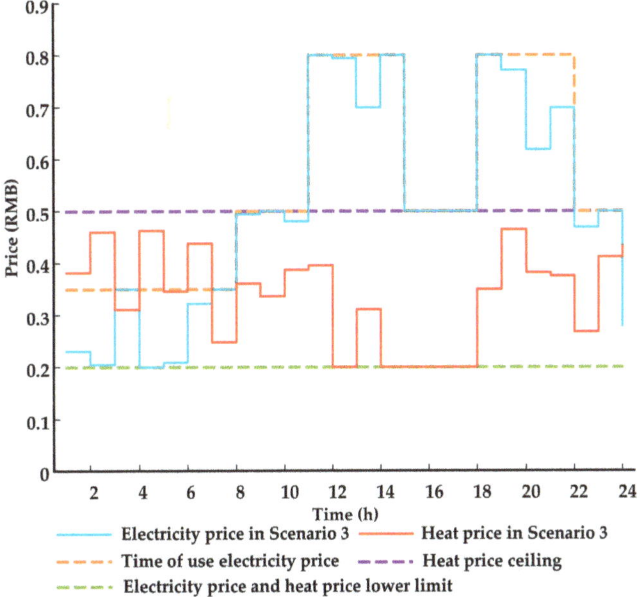

Figure 4. The price of electricity and heat that the operator trades with users.

The electricity price trend set by the community operator basically follows the electricity price in the energy market, and the peak electricity price is in the peak energy use period, which is in line with the actual situation. The heat price set by the community operator fluctuates within the limit of the heat price. Judging from the electricity and heat prices traded by the community operator and the users in Figure 4, the average price of electricity set by the community operator in Scenario 3 is 0.5 RMB/(kWh), the average price of electricity in the energy market is 0.56 RMB/(kWh). The average price of heat within the community is 0.33 RMB/(kWh), and the average price of heat in the energy market is 0.35 RMB/(kWh). Based on the above analysis, the average electricity price and average heat price set by the community operator are 10.7% and 5.7% lower, respectively, than in the market. In Figure 4, in Scenario 3, the electricity and heat prices traded between the community operator and the users can protect the benefit of the users and promote energy transactions between the users and the community operator.

Figure 5 shows the results of electricity and gas transactions between the community operator and the energy market in different scenarios. Taking the peak period of electricity price as an example, the following analysis is made: from Figure 5, it can be seen that in Scenarios 2 and 3, compared with Scenario 1, the community operator purchases less electricity from the power grid during the peak period of electricity price, which effectively reduces the power supply pressure on the large power grid, indicating that user interaction can indirectly participate in the power market and reduce the peak power consumption of the large power grid. Compared with Scenario 2, Scenario 3 considers the convertible load, and part of the electricity load demand of the users is supplied by natural gas during the peak electricity price period. The energy market has the least amount of electricity purchased and the largest amount of gas during the corresponding period.

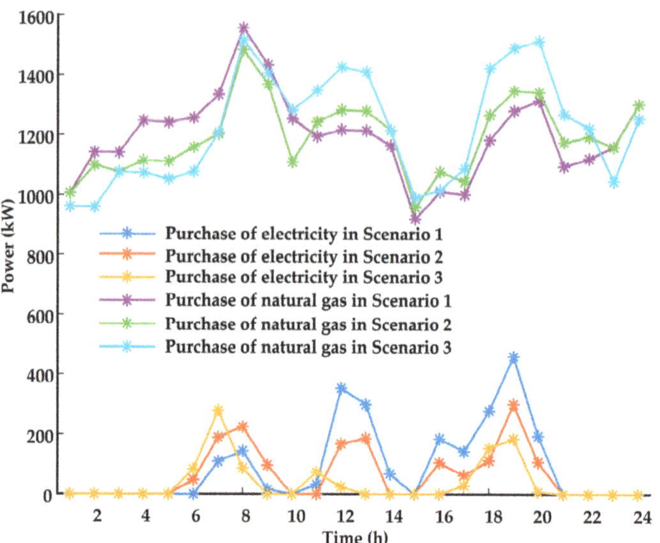

Figure 5. Electricity and natural gas trading results between the community operator and the energy market in different scenarios.

4.3.2. Analysis of Energy Use Strategies for User Interaction

Figure 6 shows the results of electricity and natural gas transactions between the community operator and the users in different scenarios. Scenario 1 does not consider user interaction, and its electricity and natural gas loads are all original loads. As shown in Figure 6, in both Scenarios 2 and 3, considering user interaction, the electrical load peak smoothed to varying degrees. Scenario 1 has the highest electrical load peak, followed by Scenario 2. Scenario 3 adopts the strategies of this paper, with the smallest electrical load peak and the largest gas load fluctuation.

Considering the limitations of space, this paper will focus on the analysis of the interactive energy use strategies of users' electricity–gas convertible load. In Scenario 3, the amount of the convertible load interaction, including consumer psychology, is shown in Figure 7, which shows the relationship between the users' actual convertible load response and the equivalent energy price difference. This is consistent with the optimized results in Figure 6. For users, when the electricity price is lower than the equivalent natural gas price (the natural gas price multiplied by the electricity–gas conversion coefficient). For example, to meet the same demand for hot water, the electricity cost of the users is lower than the natural gas cost. When the natural gas price is high, the cost of electricity is higher than the cost of natural gas. Therefore, when the electricity price is lower than the equivalent natural gas price, users replace part of the convertible natural gas load with electricity, and when the electricity price is higher than the equivalent natural gas price, users replace part of the convertible electricity load with natural gas. Compared with Scenario 2, the convertible load is considered in Scenario 3, and the electricity load curve during the peak period of the electricity price during the periods of 11:00~14:00 and 18:00~21:00 is lower than that of Scenario 2. In the low electricity price period from 1:00 to 7:00, the power load curve of Scenario 3 is higher than that of Scenario 2. After considering the convertible load, the electrical load curve of the users in Scenario 3 is smoother, and the outline of the electric load curve is optimized.

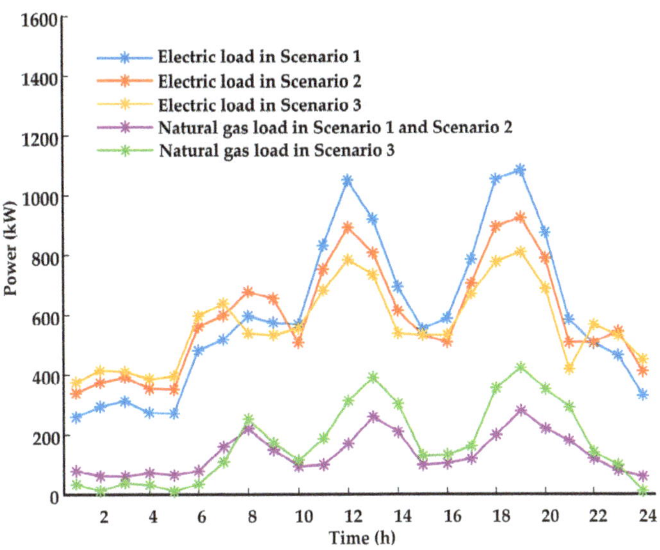

Figure 6. Electricity and natural gas loads of users in different scenarios.

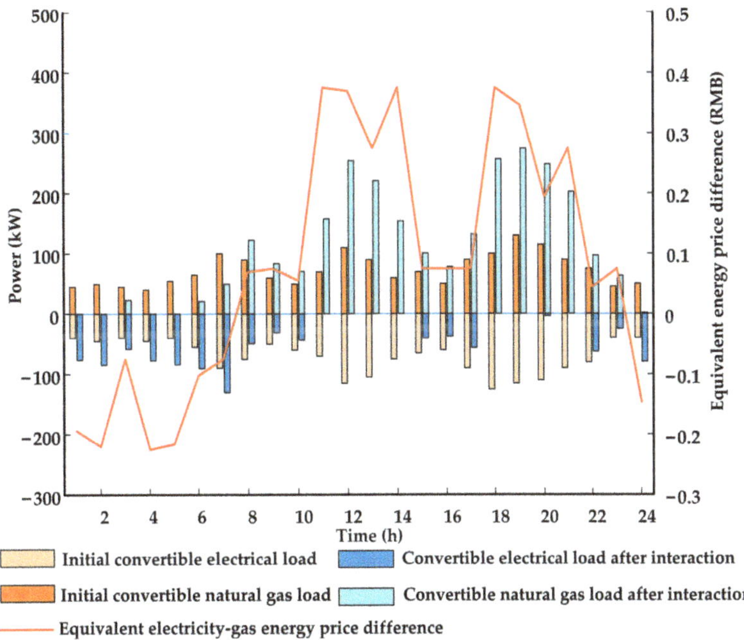

Figure 7. The relation between the convertible load and the equivalent energy price difference.

When the users' heat load does not participate in the interaction, that is, the original heat load (Scenario 1), the indoor temperature is kept at the most comfortable temperature. When considering the users' adjustable heat load, the indoor temperature is kept in a suitable range (Scenario 3 is used as an example). The heat load and indoor temperature of Scenarios 1 and 3 are shown in Figure 8.

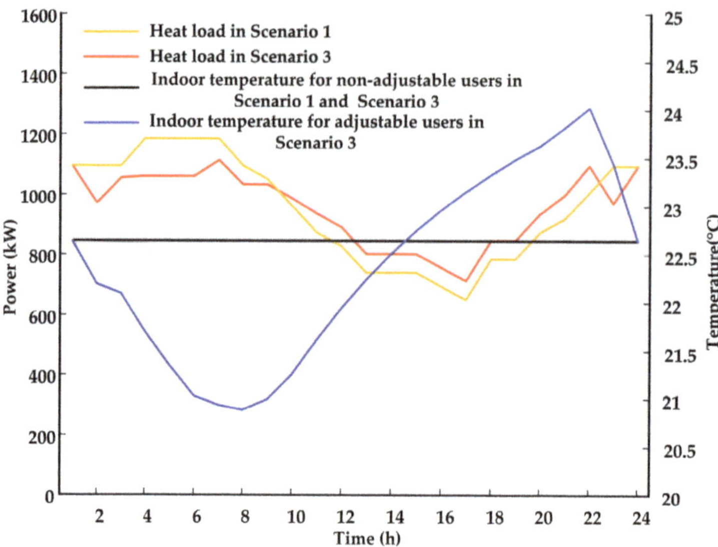

Figure 8. Heat load and indoor temperature in Scenarios 1 and 3.

According to Figure 8, the indoor temperature of users in Scenarios 1 and 3 cannot be adjusted at room temperature and is always maintained at about 22.6 degrees Celsius. The indoor temperature of users in Scenario 3 can be adjusted to maintain the indoor temperature within 20.5 to 24.5 degrees Celsius, ensuring the users' thermal comfort. From time period 1 to time period 9, the heat load provided by Scenario 3 is lower than the original heat load, and coupled with heat loss, the temperature decreases. At the same time, to keep the temperature of the first and last sections consistent, the heat load adjustment in the last few periods is relatively large.

4.3.3. Cost-Benefit Analysis of the Community Operator and Users

To show the superiority of the model and trading strategy proposed in this paper, the profit and profit rate of the community operator, the cost of the users, and the value-added benefit are compared in different scenarios. The comparison results are shown in Table 1.

Table 1. Comparative Study 1.

	Compare Items	Scenario 1	Scenario 2	Scenario 3
Community operator	Cost (RMB)	13,940	12,486	11,976
	Profit (RMB)	4459	4329	4720
	Profit margin (%)	31.9	34.7	39.4
	Renewable energy utilization (%)	88	93	96
Community users	Total cost (RMB)	18,399	16,815	16,690
	Value-added benefit	15,121	16,993	17,412

Overall, it can be seen from Table 1 that, compared with Scenarios 1 and 2, Scenario 3 is the establishment of the multi-energy prices within the community under the model proposed in this paper, the profit and profit margin of the operator, the value-added benefit of the users' energy use, and the renewable energy utilization have increased significantly.

The specific analysis is as follows: In terms of on-site consumption of renewable energy, the on-site renewable energy utilization in Scenario 2 and Scenario 3 has increased by more than 5% compared with Scenario 1. Scenario 3 has the highest renewable energy utilization, followed by Scenario 2 and Scenario 1. Therefore, user interaction can improve on-site renewable energy utilization. In terms of the community operator's profit and the energy value-added benefit of the users, compared with Scenario 1, Scenario 2 reduces the community operator's profit by 130 RMB. The cost of the users' energy use has been reduced by 1584 RMB, and the value-added benefit of the users' energy use has been greatly improved. This is because Scenario 1 is sold to the users at the market energy prices, the retail energy prices are relatively high, and the users are not given preferential energy prices. Scenario 2 considers user interaction and gaming, the retail price of energy in the community is determined, and the energy prices are constrained within an appropriate range, which can greatly improve user satisfaction with energy use. Compared with Scenario 2, the community operator's profit in Scenario 3 has increased by 391 RMB. After converting the load, the operator reduces the penalty cost of renewable energy curtailment and the power supply cost during peak power consumption. For the users, the total cost is reduced by 125 RMB after considering the convertible load. At the same time, the value-added benefit of the users' energy use in Scenario 3 compared with Scenario 2 has been improved, and the satisfaction with energy use has been further improved. Compared with Scenarios 1 and 2, Scenario 3 has the highest profit for the community operator, the lowest total energy use cost for the users, and the highest value-added benefit. Although the community operator in Scenario 3 has the lowest turnover, it has the lowest cost. From the profit side, the profit in Scenario 3 is 5.9% and 9% higher than that in Scenario 1 and Scenario 2, respectively. In terms of profit margin, Scenario 3 has the highest profit margin, which is 7.5% and 4.7% higher than Scenario 1 and Scenario 2, respectively. Meanwhile, the value-added benefit of the users in Scenario 3 is 15.2% and 2.5% higher than that in Scenario 1 and Scenario 2, respectively. Scenario 3 can maximize the benefits of the community operator and the users. The community operator can actively guide user interaction during pricing by optimizing retail energy prices, improving the users' value-added benefit of energy use, and at the same time increasing the community operator's own profit and making the users more satisfied with the services provided by the community operator to consolidate and expand the user base and provide other value-added services. To sum up, Scenario 3 can consider the benefits of the community operator and the users and achieve a win–win situation.

To further clarify the work of this paper, Scenario 4 is added: a centralized optimization method is adopted to optimize the maximum profit of a community operator with a single objective [9], considering user interaction, but not considering the convertible load. The comparison results are shown in Table 2.

Table 2. Comparative Study 2.

Compare Items		Scenario 3	Scenario 4
Community operator	Cost (RMB)	11,976	13,153
	Profit (RMB)	4720	4957
	Profit margin (%)	39.4	37.6
Community users	Total cost (RMB)	16,690	-
	Value-added benefit	17,412	-
The optimization method		The master–slave game	The centralized optimization
Whether user interaction is considered		√	√
Whether the convertible load is considered		√	×
Whether the retail energy prices have been optimized		√	×
Whether a win–win situation has been achieved		√	×

As can be seen from Table 2, Scenario 4 adopts the centralized optimization with the single goal of maximizing the profit of the community operator, without considering the interests of the users. From the optimization results, the profit of the community operator in Scenario 4 is higher than that in Scenario 3 because the community operator does not offer preferential energy prices to the users and fails to balance the interests of users in Scenario 4.

Meanwhile, the cost of the community operator in Scenario 3 is lower than that in Scenario 4, and the profit margin of the community operator in Scenario 3 is 1.8% higher than that in Scenario 4. In Scenario 3, the community operator can optimize the energy prices and balance the value-added benefit of the users through appropriate profit sharing. Although the profit is reduced, it can improve the users' satisfaction with the use of energy and improve the profitability of the community operator. In terms of user interaction, Scenario 3 considers the convertible load, which can enrich the way users use energy. Under the optimization method of the master–slave game in Scenario 3, the community energy prices can be optimized, and the win–win situation between the community operator and the users can be realized.

Through the above comparative analysis, using the multi-energy transaction decision optimization model of the community operator and the users considering user interaction proposed in this paper, it is possible to formulate reasonable retail energy prices for the community, guide the users to interact, and improve the community operator's profit and value-added benefit of the users' energy use to achieve a win–win situation.

5. Conclusions

The master–slave game model constructed in this paper describes the energy transaction between the community operator and the users and proposes the optimization model of the multi-energy transaction decision between the community operator and the users. The upper model considers the maximum profit of the community operator, while the lower model aims for the maximum value-added benefit of the users. The model is effectively analyzed by an example, and the relevant conclusions are as follows:

(1) The retail energy prices of the community determined by the decision in this paper are reasonable and acceptable to the users. The average price of electricity and the average price of heat set by the community operator are 10.7% and 5.7% lower, respectively, than the market, which protects the interests of the users. The model is extensible. By modifying the corresponding model, more user groups can be promoted. For different countries and regions, the model of the upper community operator can be modified according to the actual situation, including the type of equipment and the type of renewable energy, which can be adjusted according to the needs. The lower user model can determine the types of user interaction load (including reducible, transferable, transferable, etc.) according to the living habits of residents in different countries and regions. At the same time, the lower model has variable data related to the users and can be applied to residential communities with different energy preferences.)

(2) With the continuous improvement of the user side equipment, convertible load becomes possible. Users can choose appropriate energy modes to meet their energy needs according to different energy prices. The refined user interaction model that considers energy conversion constructed in this paper can reduce user costs.

(3) The optimization model of the multi-energy transaction decision between the community operator and the users proposed in this paper considers the energy conversion on the user side, which can not only improve the profit of the community operator, but also increase the value-added benefit of energy use and realize a win–win situation for the community operator and the users. Using the strategy proposed in this paper to set the community prices increases the community operator's profit and profit margin by 5.9% and 7.5%, respectively, compared to using market energy prices directly. At the same time, the value-added benefit to users also increases by 15.2%. In addition, user interaction can indirectly reduce the peak value of the grid, which is beneficial to grid security.

The model established in this paper mainly formulates the retail prices of community energy from the dimensions of the community and the users, ignoring the energy connection between the community and the community. At the same time, the impact of load and renewable energy uncertainty is ignored. The next step will continue to study the energy transaction strategies between the community and the community, and the impact of uncertainty. In terms of the community and the community, a single community may have an energy surplus or a shortage at some time. Multi-communities can trade surplus or shortage energy according to a certain mode to realize the efficient use of resources. Uncertainties also affect the decisions of the community operator. We will continue to study these two aspects in future research.

Author Contributions: Conceptualization and methodology, Y.L. and X.W.; simulation and analysis, Y.L.; investigation, X.W.; data curation, Y.L.; writing—original draft preparation, Y.L.; writing—review and editing, Y.L. and X.W.; supervision, X.W.; literature research, Y.L. All authors have read and agreed to the published version of the manuscript.

Funding: This research was supported by the Natural Science Foundation of Xinjiang Uygur Autonomous Region under Grant 2020D01C031.

Institutional Review Board Statement: Not applicable.

Informed Consent Statement: Not applicable.

Data Availability Statement: Not applicable.

Conflicts of Interest: The authors declare no conflict of interest.

Appendix A

Table A1. The equipment parameters.

Equipment	Parameter Type	Parameter Value
CHP	Rated Capacity	300 kW
	Minimum output power	100 kW
	Electrical efficiency fitting coefficient	$\alpha_{CHP,0} = 0.09$, $\alpha_{CHP,1} = 0.44$, $\alpha_{CHP,2} = -0.14$, $\alpha_{CHP,3} = -0.11$, $\alpha_{CHP,4} = 0.06$
	Thermoelectric ratio fitting coefficient	$\alpha_{\psi,0} = 3.82$, $\alpha_{\psi,1} = -5.84$, $\alpha_{\psi,2} = 3.6$
	Operation and maintenance cost	0.04 RMB/(kWh)
EHP	Rated Capacity	200 kW
	Thermal efficiency fitting coefficient	$\alpha_{EHP,0} = 2.61$, $\alpha_{EHP,1} = 0.36$, $\alpha_{EHP,2} = 0.026$
	Operation and maintenance cost	0.06 RMB/kWh
GB	Rated Capacity	1000 kW
	Thermal efficiency fitting coefficient	$\alpha_{GB,0} = 0.81$, $\alpha_{GB,1} = 0.13$
	Operation and maintenance cost	0.02 RMB/kWh
Electricity storage	Rated Capacity	500 kW
	Charge/Discharge efficiency	0.98
	Attrition rate	0.02
	Operation and maintenance cost	0.01 RMB/kWh
Heat storage	Rated Capacity	500 kW
	Charge/Discharge efficiency	0.95
	Attrition rate	0.02
	Operation and maintenance cost	0.01 RMB/kWh

Table A2. Data parameters for users.

Parameter	Meaning	Value	Parameter	Meaning	Value
$\pi_{con,min}^{e,g}$	Dead threshold for electrical energy conversion	0	$\pi_{con,max}^{e,g}$	Saturation value of electrical energy conversion	0.2
$\pi_{con,min}^{g,e}$	Dead threshold for natural gas conversion	0	$\pi_{con,max}^{g,e}$	Saturation value of natural gas conversion	0.15
f_1^e	First power coefficient of electrical energy preference	1.5	f_2^e	Quadratic coefficient of electrical energy preference	0.0009
f_1^h	First power coefficient of thermal energy preference	1.1	f_2^h	Quadratic coefficient of thermal energy preference	0.0011
f_1^g	First power coefficient of natural gas preference	1.2	f_2^g	Quadratic coefficient of natural gas preference	0.001
N_1	Number of the users with adjustable heating temperature	500	N_2	Number of the users with non-adjustable heating temperature	300
$L_{tran,max}^{e,out}$	Maximum load that can be transferred out	80 kW	$L_{tran,max}^{e,in}$	Maximum load that can be transferred in	80 kW
$t_{adj,max}^{e}$	Maximum duration of electrical load reduction	4 h	T_{set}	The most comfortable indoor temperature	22.6 °C

Table A3. Time of use electricity prices.

Period	Market Electricity Price (RMB/kWh)
Valley period: 01:00—07:00	0.35
Normal period: 08:00—10:00; 15:00—17:00; 22:00—24:00	0.5
Peak period: 11:00—14:00; 18:00—21:00	0.8

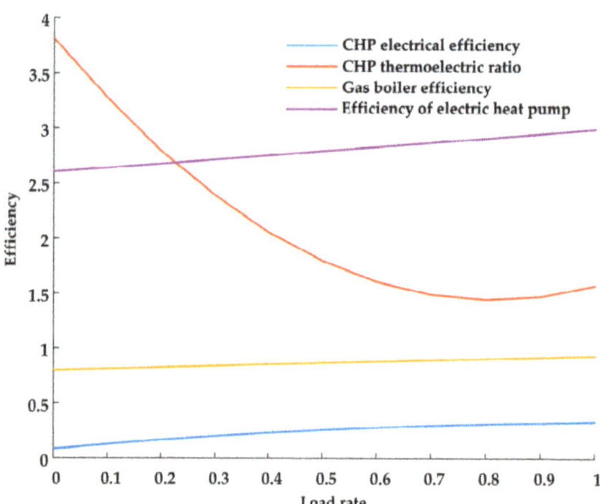

Figure A1. Fitting curve of equipment efficiency.

References

1. Bai, L.; Li, F.; Cui, H.; Jiang, T.; Sun, H.; Zhu, J. Interval optimization based operating strategy for gas-electricity integrated energy systems considering demand response and wind uncertainty. *Appl. Energy* **2016**, *167*, 270–279. [CrossRef]
2. Wang, L.; Hou, C.; Ye, B.; Wang, X.; Yin, C.; Cong, H. Optimal Operation Analysis of Integrated Community Energy System Considering the Uncertainty of Demand Response. *IEEE Trans. Power Syst.* **2021**, *36*, 3681–3691. [CrossRef]
3. Gokcek, T.; Sengor, I.; Erdinc, O. A Novel Multi-Hierarchical Bidding Strategies for Peer-to-Peer Energy Trading among Communities. *IEEE Access* **2022**, *10*, 23798–23807. [CrossRef]

4. Mediwaththe, C.P.; Stephens, E.R.; Smith, D.B.; Mahanti, A. Competitive Energy Trading Framework for Demand-Side Management in Neighborhood Area Networks. *IEEE Trans. Smart Grid.* **2018**, *9*, 4313–4322. [CrossRef]
5. Li, J.; Tian, L.; Cheng, L.; Guo, J. Optimization Planning of Micro-energy System Considering the Characteristics of Variable Operating Conditions (1) Basic Model and Analysis. *Autom. Electr. Power Syst.* **2018**, *42*, 18–26+49.
6. Chen, W.; Mu, Y.; Jia, H.; Wei, W.; He, W. Optimization Scheduling Method for Regional Integrated Energy System Considering Equipment Variable Operating Condition Characteristics. *Power Syst. Technol.* **2021**, *45*, 951–958.
7. Wang, Y.; Ma, Y.; Song, F.; Ma, Y.; Qi, C.; Huang, F.; Xing, J.; Zhang, F. Economic and efficient multi-objective operation optimization of integrated energy system considering electro-thermal demand response. *Energy* **2020**, *205*, 118022. [CrossRef]
8. Guo, Z.; Zhang, R.; Wang, L.; Zeng, S.; Li, Y. Optimal Operation of regional integrated energy system considering demand response. *Appl. Therm. Eng.* **2021**, *191*, 116860. [CrossRef]
9. Liu, R.; Yang, T.; Sun, G.; Lin, S.; Muyeen, S.M.; Ma, T. Optimal Dispatch of Community Integrated Energy System Considering Comprehensive User Satisfaction Method. *IEEJ Trans. Electr. Electron. Eng.* **2022**. [CrossRef]
10. Miao, B.; Lin, J.; Li, H.; Liu, C.; Li, B.; Zhu, X.; Yang, J. Day-Ahead Energy Trading Strategies of Regional Integrated Energy System Considering Energy Cascade Utilization. *IEEE Access* **2020**, *8*, 138021–138035. [CrossRef]
11. Li, Y.; Feng, C.; Wen, F.; Wang, K.; Huang, Y. Energy Pricing and Management of Park Energy Internet Including Electric Vehicles and Electricity-to-Gas Conversion. *Autom. Electr. Power Syst.* **2018**, *42*, 192–196.
12. Wei, W.; Liu, F.; Mei, S. Energy Pricing and Dispatch for Smart Grid Retailers Under Demand Response and Market Price Uncertainty. *IEEE Trans. Smart Grid.* **2015**, *6*, 1364–1374. [CrossRef]
13. Paudel, A.; Chaudhari, K.; Long, C.; Gooi, H.B. Peer-to-Peer Energy Trading in a Prosumer-Based Community Microgrid: A Game-Theoretic Model. *IEEE Trans. Ind. Electron.* **2019**, *66*, 6087–6097. [CrossRef]
14. Ma, L.; Liu, N.; Zhang, J.; Wang, L. Real-Time Rolling Horizon Energy Management for the Energy-Hub-Coordinated Prosumer Community from a Cooperative Perspective. *IEEE Trans. Power Syst.* **2019**, *34*, 1227–1242. [CrossRef]
15. Fu, Y.; Li, H. Energy Management and Pricing Strategy of Multi-energy Supply Service in Parks Considering User Utility Evaluation. *Modern Electric Power.* **2022**, *39*, 317–328.
16. Fleischhacker, A.; Corinaldesi, C.; Lettner, G.; Auer, H.; Botterud, A. Stabilizing Energy Communities Through Energy Pricing or PV Expansion. *IEEE Trans. Smart Grid.* **2022**, *13*, 728–737. [CrossRef]
17. Anoh, K.; Maharjan, S.; Ikpehai, A.; Zhang, Y.; Adebisi, B. Energy Peer-to-Peer Trading in Virtual Microgrids in Smart Grids: A Game-Theoretic Approach. *IEEE Trans. Smart Grid.* **2020**, *11*, 1264–1275. [CrossRef]
18. Wei, F.; Jing, Z.X.; Wu, P.Z.; Wu, Q.H. A Stackelberg game approach for multiple energies trading in integrated energy systems. *Appl. Energy* **2017**, *200*, 315–329. [CrossRef]
19. Li, P.; Wang, Z.; Wang, N.; Yang, W.; Li, M.; Zhou, X.; Yin, Y.; Wang, J.; Guo, T. Stochastic robust optimal operation of community integrated energy system based on integrated demand response. *Int. J. Electr. Power Energy Syst.* **2021**, *128*, 106735. [CrossRef]
20. Ruan, W.; Wang, B.; Li, Y.; Yang, S. Research on User Response Behavior Under Peak and Valley Time-of-use Electricity Prices. *Power Syst. Technol.* **2012**, *36*, 86–93.
21. Geidl, M.; Koeppel, G.; Favre-Perrod, P.; Klockl, B.; Andersson, G.; Frohlich, K. Energy hubs for the future. *IEEE Power Energy Mag.* **2007**, *5*, 24–30. [CrossRef]
22. Wang, Y.; Zhang, N.; Kang, C.; Kirschen, D.S.; Yang, J.; Xia, Q. Standardized Matrix Modeling of Multiple Energy Systems. *IEEE Trans. Smart Grid.* **2019**, *10*, 257–270. [CrossRef]
23. Cui, Y.; Guo, F.; Fu, X.; Zhao, Y.; Han, C. Coordinated and optimized source-charge dispatch of an integrated energy system that promotes wind power consumption through energy-supply conversion. *Power Syst. Technol.* **2022**, *46*, 1437–1447.
24. Li, Y.; Zhang, Z.; Yang, L.; Tang, D.; Dong, C.; Ji, C. A Decision-making Model for Incremental Power Distribution and Sales Companies Considering Collaborative Optimization of Source, Load and Storage. *Autom. Electr. Power Syst.* **2021**, *45*, 125–132.
25. Giannelos, S.; Konstantelos, I.; Strbac, G. Option Value of Demand-Side Response Schemes Under Decision-Dependent Uncertainty. *IEEE Trans. Power Syst.* **2018**, *33*, 5103–5113. [CrossRef]
26. Wu, C.; Gu, W.; Xu, Y.; Jiang, P.; Lu, S.; Zhao, B. Bi-level optimization model for integrated energy system considering the thermal comfort of heat customers. *Appl. Energy* **2018**, *232*, 607–616. [CrossRef]
27. Sun, Y.; Zhang, C.; Li, Z.; Zhang, X.; Li, F. Flexible Load Multi-power-level Regulation Strategies Considering multi-regional User Differentiated PMV. *Proc. CSEE* **2021**, *41*, 7574–7586.
28. Mu, Y.; Chen, W.; Yu, X.; Jia, H.; Hou, K.; Wang, C.; Meng, X. A double-layer planning method for integrated community energy systems with varying energy conversion efficiencies. *Appl. Energy* **2020**, *279*, 115700. [CrossRef]
29. Zhao, H.; Miao, S.; Li, C.; Zhang, D.; Tu, Q. Research on the Optimal Operation Strategies of the Integrated Energy System in the Park Considering the Coupled Response Characteristics of Cooling, Heating and Electricity Demand. *Proc. CSEE* **2022**, *42*, 573–589.
30. Maharjan, S.; Zhu, Q.; Zhang, Y.; Gjessing, S.; Basar, T. Dependable Demand Response Management in the Smart Grid: A Stackelberg Game Approach. *IEEE Trans. Smart Grid.* **2013**, *4*, 120–132. [CrossRef]
31. Liu, R.; Li, Y.; Yang, X.; Li, Y.; Sun, G.; Shi, S. Two-stage Optimal Scheduling of Community Integrated Energy System Considering Demand Response. *Acta Energ. Sol. Sin.* **2021**, *42*, 46–54.

32. Li, L. Development and Problems of Residential Centralized Cooling (Heating) in Hefei. *Anhui Archit.* **2021**, *28*, 79–80.
33. Zhao, Y.; Gao, H.; Wang, Z.; Zhang, R.; Zhong, L.; Li, Y.; Liu, J. Optimization Decision of Multi-energy Trading Game for Business Park Operators Considering User Power Substitution. *Power Syst. Technol.* **2021**, *45*, 1320–1331.

processes

MDPI

Article

Optimization Strategy of Hybrid Configuration for Volatility Energy Storage System in ADN

Guoping Lei [1], Yinhua Huang [1], Nina Dai [1], Li Cai [1], Li Deng [1], Shenghao Li [2] and Chao He [1,*]

[1] School of Electronic and Information Engineering, Chongqing Three Gorges University, Chongqing 404100, China
[2] School of Big Data and Internet of Things, Chongqing Vocational Institute of Engineering, Chongqing 402260, China
* Correspondence: hechao@sanxiau.edu.cn; Tel.: +86-158-7059-6170

Abstract: This study aims to address the issues of volatile energy access to the active distribution network (ADN), which are the difficulty of frequency regulation, the increased voltage deviation of the ADN, the decrease in operational security and stability, etc. In this study, a two-stage majorization configuration model is established to identify and understand how volatility energy affects a hybrid energy storage system (HESS). The ADN and HESS with lead-acid batteries and supercapacitors (SC) are examined using day forecast data for wind, solar, and load. In this planning stage, the integrated cost, network loss, and node voltage deviation are considered as optimal objectives in a multi-objective optimization model, while the revised multi-objective optimization particle swarm approach is used to solve the initial value of capacity configuration. In the operation stage, optimizing objectives like wind output power fluctuations, the frequency deviation of HESS is used to solve the modified value of the configuration capabilities of the SC, and the output of different types of units in ADN is further optimized by the quantum particle swarm with the addition of a chaotic mechanism. The simulation study is conducted to determine the best configuration result based on case 33 node examples, and the simulation results demonstrate the model's viability.

Keywords: ADN; HESS; operation strategy; optimal configuration; frequency regulation

Citation: Lei, G.; Huang, Y.; Dai, N.; Cai, L.; Deng, L.; Li, S.; He, C. Optimization Strategy of Hybrid Configuration for Volatility Energy Storage System in ADN. *Processes* **2022**, *10*, 1844. https://doi.org/10.3390/pr10091844

Academic Editors: Haoming Liu, Jingrui Zhang and Jian Wang

Received: 22 August 2022
Accepted: 5 September 2022
Published: 13 September 2022

1. Introduction

Under the background of the worldwide "Carbon Double", the development of a series of volatile energy [1], like tidal energy and solar energy, has received unprecedented attention. Under the pressure of a high proportion of instability volatility energy consumption, multiple countries have put forward supporting development policies of "volatility energy + ESS", and the significance of energy storage devices for the heavy penetration of volatility energy sources is totally mirrored. The optimal configuration of ESS incorporates a direct effect on the active control ability of ADN, which makes the ADN preferable to the traditional distribution network (TDN), and realizes the volatile energy interoperability between the grid and the electricity consumption side [2,3].

At present, the configuration strategy of the ESS of the distribution network has been observed in many studies at home and abroad. Wang et al., according to the data from load-side transformers and solar power, established an energy storage capacity allocation scheme with optimal economic efficiency based on intelligent algorithms and energy storage allocation strategies for customer power consumption characteristics [4]. Chen et al. established a model with the highest wind-storage combined system power sales revenue as the optimization objective and used the Ant-Lion algorithm to solve the optimal allocation scheme for wind generation (WG) cluster power backup and energy storage power and capacity [5]. Using dynamic solar planned output data as a constraint and the maximum average annual benefit over the life cycle as the optimization objective, Hong et al. used particle swarm algorithms (PSO) and time-series simulation calculations to solve for the best configuration of ESS [6]. They also considered the energy storage investment cost, operation and maintenance, penalty costs for deviating from the planned output, and revenue from PV plants. Liu et al. analyzed the influence of fluctuations of load power on the distribution network and proposed a model predictive control-based optimization strategy for energy storage allocation and scheduling with the goal of economic efficiency of user-side energy storage operation [7].

However, the above models only formulate the configuration of the ESS in terms of operational economics and none of them take into account the dynamic characteristics of the ESS. Shi et al. analyzed the characteristics of historical wind and solar power output fluctuations at two durations of 15 min and 10 min and studied the capacity allocation strategies of ESSs based on smoothing energy output fluctuations and participating in system frequency regulation [8]. Wu et al. analyzed the output characteristics of the combined power generation farms with wind and solar, and proposed a project comparing the stabilization index and smoothing effect evaluation index to analyze the filtering effects of the sliding average method and least-squares procedure, to decide the output power level of ESS configuration [9]. Wang et al., resting on the historical information characteristics of the WG and PV, proposed a capacity optimization configuration method based on the analysis of the wind and PV output volatility under different capacity allocation schemes to guarantee the chance that the system output change rate satisfies the maximum requirements [10]. In Ref. [11], the ADN energy storage operating approach to smooth out the system's power fluctuation is suggested. An ESS configuration scheme is structured with fixed expenses and operating expenses in the cycle as the optimization targets, and the dynamic programming arithmetic is needed to calculate the energy storage installation capacity, power, and installation location. The above model takes into consideration the dynamic characteristics of the ESS and additionally smoothed out the volatility to a definite extent; however, the improvement of the configuration results is not obvious when solely one layer model is employed for designing.

In Ref. [12], the authors described the design of a two-level estimate model for allocating storage capacity. The outer layer determines the in-out power and capacity of the ESS with the calculated goal of minimizing the expense to invest in the storage system, the inner layer determines the charging and discharging power of the ESS to minimize the system transit line's power fluctuation, and a probabilistic approach to multiple scenarios is adopted to calculate the conclusions of the ESS allocation. In Ref. [13], a HESS two-layer planning scheme on account of the operational life span in the operation phase was constructed. In the upper layer, with the objectives of the lowest investment cost, the linear programming algorithm is adopted to estimate the total action domain of HESS, which provides a reference range for the actual operation of HESS and formulates the energy storage operation strategy considering the storage charging and discharging capacity; the lower layer takes the maximum operational life span of the battery during the operation phase as the objective function, and the PSO algorithm is taken to calculate the best configuration of battery and supercapacitor capacity. The works of Refs. [11,12] are based on the use of ESSs within the distribution network for double-layer configuration, purely with the support of the distribution network; however, they did not consider that the role of WG in the configuration of the ESS is the existence of a negative correlation, and provide no analysis on the aspects of wind generation concerned in system frequency.

With the increase of penetration power of wind and PV, the proportion of conventional generating sets is gradually reduced, and the power grid inertia and FM intensity are constantly reduced, which can have an effect on the security and stability of operation within the ADN with comparatively high volatile energy proportion [14]. To take full advantage of the ESS and cut down the cost, this paper takes into account the optimal configuration of the ESS on the ADN side and the energy side, and considers the investment to study the optimal configuration of the HESS of lead-acid batteries and supercapacitors with the idea of "integrated planning of energy storage capacity configuration and dispatching strategy" [15]. It is also contrasted with various battery types used for energy storage and serves as a guide for user-side energy storage designs. Based on the initial values of capacity and power provided by the traditional energy storage allocation method, an operational strategy for volatility energy utilization value enhancement is introduced at the planning level, and an optimal scheduling strategy to take into account the system frequency deviation is introduced at the dispatching level, using the output of WG and ESSs to take part in FM to correct the configuration of planning level supercapacitors energy storage capacity's initial values. The contributions of this paper are summarized as follows:

(1)   The problem of the impact of fluctuating energy output on the configuration of energy storage systems is analyzed, laying the foundation for the subsequent development of operational dispatching strategies based on equivalent load curves.
(2)   A hybrid energy storage system using lead-acid batteries and supercapacitors is utilized to diversify the types of energy storage and expand the scope of optimization.
(3)   By considering both the distribution grid side and the energy side, a two-tier energy system optimization strategy with joint participation of wind storage in system frequency regulation is proposed. Through day-ahead optimization and intra-day correction multi-timescale hybrid energy storage configuration optimization, the distribution grid economy and renewable energy utilization are improved.

2. Impacts of Volatility Energy Power on ESS and Mathematical Model

2.1. Analysis of the Impact of Volatility Energy Power on ESS

In this paper, volatility energy mainly adopts WG generation and PV. The sum of actual load and negative load (each power generation) is taken as the equivalent load of volatility energy access to ADN, and the period of charging and discharging of HESS is segmented by the extremal variation of the equivalent load figure.

Equation (1) presents the specific calculation procedure.

$$P_{e,load}(t) = P_{load}(t) - P_{WG}(t) - P_{PV}(t) \tag{1}$$

where $P_{e,load}(t)$, $P_{load}(t)$, $P_{WG}(t)$ and $P_{PV}(t)$ illustrate the equivalent load, realistic load, WG power, and PV power in period t, respectively.

Currently, the fluctuation of photovoltaics often takes place in intervals of less than 1 min; when considering how to smooth out fluctuations in PV power, control using the energy storage system's output is frequently used [16]. When PV power fluctuations do not exceed the maximum permissible power of the HESS, the HESS's power is often employed to smooth these power variations while keeping the PV converter operating in MPPT mode. To put it another way, downward power fluctuations are tamed by discharging (when the power value of HESS greater than 0), and upward power variations are tamed by charging (when HESS power is negative). The details of the coordinated control strategy are discussed in Section 3. Wind power and load fluctuations follow the same pattern.

The variations in PV power throughout a minute is discussed in this paper. The power fluctuations are the interval size between the utmost and minimum power values measured at the purpose of common coupling over the course of 1 min, as described in Figure 1.

The comparison of typical daily PV output and wind output curve and actual load curve in a certain place is shown in Figure 2.

From Figure 1, it is obvious that the peak period of PV output is 9:00–14:00, while the typical daily load curve peaks at around 12:00 and 20:00, indicating that the peak period of PV and the peak period of the load curve during the day coincide. Thus, the peak-to-valley's distances of the equivalent load curve will be curtailed after the superposition of PV output with realistic load, and PV power generation is positively correlated with the peak-to-valley difference. On the contrary, the peak period of wind and the valley period of load curve at night are similar; the peak-to-valley's distances after the superposition of wind output and realistic load will be increased, and its influence is negatively correlated.

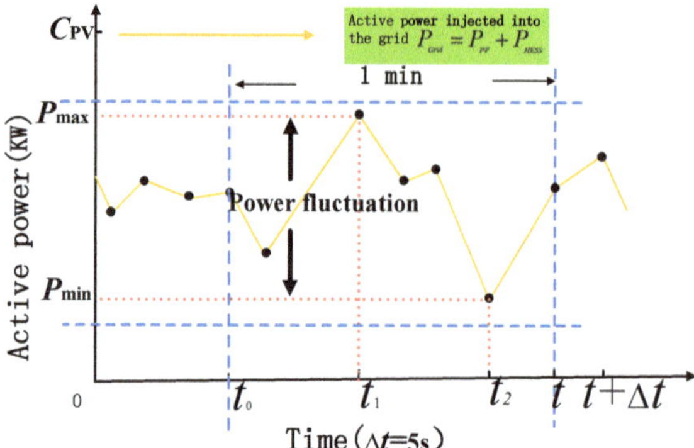

Figure 1. PV power fluctuations in one minute.

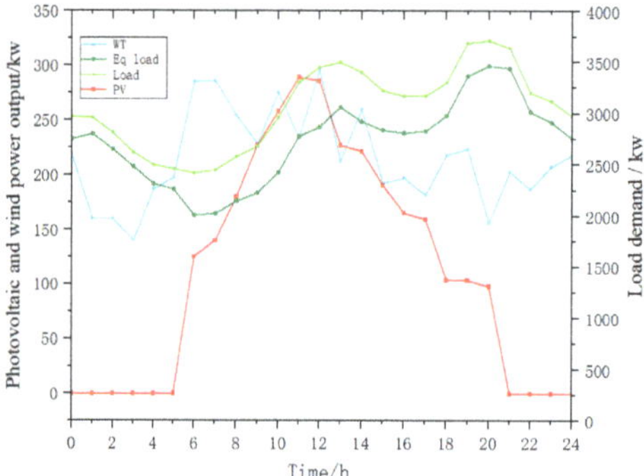

Figure 2. Comparison of the load curve and the volatile energy output.

To sum up, PV output reduces the necessity of energy storage configuration, and wind output increases the necessity. While the HESS is configured with the equivalent load curve, the access of volatility energy will realize peak cutting and valley filling and affect the HESS's operation strategy.

2.2. Mathematical Model of ESS

The usage of energy storage devices can help to reduce network losses and power quality fluctuations [17] that are brought on by unstable energy sources linked to ADN as well as some of the energy consumption and utilization rate fluctuations. It is challenging to fulfill this need with a single kind [18] of energy storage device, though. The double-layer planning model established in this paper can fully utilize the complementary characteristics of lead-acid batteries and supercapacitors [19] to effectively extend the service life of the system, save cost, improve the overall performance of energy storage, and solve the problem to the greatest extent. This is demonstrated by the low frequency of lead-acid battery charging and discharging and the high frequency of supercapacitor charging and discharging [20].

It is mathematically modeled from the State of Charge (SOC) and the charging/discharging power.

$$SOC(t) = \begin{cases} (1-\eta)SOC(t-1) + \frac{P_c(t) \bullet \Delta t \bullet \gamma}{E_e}, charge \\ (1-\eta)SOC(t-1) + \frac{P_d(t) \bullet \Delta t}{E_e \bullet \lambda}, discharge \end{cases} \tag{2}$$

where $SOC(t)$ evaluates the SOC level in period t, η delineates the loss rate of remaining power per hour, $P_c(t)$ and $P_d(t)$ clarify charging and discharging power, γ and λ are charging and discharging efficiencies, E_e is the rated capacity, Δt is the sampling interval, and the value of this paper is 1 h.

3. A Double-Layer Multi-Objective Optimization Model

This study uses the two-layer decision optimization model to solve the ESS configuration scheme. The two-layer model can comprehensively consider the problems of the configuration of ESS and various problems in the and.

3.1. Planning Layer Optimization Model

According to the load situation of the ADN, the maximum capacity value of the ESS is initially calculated, the day forecasts of wind, solar, and load are used to carry out the preliminary planning of the ESS, and an optimization model is built with the combined cost, network loss, and nodal voltage deviation as the optimal objectives.

3.1.1. Objective Functions

- Comprehensive cost of a full-day life cycle;

$$minF_1 = C_{inv} + C_{run} + C_{PV} + C_{WG} + C_{buy} + C_{ploss} \tag{3}$$

where F_1 introduces the daily comprehensive expenses of HESS, C_{inv} and C_{run} are the daily investment expenses and daily operation and maintenance expenses of HESS, C_{PV}, C_{WG} represent the operation and maintenance expenses of PV and wind farms, C_{ploss} describes the network loss expenses, and C_{buy} describes the daily power purchase expenses, which to some extent characterizes the ability of "Peak cut and fulfill valleys". The calculation formulas of each component are as follows:

$$C_{inv} = \sum_{j=1}^{N} \frac{\tau(1+\tau)^l \left(c_p P_{ess} + c_e E_e\right)}{24\left((1+\tau)^l - 1\right)} \bullet T$$

$$C_{run} = \sum_{j=1}^{N} \int_0^T \left(c_{om} P_{ess}(t)\right) dt$$

$$C_{PV} = \int_0^T \left(c_{PV} P_{PV}(t)\right) dt \tag{4}$$

$$C_{WG} = \int_0^T \left(c_{WG} P_{WG}(t)\right) dt$$

$$C_{buy} = \int_0^T m_g \left(P_{LAB,c}(t) + P_{SC,c}(t)\right) dt$$

$$C_{ploss} = \int_0^T m_a \left(P_{ploss}(t)\right) dt$$

where N indicates energy storage units' amount, P_{ess} indicates the HESS's power rating, c_p and c_e evaluate power and capacity cost coefficients, c_{om}, c_{PV}, and c_{WG} denote HESS's, PV's, and WG's operation and maintenance cost factor, $P_{ess}(t)$ indicates the actual power level of the HESS in period t, $P_{LAB,c}(t)$ and $P_{SC,c}(t)$ describe the charge powers of lead-acid batteries and supercapacitors, m_g and m_a evaluate unit electricity price and unit network loss cost, and $P_{ploss}(t)$ denotes the network active loss power.

- Network loss

$$minF_2 = \sum_{t=1}^{T} \sum_{i=1}^{I} P_{loss,i,t} \tag{5}$$

where F_2 delineates the network loss for 24 h, T defines the dispatching time, I is the nodes' amount of the ADN, and $P_{loss,i,t}$ denotes the power loss of line i at time t.

- Node voltage deviation

$$minF_3 = \sum_{i=1}^{I} |V_i - V_N| \tag{6}$$

where V_N denotes the node's rated voltage, V_i is the voltage on node i, and F_3 is smaller, meaning the node voltage is more stable.

3.1.2. Constraint Condition

- SOC of energy storage constraint;

To avoid over-charging and over-discharging, SOC has a certain range limit, which cannot be fully discharged or fully charged.

$$SOC_{\min} \leq SOC(t) \leq SOC_{\max} \tag{7}$$

where $SOC_{\min}$ and $SOC_{\max}$ are the minimum charge level and maximum residual charge level, respectively.

- Node voltage constraints

$$U_{i,\min} \leq U_i \leq U_{i,\max}(i = 1,2,3\cdots,I) \tag{8}$$

where $U_{i,\min}$ and $U_{i,\max}$ are the minimum and maximum voltages at node i, respectively.

- Branch circuit current constraints

To guarantee the HESSs can operate well and stably.

$$I_i \leq I_{i,\max}(i = 1,2,3\cdots,m) \tag{9}$$

where $I_{i,\max}$ is the upper limit of the current in the i-th branch, and m is the number of branches.

- Power balance constraints

$$\begin{cases} P_{G,i} - P_{N,i} = U_i \sum_{j=1}^{N} U_j \left(X_{ij} \cos \phi_{ij} + Y_{ij} \sin \phi_{ij} \right) \\ Q_{G,i} - Q_{N,i} = U_i \sum_{j=1}^{N} U_j \left(X_{ij} \sin \phi_{ij} - Y_{ij} \cos \phi_{ij} \right) \end{cases} \tag{10}$$

where $P_{G,i}$ and $Q_{G,i}$ indicate the power output of active and reactive to the power supply at nodes i, $P_{N,i}$ and $Q_{N,i}$ indicate the power output of active and reactive at nodes i, U_i and U_j are the voltage amplitude at nodes i and j, X_{ij} and Y_{ij} evaluate real and virtual parts of the node-admittance matrix elements, and Φ_{ij} evaluates the voltage angular phase difference of nodes i and j.

3.2. Operation Layer Optimization Model

Due to the access of WG and load, there is frequency fluctuation in the distribution network system. This study adds 120 MW of WG in the operation layer to adjust the energy storage system output, using the energy storage output to suppress the fluctuation [21,22]; 5% of the WG output is used for system frequency regulation. When the fluctuation frequency range exceeds 0.2 Hz, the whole capacity of WG is added to the distribution network; otherwise, the energy storage capacity calculated under the optimal strategy is used. In turn, the value of the additional capacity to the operation layer needed for the supercapacitors is calculated.

3.2.1. Objective Function

- Minimal fluctuations in WG output

$$\min F_4 = \sqrt{\frac{1}{n}\sum_{t=1}^{n}\left[P_{WG}(t) - \overline{P}_{WG}\right]^2} \tag{11}$$

where $\overline{P}_{WG}$ is the average active power of the all-day life cycle.

- Minimal system frequency deviation

$$\min F_5 = \frac{f(t) - f_e}{f_e} \bullet 100\% \tag{12}$$

where $f(t)$ is system frequency at time t, and f_e is system-rated frequency.

3.2.2. Constraint Condition

- Charge and discharge power constraint

$$-P_{LAB,c,\max} \leq P_{LAB}(t) \leq P_{LAB,d,\max} \tag{13}$$

$$-P_{SC,c,\max} \leq P_{SC}(t) \leq P_{SC,d,\max} \tag{14}$$

where $P_{LAB,c,\max}$ and $P_{LAB,d,\max}$ reflect the lead-acid batteries' charging/discharging powers crest values, $P_{SC,c,\max}$ and $P_{SC,d,\max}$ reflect the supercapacitors' charging/discharging powers crest values.

- Charge and discharge times constraint

The life span of energy storage units increase and the costs reduce by reducing the number of charging and discharging occurrences during operation.

$$\begin{array}{c} 0 \leq x \leq N \\ 0 \leq y \leq M \end{array} \tag{15}$$

where $x, y, N,$ and M are the number of charging and rated charging of lead-acid batteries and supercapacitors, respectively.

4. Scheduling Strategy and Solution Algorithm

4.1. Scheduling Strategy for Energy Storage Systems

The two components of the scheduling strategy are as follows: the division of continuous charging and discharging periods following the "time-of-day tariff" [23]. The segmentation of charging and discharging periods is used to determine the power of the ESS to charge and discharge in each period.

① HESS is configured according to the load curve, charging at the curve trough, and discharging at the peak. To improve the utilization of the energy storage system, for the flat tariff period, if the period before and after it is a high tariff period then the charging time is T_e; if both the preceding and following periods are low tariff periods, then the discharge time is T_d. The charging/discharging periods are distinguished on account of the time-of-day tariff strategy, and the high and low electricity price periods corresponding to the charging and discharging periods; $T_{e,1}$ and $T_{e,2}$ are the charging periods and $T_{d,1}$, $T_{d,2}$ and $T_{d,3}$ are the discharging periods. The results of the charging and discharging time periods are delineated in Figure 3. The times of 4:00–9:00 and 15:00–19:00 are low electricity prices, while 10:00–14:00 and 20:00–3:00 are high electricity prices, in the known charging and discharging period, considering the SOC.

② The variable power charging/discharging mode is adopted to determine the power values of multiple ESSs when charging and discharging. The specific process is as follows.

The smaller the equivalent load within Δt, the more energy storage charging is required. The equivalent load values within Δt for each sampling interval of period T are sorted in order from smallest to largest, and the size of the charging power level of the HESS within Δt corresponding with the equivalent load is determined, respectively. To make the fluctuation of the equivalent load curve of the HESS after charging as small as possible (except for the HESS within Δt with the smallest equivalent load, which is charged by the maximum power), the storage system is charged at a variable power less than the maximum power. The calculation is shown in Equation (16).

Charging:

$$P_c(t) = \begin{cases} \alpha \bullet (P_{c,\max} + P_{L,\min} - P_L(t)), & P_L(t) < (P_{c,\max} + P_{L,\min}) \\ P_{c,e} & , P_L(t) \geq (P_{c,\max} + P_{L,\min}) \end{cases} \tag{16}$$

For each determined charging power, the *SOC* also increases, and undetermined sampling intervals are charged at zero power until the power magnitude of all sampling intervals is determined and all charging power values for the HESS are output. In contrast, the *SOC* decreases during the discharging process.

The process of determining the magnitude of the discharge power is similar to the above process. The differences are the larger the equivalent load in Δt, the greater the need for energy storage discharge, and in order of equivalent load from largest to smallest to determine its corresponding the size of the discharge power of the HESS within Δt. Except for the Δt with the largest equivalent load, the HESS is discharged by the maximum power value; during other Δt the HESS is discharged at a variable power that is less than the maximum power. The calculation is shown in Equation (17).

Discharging:

$$P_c(t) = \begin{cases} \beta \bullet (P_L(t) + P_{d,\max} - P_{L,\max}), & P_L(t) > (P_{L,\max} - P_{c,\max}) \\ P_{d,e} & , P_L(t) \leq (P_{L,\max} - P_{c,\max}) \end{cases} \tag{17}$$

where $P_{c,\max}$, $P_{d,\min}$, $P_{c,e}$, $P_{d,e}$ are charging/discharging powers in period t, maximum charging power, minimum discharge power, and rated charging/discharging power, respectively. $P_L(t)$, $P_{L,\min}$, $P_{L,\max}$, α, β are large equivalent load values during the sampling period t, minimum, and maximum, equivalent load values at the sampling interval, and charging and discharging power weights, respectively.

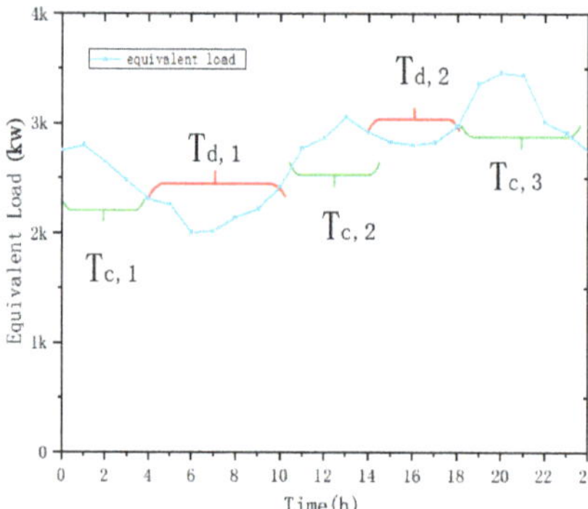

Figure 3. The strategy of charging and discharging.

4.2. The Computational Flow of Multi-Objective Chaotic Particle Swarm Algorithm

A mathematical optimization methodology for dealing with multi-layer analytical processes is called Chaos Particle Swarm Optimization (CPSO). According to how well it fits its surroundings, each particle is gradually shifted to a better location. After solving each sub-step or step's requirement's part-optimal solution in the correct order, the optimal prescription from the set of local optima is then employed as the optimization's final output.

Therefore, a modified Chaos Particle Swarm Optimization (MCPSO) is used to solve this problem. The chaotic property is used to improve the diversity of the population and the ergodicity of the particle search, and the inclusion of chaotic states into the optimization variables gives the particles the ability to search continuously. The specific flow chart is presented in Figure 4.

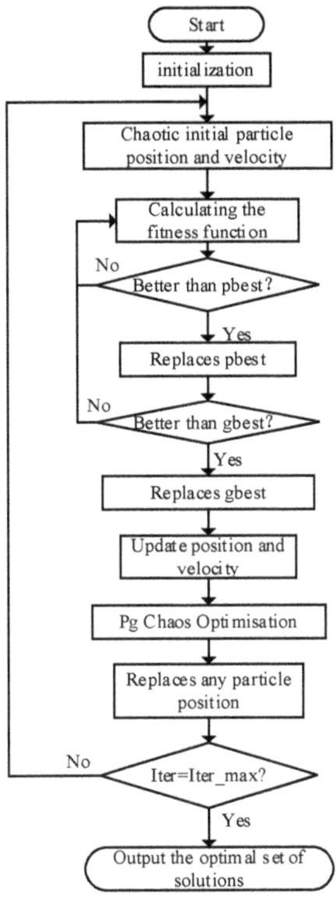

Figure 4. An MCPSO flow chart for solving the optimal HESS configuration.

The detailed operations are shown as follows.

Step 1: The maximum allowable times of iterations the range of fitness error values, and the algorithm-related parameters: inertia weights, and learning factors are initialized.

Step 2: Chaotic initialization of particle positions and velocities are determined.

(1) An n-dimensional vector $x_1 = (x_{11}, x_{12}, \cdots, x_{1n})$ between [0,1] is randomly generated, using the Logistic chaotic system equation by Equation (18) to obtain N vectors $x_1, x_2, \cdots, x_N$.

$$x_{n+1} = \partial x_n(1 - x_n), n = 0, 1, 2, \cdots \tag{18}$$

(2) After calculating the fitness function for all particles, Z initial velocities are generated at random from Y initial populations by choosing the Z initial solutions with the best performance.

Step 3: pBest is set as the new position if the particle fitness is greater than the individual extreme.

Step 4: The global extreme gBest is set to the new position if the particle fitness is greater than it.

Step 5: Dynamically update learning factor.

(1) Take the average value of the particle adaptation value.

(2) The particle adaptation value is compared with the average value. When the average value is more than the adaptation, the learning effect maximum value is taken. Otherwise, the learning factor is solved by using Equation (19).

$$w = w_{\min} + \frac{w_{\max} - w_{\min}}{\bar{x} - x_1}(x_i - x_1) \tag{19}$$

where $w_{\min}$, $w_{\max}$ are the learning factor's minimum and maximum values, and x_1, x_i, $\bar{x}$ are the average of the adaptation values of the 1st and ith particle, and the population adaptation values.

Step 6: Redefine the particles' positions and velocities.

Step 7: Chaos optimization to get the best position.

Calculate the adaptation value for each feasible solution experienced by the chaotic variables in the original solution space, and select the feasible solution with the best performance. Map the vectors in the optimal position to the definition domain of the Logistic equation [0,1], iterate with the Logistic equation to generate a sequence of chaotic variables, then return the generated sequence of chaotic variables to the original solution space through the inverse mapping.

Step 8: Substitute for any one particle's position present in all particles with p^*.

Step 9: The search terminates and the global optimal position is output if the halting condition is met. If not, go back to Step 3.

5. Case Study

5.1. Basic Parameters for the Case

In this research, the modified case 33 node examples system was used as an arithmetic example, and 300 kw WG and 300 kw PV were added to nodes 19 and 26. The system structure is delineated in Figure 5 for the HESS configuration. The WG output, PV output, and load curves for a typical day at a site are described in Figure 2.

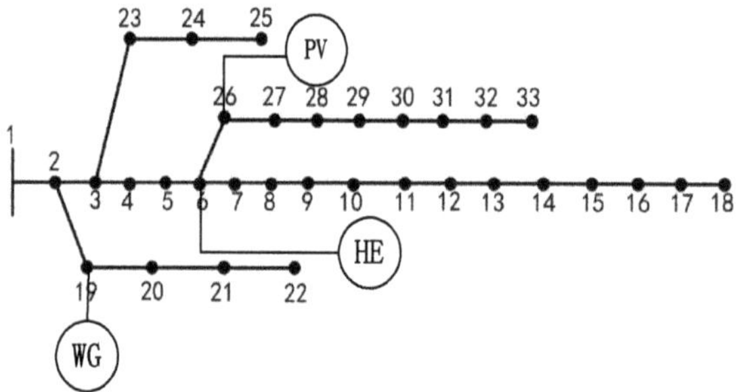

Figure 5. Example system of PV, WG, and HESS access in ADN.

A time-of-day tariff was set, with a tariff of RMB 1.0/kWh during peak hours (HESS discharging periods), RMB 0.35/kWh during low hours (HESS charging periods), and a flat tariff of RMB 0.55/kWh for the rest of the day. The parameters related to supercapacitor and lead-acid battery units are expressed in Table 1.

5.2. Analysis of the Impact of Energy Storage System Access Nodes

As shown in Figure 6, when the same capacity (400 kW) energy storage device was connected to different nodes, the voltage stability and minimum voltage difference were obvious. In nodes 8 to 18 and nodes 29 to 33 access, the node voltage was lower but the corresponding voltage stability index was also not high, thus the voltage lifting effect was not obvious. Therefore, it is not conducive to voltage safety and stability. Thus, if only the voltage stability indicators are considered, the nodes in Table 2 can be connected.

Assuming access to one of the nodes first, the size of the active network loss of the energy storage device at different access points of the network-wide 33 nodes is derived, as shown in Figure 7. As can be seen from the figure, the network-wide active network loss values are relatively low at nodes 1, 2, 6, 19–22, 28, and nodes 31–33, so it is possible to choose between these nodes.

Table 1. Related parameters of the energy storage unit.

Parameters	Lead-Acid Batteries	SC
SOC_{min}	0.4	0.1
SOC_{max}	0.8	0.9
charging and discharging efficiency (%)	98	98
capacity cost factor (RMB/kW)	1500	2400
capacity maintenance cost factor (RMB/year)	0.045	0.015
service life (year)	10	20
power cost factor (RMB/MW)	300	300
initial volume of SOC	0.4	0.1
discount rate (%)	10	10
power factor (%)	98	98

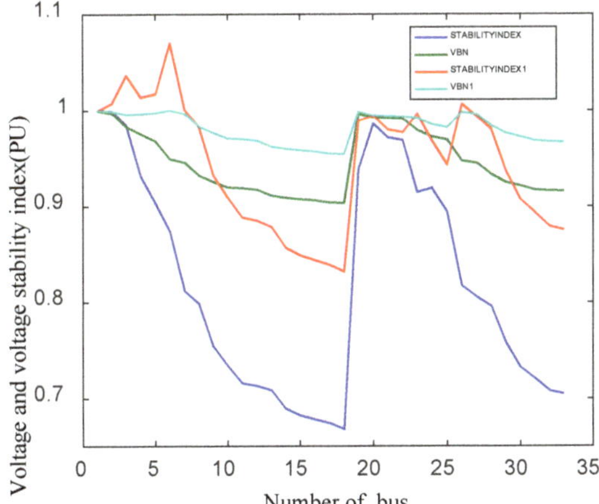

Figure 6. Voltage stability index for different access points.

Table 2. Effect of different access points on grid voltage and stability. VBN: Voltage of the branch node.

Access Nodes	Stability Index	VBN
1	1	1
2	0.99	0.99
3	0.98	0.98
20	0.98	0.99
21	0.97	0.99
22	0.96	0.99

After considering both the stability index curve and the active network loss curve, the energy storage device is connected to the above nodes, the tidal current calculation is carried out, and finally node 6 and node 30 are selected. Due to the three branches of the IEEE 33 node distribution system, the position of node 6 can be made the Interaction of energy, generating information faster and more economically secure, which is the interaction between the wind generation systems, PV systems, and loads of the individual nodes.

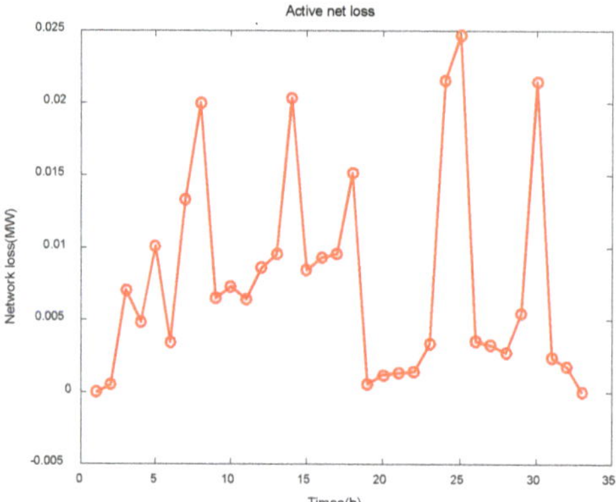

Figure 7. Effect of different access points on active power network loss.

5.3. Interpretation of Result

5.3.1. Capacity Configuration Results and Economic Analysis

This paper takes a comparison under three scenes and thus judges the validity and reliability of this study. Scene 1 is a single layer multi-objective improved particle swarm optimization algorithm for configuration, without consideration of lower layer optimization. In scene 2, a double-layer optimization configuration model, the upper layer is optimized by a multi-objective chaotic particle swarm algorithm but does not consider that the operation layer wind power does not participate in the impact of system frequency modulation on the configuration. Scene 3 is the proposed solution of this study.

Table 3. Configuration results in the three scenarios (KW).

Parameters	Scene 1	Scene 2	Scene 3
capacity of lead-acid batteries	464.55	425.32	157.34
capacity of supercapacitors	1100	926.93	695.64
correction of supercapacitors	0	5.25	5.95

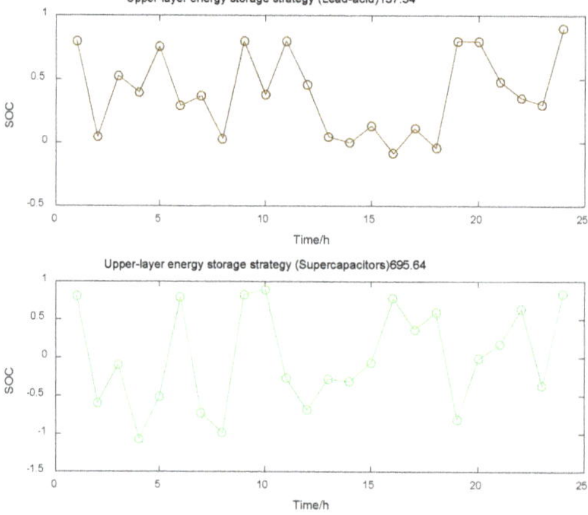

Figure 8. Charging and discharge strategy of the HESSs in Scene 3.

In addition to lead-acid batteries, other types of hybrid batteries such as Li-Ion batteries and NaS batteries were also tested, demonstrating that the simulation results are rather different. This was done to diversify the forms of energy storage and widen the scope of optimization.

Table 4. The cost profiles of the configuration schemes (RMB).

Parameters	Scene 1	Scene 2	Scene 3
Investment costs	3.34×10^6	2.86×10^6	1.91×10^6
operation and maintenance costs	201.62	176.61	84.23
network loss costs	3.49	3.27	2.67
wind and PV operation and maintenance costs	80.34	80.34	80.34
power purchase costs	52.27	50.25	45.29
correction costs	0	71.14	61.06
total costs	3,337,137.73	2,934,047.53	1,844,656.99

From Tables 3 and 4 it can be concluded that:

(1) In Scene 1, a hybrid energy storage equipment is added to the system, and although the operation layer energy storage dispatching strategy does not take into account the system frequency deviation, it has a certain soothing effect on the equivalent load curve, achieving a certain effect of "Peak cut" and optimizing the operation of the grid.

(2) Scene 2 is based on Scene 1, using a double-layer planning model, with only the lower layer of ESS taking part in the system's FM. Therefore, the lower layer is used to correct the capacity of supercapacitors, with a correction value of 5.2456 kw, reducing the total cost by RMB 403,090.20. The cost reduction rate is about 12.08%, which achieves integrated planning of capacity dispatch and further improves the effect of peak and valley reduction.

(3) Scene 3 is an optimized configuration of the HESS based on a double-layer planning model, with WG added to the lower layer to participate in system FM, correcting the supercapacitor capacity value of the improvement. The charging and discharging strategy for hess in Scene 3 is introduced in Figure 8. The total costs of Scene 3 relative to Scene 1 and Scene 2 are RMB 1,492,480.74 and RMB 1,089,390.54 saved, respectively. The reduction rates are approximately 44.72% and 37.13%, with a total cost reduction while the effect of network loss optimization is also more obvious.

Table 5. Comparison of the three battery storage costs in Scene 3 (RMB).

Parameters	NaS	Li-Ion
Investment costs	3.40×10^6	3.75×10^6
operation and maintenance costs	249.38	343.32
network loss costs	3.26	3.36
wind and PV operation and maintenance costs	61.60	61.60
power purchase costs	50.93	49.93
correction costs	63.12	56.97
total costs	3,399,028.28	3,748,015.17

The analysis of lithium batteries and sodium-sulfur batteries in Scene 3 of this paper reveals that the life-cycle costs of lithium batteries are RMB 3,747,958.20 and the life-cycle costs of sodium-sulfur batteries are RMB 3,398,965.16, like Table 5, which leads one to the conclusion that lead-acid batteries are more cost-effective than other energy storage batteries because their price per unit capacity and power are lower.

5.3.2. Network Loss Analysis after Optimization

The net loss can be greatly improved after the hybrid energy storage device in node 6 is delineated in Figure 9. The network loss before the configuration optimization is 4.72 MW, after is 4.19 MW, and the net loss is reduced by about 0.53 MW. At node 14, the network loss reduction is the largest, at about 0.045 MW.

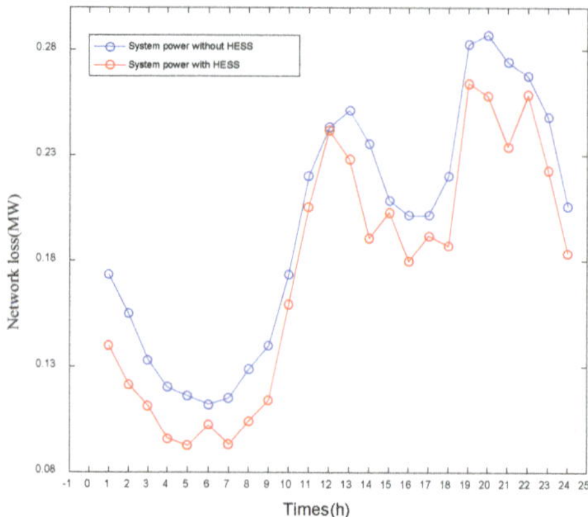

Figure 9. Network loss comparison after adding HESS to ADN.

5.3.3. Optimum Voltage Analysis of Distribution Network

After energy storage optimization, the minimum voltage values of multiple nodes of the distribution network system are increased, and the minimum voltage increase at node 11 is the most obvious, which is described in Figure 10. To some extent, it can be explained that the energy storage system configuration has significantly improved the voltage of the network.

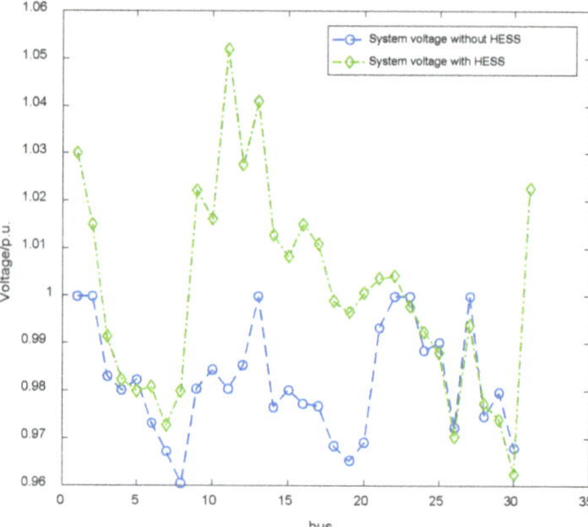

Figure 10. Optimized voltage comparison curve after distribution network.

5.3.4. Analysis of the Frequency Bias after Optimization

Under normal frequency fluctuations (within the range of [−0.2,0.2]), the wind generator is effectively adjusted according to the reserved 20% output margin. In emergencies with large fluctuations (outside the range of [−0.2,0.2]), the maximum capacity of the wind generator is used to adjust the system frequency and the load side is demand-responding, with an adjustment factor within the rated value of 1.5.

The WG output during the system frequency modulation is displayed in Figure 11. The supercapacitor energy storage outputs during system frequency modulation are described in Figure 12.

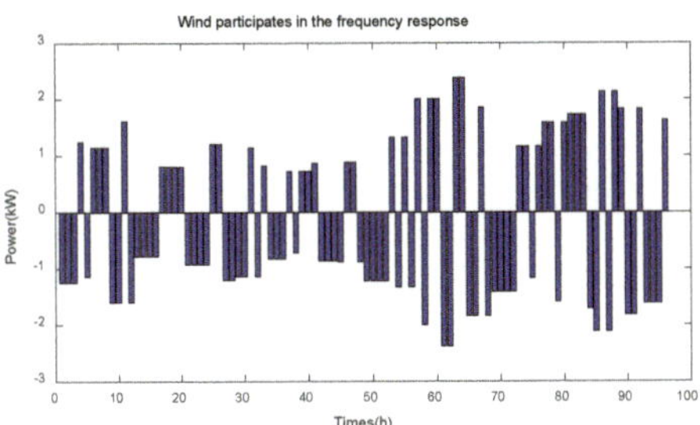

Figure 11. The output of WG participation in frequency regulation.

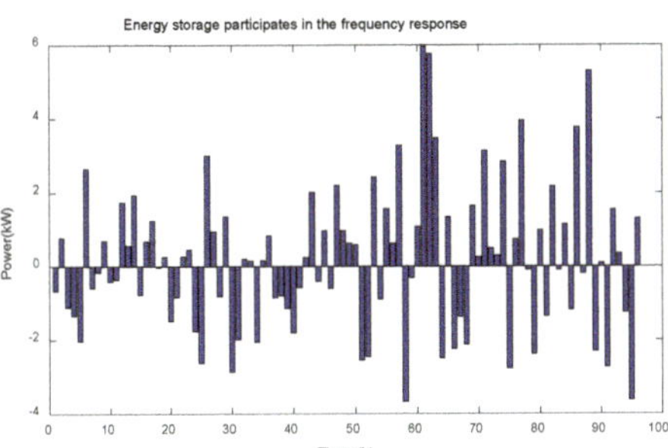

Figure 12. The output of HESS participation in frequency regulation.

The frequency fluctuation range after optimization is obviously reduced by extending the time for system frequency regulation to 96 h, which is described in Figure 13, with the frequency fluctuating within $-0.032\sim0.042$ Hz before optimization and within $-0.024\sim0.016$ Hz after optimization. To some extent, the frequency fluctuations of the ADN are abated.

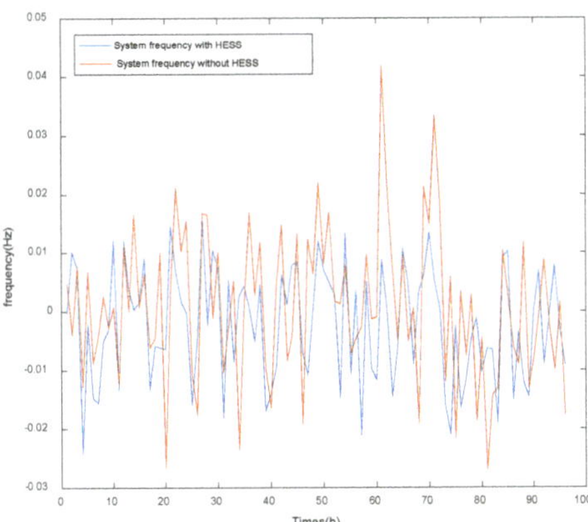

Figure 13. Comparison of frequency response before and after each period.

6. Conclusions

This paper studies the effect of energy storage charging/discharging tactics and WG participation in frequency modulation on HESS configuration and operation in ADN containing volatile energy sources, which is solved in MATLAB using a dynamic chaotic particle swarm algorithm. Through the simulation analysis of the improved case 33 nodes power distribution system, the three conclusions are obtained.

(1)   The first layer is developed according to the equivalent load curve and proposes a charging and discharging strategy for the HESS considering the "time-of-day tariff", while the second layer adopts WG and HESS to suppress fluctuations in the operation strategy, which can achieve better economic results and "Peak cut and fulfill valleys" with less investment and operation costs.

(2)   Optimizing the configuration of the HESS with "integrated planning of the configuration capacity and dispatching strategy" and establishing a mathematical model for the optimal configuration of the capacity ensures that research of energy storage configuration can be more reasonably and accurately grasped, and the risk of over-investment or under-investment can be reduced.

(3)   Consideration of the dynamic characteristics of HESS operation, which can achieve the goal of smoothing fluctuating energy's power fluctuation, as well as improving the voltage quality of the distribution network and reducing network losses, which is of more practical significance.

Author Contributions: Conceptualization, G.L., Y.H. and C.H.; methodology, G.L.; software, Y.H.; validation, G.L., L.C. and N.D.; formal analysis, Y.H. and L.D.; investigation, L.D. and S.L.; resources, G.L.; writing—review and editing, Y.H. All authors have read and agreed to the published version of the manuscript.

Funding: This research was funded by the Natural Science Foundation Project of Chongqing Science and Technology Commission (grant numbers cstc2021jcyj-msxmX0301, cstc2020jcyj-msxmX0034, 2022NSCQ-MSX4086, and 2022NSCQ-MSX3846), the Science and Technology Research Program of Chongqing Municipal Education Commission (grant numbers KJZD-K202101202 and KJZD-K202103401), and the Wanzhou District Innovation and Entrepreneurship Demonstration Team (No.2021-07).

Institutional Review Board Statement: Not applicable.

Informed Consent Statement: Not applicable.

Data Availability Statement: Not applicable.

Conflicts of Interest: The authors declare no conflict of interest.

References

1. Paul, S.; Dey, T.; Saha, P.; Dey, S.; Sen, R. Review on the development scenario of renewable energy in different country. In Proceedings of the 2021 Innovations in Energy Management and Renewable Resources (52042), Kolkata, India, 5–7 February 2021; pp. 1–2.
2. Mokryani, G. Energy Storage Systems in Future Distribution Networks. In *Future Distribution Networks: Planning, Operation, and Control*; AIP Publishing: Melville, NY, USA, 2022. [CrossRef]
3. Guo, X.; Chen, Q.; Liang, W.; Wang, S.; Wang, Y.; Qin, Y. Intelligent Optimization Algorithm of Active Distribution Network Based on Load Storage Cooperative Optimization of Source Network. In Proceedings of the 2021 IEEE 4th International Conference on Automation, Electronics and Electrical Engineering (AUTEEE), Shenyang, China, 19–21 November 2021; pp. 355–360.
4. Wang, H.; Ma, C.; Sheng, W.; Sha, G.; Zhao, C.; Liu, N.; Duan, Q. Energy Storage Configuration Optimization Strategy for Islanded Microgrid Interconnection Based on Energy Consumption Characteristics. *Discrete Dyn. Nat. Soc.* **2021**, *2021*, 2826670. [CrossRef]
5. Chen, T.; Yang, J.; Jiang, W.; Sui, Z.; Wang, Y. Optimal configuration of energy storage capacity in wind farms based on cloud energy storage service. *IET Renew. Power Gener.* **2022**, *16*, 211–222. [CrossRef]
6. Mao, M.; Hong, J.; Zhang, L. Energy storage optimization configuration method considering conditional forecast error correction. *Taiyangneng Xuebao* **2021**, *42*, 410–416.
7. Liu, Y.; Liu, Q.; Guan, H.; Li, X.; Bi, D.; Guo, Y.; Sun, H. Optimization Strategy of Configuration and Scheduling for User-Side Energy Storage. *Electronics* **2022**, *11*, 120. [CrossRef]
8. Shi, X.; Shi, X.; Dong, W.; Zang, P.; Jia, H.; Wu, J.; Wang, Y. Research on energy storage configuration method based on wind and solar volatility. In Proceedings of the 2020 10th International Conference on Power and Energy Systems (ICPES), Chengdu, China, 25–27 December 2020; IEEE: New York, NY, USA, 2020; pp. 464–468.
9. Wu, T.; Shi, X.; Li, L.; Zhou, C.; Zhou, H.; Su, Y. A capacity configuration control strategy to alleviate power fluctuation of hybrid energy storage system based on improved particle swarm optimization. *Energies* **2019**, *12*, 642. [CrossRef]
10. Wang, D.; Zhao, Y.; Tao, Q.; Xue, J.; Ye, J. Research on Planning and Configuration of Multi-objective Energy Storage System Solved by Improved Ant Colony Algorithm. In Proceedings of the 2018 China International Conference on Electricity Distribution (CICED), Tianjin, China, 17–19 September 2018; pp. 2279–2283.

11. Gong, Q.; Wang, Y.; Fang, J.; Qiao, H.; Liu, D. Optimal configuration of the energy storage system in ADN considering energy storage operation strategy and dynamic characteristic. *IET Gener. Transm. Distrib.* **2020**, *14*, 1005–1011. [CrossRef]
12. Li, J.; Xue, Y.; Tian, L.; Yuan, X. Research on optimal configuration strategy of energy storage capacity in grid-connected microgrid. *Prot. Control Mod. Power Syst.* **2017**, *2*, 35. [CrossRef]
13. Yao, M.; Cai, X. Energy storage sizing optimization for large-scale PV power plant. *IEEE Access* **2021**, *9*, 75599–75607. [CrossRef]
14. Song, J.; Hu, C.; Su, L. Distributed Wind Power and Photovoltaic Energy Storage Capacity Configuration Method under Big Data. In Proceedings of the 2021 International Conference on Electronic Communications, Internet of Things and Big Data (ICEIB), Yilan, Taiwan, 10–12 December 2021; pp. 318–322.
15. Ma, T.; Gu, H.; Zha, Z.; Li, G.; Lou, H. A Two-Layer Optimization Model for Energy Storage Configuration in the Distribution Network. In *IOP Conference Series: Earth and Environmental Science*; IOP Publishing: Bristol, UK, 2021; Volume 647, p. 012012.
16. Ma, W.; Wang, W.; Wu, X.; Hu, R.; Tang, F.; Zhang, W. Control Strategy of a Hybrid Energy Storage System to Smooth Photovoltaic Power Fluctuations Considering Photovoltaic Output Power Curtailment. *Sustainability* **2019**, *11*, 1324. [CrossRef]
17. Wang, P.; Zhang, F.; Chen, Q. Bi-level optimal configuration of hybrid energy storage for wind farms considering battery life. *J. Phys. Conf. Ser.* **2022**, *2247*, 012005. [CrossRef]
18. Wang, Y.; Song, F.; Ma, Y.; Zhang, Y.; Yang, J.; Liu, Y.; Zhang, F. Research on capacity planning and optimization of regional integrated energy system based on hybrid energy storage system. *Appl. Therm. Eng.* **2020**, *180*, 115834. [CrossRef]
19. Liu, C. *Optimal Configuration and Energy Management of Hybrid Energy Storage System with Lithium Battery and Supercapacitor*; University of Science and Technology of China: Hefei, China, 2020.
20. Shen, Y.; Hu, W.; Liu, M.; Fang, Y.; Hong, X. Energy storage optimization method for microgrid considering multi-energy coupling demand response. *J. Energy Storage* **2022**, *45*, 103521. [CrossRef]
21. Liu, H.; Liu, Y.; Zhang, C.; Sun, L.; Wu, X. Configuration of an Energy Storage System Considering the Frequency Response and the Dynamic Frequency Dispersion. *Front. Energy Res.* **2022**, *9*, 807763. [CrossRef]
22. Liu, D.; Jin, Z.; Chen, H.; Cao, H.; Yuan, Y.; Fan, Y.; Song, Y. Peak Shaving and Frequency Regulation Coordinated Output Optimization Based on Improving Economy of Energy Storage. *Electronics* **2022**, *11*, 29. [CrossRef]
23. Liu, Y.; Wang, L.; Zeng, Z.; Bei, Y. Optimal charging plan for electric bus considering time-of-day electricity tariff. *J. Intell. Connect. Veh.* **2022**, *5*, 123–137. [CrossRef]

Article

Cost Analysis of Synchronous Condenser Transformed from Thermal Unit Based on LCC Theory

Chenghao Li [1], Mingyang Liu [1], Yi Guo [2], Hanqing Ma [2], Hua Wang [1] and Xiaoling Yuan [2,*]

[1]   State Grid Henan Electric Power Research Institute, Zhengzhou 450052, China
[2]   College of Energy and Electrical Engineering, Hohai University, Nanjing 211100, China
*   Correspondence: lingx@hhu.edu.cn

Citation: Li, C.; Liu, M.; Guo, Y.; Ma, H.; Wang, H.; Yuan, X. Cost Analysis of Synchronous Condenser Transformed from Thermal Unit Based on LCC Theory. *Processes* **2022**, *10*, 1887. https://doi.org/10.3390/pr10091887

Academic Editor: Davide Papurello

Received: 31 August 2022
Accepted: 15 September 2022
Published: 17 September 2022

Abstract: With the development of large-scale renewable energy consumption and multi-infeed high voltage direct current (HVDC) systems, the demand of a system for the synchronous condensers with a strong dynamic reactive power support capacity and a strong short-time overload capacity is increasing. Meanwhile, with the reuse of a large number of retired thermal units, the most practical and economic way is to transform thermal units into synchronous condensers. The cost difference in the life-cycle of the synchronous condenser transformed from a thermal unit (SCTTU) and the newly established synchronous condenser (NESC) is a key factor that affects the decision-making and construction of the transformation from thermal unit to synchronous condenser. However, the life-cycle cost (LCC) of the synchronous condenser transformed from a thermal unit and the newly established synchronous condenser contains many uncertain factors, which affect the accuracy of the LCC estimation value. In order to quantify the impact of the blind information on the cost of the synchronous condenser station, blind number theory is introduced to establish the blind number model of the LCC of the synchronous condenser transformed from a thermal unit and the newly established synchronous condenser. Additionally, the LCC of the NESC and SCTTU with a different life-cycle under the capacity of 2×300 MVar are estimated. The results show that the cost of the SCTTU with a long service life of more than 15 years is significantly lower than that of the NESC and, thus, the SCTTU has better economic performance. The economic performance of the SCTTU with a life-cycle of less than 15 years is not better than that of the NESC. Compared with the traditional calculation method of a single cost value, the blind number model can obtain the possible distribution interval of LCC and the reliability of the corresponding interval, which makes the estimation results more valuable for practical engineering reference.

Keywords: synchronous condenser transformed from thermal unit (SCTTU); newly established synchronous condenser (NESC); life-cycle cost (LCC); blind number theory

1. Introduction

Building a novel power system with renewable energy as the main body is an important way for China to achieve the goal of "carbon peaking and carbon neutrality". With the State Grid Corporation further accelerating the construction of a novel power system and fully promoting the realization of the "dual carbon" goal, the dominant position of novel energy power generation with "weak support" will become increasingly prominent. The output space of traditional thermal power units with "strong support" will be limited, the reactive power demand of the power system will rise sharply, and the voltage stability will face great challenges [1–3]. In order to increase the proportion of dynamic reactive power supply, optimize the utilization rate of thermal power units, and improve the stable operation level of the power grid, the new large capacity synchronous condenser has been widely used in HVDC transmission and reception terminals in recent years, and has played an important role in suppressing the DC commutation failure and improving the voltage stability of the system [4]. Up to now, 47 new large-capacity 300 MVar synchronous condensers have been built, and 39 have been put into operation in China. However, the new large capacity synchronous condenser faces a series of problems, such as high cost, high operation and maintenance costs, and great construction difficulties.

Compared with the NESC, there are two advantages in transforming the thermal units into synchronous condensers. One is the advantage of economic cost, and the other is the advantage of improving the utilization rate of thermal units. The SCTTU can save the investment cost of the synchronous condenser unit itself and, at the same time, the retired thermal unit can be reconstructed and reused, which will improve the utilization rate of the thermal units [5]. Therefore, power systems should make full use of the existing retired thermal units, transform them into synchronous condensers, and conduct long-term grid connected operation, which makes use of to its characteristics of strong overload capacity and fast response speed, and provides sufficient dynamic reactive power support for the power grid without occupying active space. In doing so, it is possible to reduce investment and operation costs, improve the utilization rate of thermal units, and solve the survival problems of thermal power plants. This has significant practical value and social and economic benefits [6].

At present, there have been relevant studies on synchronous condensers and synchronous condensers transformed from thermal units. With the large-scale access to HVDC transmission and new energy, the domestic research on the synchronous condensers and the synchronous condensers transformed from thermal units is gradually enriched. Jiang Zhe et al. [6] demonstrated the feasibility of the technical transformation of retired thermal units to synchronous condensers, and took Shandong power grid as an example to verify the ability of the synchronous condensers transformed from thermal units to improve the voltage stability of the power grid and the new energy grid connection characteristics. D. K. Chaturvedi [7] proposed the transformation scheme of transforming a 500 MW retired thermal unit into a 300 MVar synchronous condenser, and demonstrated the ability of SCTTU to provide dynamic compensation, improve system inertia, and improve system power quality. Karan et al. [8] proposed the steps of transforming retired thermal units into synchronous condensers to provide reactive power support for the system, so as to meet the reactive power demand of the Indian power grid under large-scale new energy access. J. An et al. [9] proposed an optimal configuration method for the conversion of thermal power units to synchronous condensers and verified the feasibility of the proposed scheme from the perspectives of technology, economy, and operation mode in combination with engineering cases. J. Kaur and N. R. Chaudhuri [10] put forward the transmission scheme from thermal units to synchronous condensers in the weak interconnection system and analyzed the supporting ability of synchronous condensers transformed from thermal units in the system. In addition, in terms of cost estimation, the China Qaidam converter station 2 × 300 MVar synchronous condenser project adopts a single value estimation method to estimate the cost of the NESC [11]. The cost estimation of the NESC of the China Jiangsu Taizhou 2 × 300 MVar synchronous condenser station is also a single value estimation method [12]. Huang Z et al. [13] proposed that the cost estimation of replacing the retired thermal power unit with new energy power station and synchronous condenser also adopts a single value estimation method. Most of the above studies focus on the aspects of technical feasibility, transformation methods, and the improvement of power grid stability after the transformation. Few economic studies are only the estimation of a single value of the cost of synchronous condensers. Therefore, there is still a lack of comprehensive analysis, both in terms of foreign and domestic circumstances, for the cost analysis of synchronous condensers transformed from thermal units.

In this paper, a cost calculation method based on LCC is proposed. At the same time, blind number theory is introduced into the cost calculation model to solve the influence of various uncertain information on the cost of the synchronous condenser project, and the distribution range of the life-cycle cost of the synchronous condenser is obtained. The life-cycle cost of the NESC and the SCTTU with different service life are compared and analyzed. Combined with the operation mode of the reactive power equipment in the market, the reactive power pricing under different operation modes of the NESC and the SCTTU is formed. The analysis results show that the life-cycle cost of the SCTTU with a long life-cycle of more than 15 years is significantly lower than that of the NESC, and the SCTTU has lower market reactive power pricing and better economy. At the same time, the cost range and the corresponding confidence of the SCTTU are obtained. Compared with the single value of the traditional prediction, these results have more engineering reference value.

The remained of this paper is organized as follows. In Section 2, the life-cycle cost model of a synchronous condenser is established. In Section 3, blind number theory is introduced, and the processing and calculation methods of blind information are introduced. In Section 4, the LCC model of the synchronous condenser based on the blind number theory is established. In Section 5, based on the existing market, a reactive power pricing model with coverage cost as the objective is constructed. In Section 6, a detailed case study proves that the proposed SCTTU is economical. In Section 7, several important conclusions are summarized.

2. Life-Cycle Cost Analysis of SCTTU

The connotation of life-cycle cost can be summarized as follows: the generalized life-cycle cost refers to all expenses incurred by all stakeholders, such as producers, consumers, and the public, in the life-cycle of the project from the perspective of the whole society. In a narrow sense, the life-cycle cost refers to the cumulative sum of the costs of the project at each stage of its life-cycle after discounting [14].

This paper is based on the life-cycle cost theory in a narrow sense, and more intuitively reflects the cost of the synchronous condenser project. According to the life-cycle stage, the life-cycle cost of the construction project mainly includes the cost of the decision-making stage, the cost of the design stage, the cost of the construction stage, the cost of the operation and use stage, and the cost of the scrapping and dismantling stage [15]. These are as follows:

(1) Cost of the decision-making stage. The cost in the decision-making stage mainly refers to the expenses incurred in the process of project planning, feasibility studies, market investigation, fundraising, scheme optimization, land acquisition, etc. [3];

(2) Cost of the design stage. The cost in the design stage accounts for a small proportion in the total investment of the project, but the design stage is an important stage in the cost control of the construction project;

(3) Cost of construction stage. The cost of this stage mainly includes labor, materials, equipment, management, and various taxes;

(4) Cost of the operation and use stage. The cost in the use stage refers to various expenses that the user needs to pay in the process of using the project, mainly including energy consumption expenses, maintenance expenses, management expenses, etc. [6];

(5) Cost of scrapping and dismantling stage. The scrapping and dismantling stage is the last stage of the project life-cycle. In this stage, the demolition of the project and the disposal of wastes are mainly carried out, and the costs are mainly demolition costs and cleaning costs.

The change in the life-cycle cost of the construction project at different stages is shown in Figure 1. Life-cycle cost analysis is an auxiliary tool for investment decision-making. Its core aim is to identify the cost items at each stage of the life-cycle, and to conduct quantitative estimation and analysis of the cost according to a certain cost estimation model and method. Finally, the Life-cycle cost of the project is obtained, and the project decision is made on this basis.

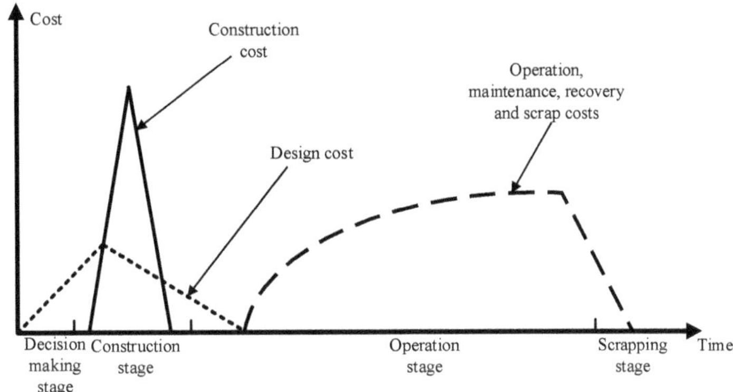

Figure 1. Diagram of cost trend at each stage of the life-cycle.

2.1. Life-Cycle Cost Analysis of SCTTU and NESC

According to the stage division of the construction project based on the life-cycle theory, combined with the engineering characteristics of the synchronous condenser, this paper brings the design decision into the construction cost. Considering that, in addition to operation and maintenance costs, overhaul and technical transformation costs, as well as market penalty costs caused by power failure, are not able to be included in operation and maintenance costs during the operation phase of the SCTTU and the NESC, they are calculated separately. The final scrapping stage is the residual value cost of the equipment. Since the residual value of the equipment is the residual value of the project, its cost calculation is negative. Therefore, the life-cycle cost of the SCTTU and the NESC can be divided into five parts.

The present value of the initial construction cost S_b (i.e., the total construction investment from year 0 to year k) is expressed as follows:

$$S_b = \sum_{t=0}^{k} \frac{S_{bt}}{(1+i)^t} \tag{1}$$

where S_{bt} is the construction investment in the year t, and i is the discount rate.

Assuming that the operation cost is generated from the year l, the present value expression of the operation and maintenance cost is as follows:

$$S_u = \sum_{t=l}^{T} \frac{S_{ut}}{(1+i)^t} \tag{2}$$

where S_{ut} is the operation and maintenance cost of the year t, and T is the service life of the synchronous condenser.

Assuming that the overhaul cost is generated from the year m, the present value expression of the overhaul and technical transformation cost is:

$$S_r = \sum_{t=m}^{T} \frac{S_{rt}}{(1+i)^t} \tag{3}$$

where S_{rt} is the overhaul and technical transformation cost of the year t.

Assuming that the penalty cost is generated from the year x, the present value expression of the penalty cost is as follows:

$$S_p = \sum_{t=x}^{T} \frac{S_{pt}}{(1+i)^t} \tag{4}$$

where S_p is the penalty cost incurred in the year t.

The present value expression of the residual value at the end of the year T is as follows:

$$S_q = S_{qt}/(1+i)^T \tag{5}$$

where S_{qt} is the residual value at the end of the year T.

Therefore, we can obtain the calculation formula of the life-cycle cost of the synchronous condenser project as follows:

$$S_{\text{LCC}} = S_b + S_u + S_r + S_p + S_q \tag{6}$$

2.2. Uncertainty Analysis of Cost in Each Stage

(1) Uncertain influencing factors of initial construction cost S_b.

Here, S_b is the costs and expenses incurred in the process of planning, design, implementation, and the completion of project construction of a synchronous condenser or a synchronous condenser transformed from a thermal unit. At present, most Chinese assets are estimated according to industry regulations, and the traditional quota is used as the pricing basis. Now, the estimation theory is relatively mature, the cost change range is not large, and the influence of uncertain factors on it is not great. Therefore, this model mainly considers the uncertain factors of operation and maintenance in the calculation. Here, S_b is determined by the budget estimate of the corresponding project and treated as a deterministic factor.

(2) Uncertain influencing factors of operation and maintenance cost S_u.

Here, S_u corresponds to the annual cycle cost, which is the cost that will occur every year in the research cycle. Most of the daily operation and maintenance costs are related to the functions and custody services of the equipment in the synchronous condenser, mainly including energy consumption costs, maintenance costs, labor costs, environmental costs, etc. The energy consumption cost refers to the energy consumption required by the operation of the project equipment. Maintenance cost refers to the cost of maintenance, overhaul, and the replacement of parts for the project equipment. Labor cost refers to the labor cost generated by the operation and management of a synchronous condenser. Environmental cost refers to the cost required to establish the equipment operation environment. According to the above analysis, among the influencing factors of S_u, personnel factors constitute the main uncertain factors, resulting in large changes in maintenance and operation management costs.

(3) Uncertain influencing factors of overhaul and technical transformation cost S_r.

Here, S_r is divided into overhaul cost and technical transformation cost. It corresponds to the non-annual cycle cost, which is not a cost that will occur every year. Overhaul cost refers to the cost of major maintenance measures that must be taken to maintain the normal operation of equipment. The cost of technological transformation is the cost incurred by introducing advanced technology, equipment, and materials to improve, update and transform the existing backward production equipment and supporting auxiliary facilities. The costs of these two parts will be affected by the environment or the development of social technology and economy, with strong uncertainty. The occurrence of overhaul cost is a random probability event, while the occurrence of technical transformation cost is a kind of unknown information and grey information, which makes it difficult to accurately estimate a certain value in actual work. This paper will introduce blind number theory to solve this problem.

(4) Uncertain influencing factors of penalty cost S_p.

Here, S_p is the power shortage cost on the demand side and the direct economic reflection of the power supply reliability level of the power grid. Its magnitude is related to the outage probability, outage duration, average outage power, and maintenance cost after outage. The empirical value shall be adopted according to the actual station operation.

(5) Uncertain influencing factors of equipment residual value S_q.

Residual value S_q is the residual value of the equipment or the whole project at the end of the analysis period. Unlike other costs mentioned above, S_q can be a positive cost or a negative value. There is a strong subjective uncertainty in the estimation of S_q, but because of its small proportion in the total cost, it is often ignored or estimated as a percentage of the initial construction investment. The uncertainty is ignored here and is estimated as a function of initial construction cost.

From the above analysis, it can be seen that S_u, S_r, and S_p are all regarded as functions of the running time of the synchronous condenser in the calculation equation (6) of the synchronous condenser LCC. This expression is inaccurate and does not conform to the actual situation. Because S_r and S_p do not occur every year, their occurrence is affected by many uncertain factors. Later, blind number theory is introduced into Equation (6) and corrected to reasonably deal with the uncertain information in the cost analysis.

3. Blind Number Reliability Model

Objective uncertainty information may be expressed in two or more forms of uncertainty. Complex information with the above four forms of uncertainty at most is called blind information. For such uncertain information, blind number theory can be used to express and process it.

3.1. Definition of a Blind Number

Let $\alpha_i \in g(I)$, where $g(I)$ is an interval grey number set, $\alpha_i \in [0,1]$, $i = 1,2\ldots,n$, $f(x)$ is a grey function defined on $g(I)$, and $f(x)$ is the following:

$$f(x) = \begin{cases} \alpha_i, & x = x_i(i = 1,2,3,\ldots,n); \\ 0, & x = other \end{cases} \tag{7}$$

When $i \neq j$, $x_i \neq x_j$, $\sum_{i=1}^{n} \alpha_i = \alpha \leq 1$, then function $f(x)$ is called a blind function.

In the expression of the blind function $f(x)$, α_i is the reliability of x_i value, α is the total reliability of $f(x)$, and n is the order of $f(x)$ [16].

3.2. Operation of Blind Numbers

Let $*$ denote four operations (add, subtract, multiply, and divide) of blind numbers, and set the blind numbers as A and B. The four operations for defining blind numbers are as follows:

$$\begin{aligned} A = f(x) &= \begin{cases} \alpha_i, & x = x_i(i = 1,2,3,\ldots,m); \\ 0, & x = other \end{cases} \\ B = g(y) &= \begin{cases} \beta_j, & y = y_j(j = 1,2,3,\ldots,n) \\ 0, & y = other \end{cases} \end{aligned} \tag{8}$$

Then, the following can be calculated:

$$\begin{array}{c} \begin{matrix} x_1 \\ \vdots \\ x_i \\ \vdots \\ x_m \end{matrix} \left[\begin{matrix} x_1 * y_1 \cdots x_1 * y_i \cdots x_1 * y_n \\ \vdots \\ x_i * y_1 \cdots x_i * y_i \cdots x_i * y_n \\ \vdots \\ x_m * y_1 \cdots x_m * y_i \cdots x_m * y_n \end{matrix} \right] \\ \ \ y_1 \quad y_i \quad y_n \end{array} \tag{9}$$

Equation (9) is called the matrix of the confidence band edge product of A with respect to B, $\alpha_1, \alpha_2, \ldots, \alpha_m$, and $\beta_1, \beta_2, \ldots, \beta_n$ are the confidence sequences of A and B, respectively. The $m * n$-order matrix is called the confidence product matrix of A on B, which is referred to as the confidence product matrix for short. The element $x_i * y_i$ in the possible value $*$matrix of A with respect to B and the element $\alpha_i * \beta_i$ in the confidence matrix of A with respect to B are called corresponding elements, and their positions are called corresponding positions [17].

If A, B, and C are blind numbers, and the operation between blind numbers satisfies the following properties [18]:

$$
\begin{aligned}
A + B &= B + A \\
A \times B &= B \times A \\
(A + B) + C &= A + (B + C) \\
(A \times B) \times C &= A \times (B \times C) \\
(A + B) \times C &= A \times C + B \times C
\end{aligned}
\tag{10}
$$

3.3. Mean Value of Blind Number

Define a and b as real numbers, and $a \leq b$, $(a+b)/2$ is the center of rational grey number $[a, b]$, denote $\odot[a, b] = (a+b)/2$, and it is a first-order unascertained rational number, then the following is true:

$$
E(f(x)) = \begin{cases} \alpha, & x = \frac{1}{\alpha}(\odot)\sum_{i=1}^{n} \alpha_i x_i \\ 0, & x = other \end{cases}
\tag{11}
$$

Here, $E(f(x))$ is the mean value of blind number $f(x)$, which reflects the average value of blind number $f(x)$ [19].

If the blind numbers $f(x)$ and $g(y)$ are known, the mean value of the blind numbers has the following properties:

$$
\begin{aligned}
E(f(x) + g(y)) &= E(f(x)) + E(g(y)) \\
E(f(x) - g(y)) &= E(f(x)) - E(g(y)) \\
E(f(x) \cdot g(y)) &= E(f(x)) \cdot E(g(y))
\end{aligned}
\tag{12}
$$

4. Life-Cycle Cost of SCTTU and NESC Based on Blind Number Theory

After collecting and sorting out the actual data of the previous synchronous condenser projects, we can know that in the actual construction of the synchronous condenser project, the estimation method and theory of the initial construction cost S_b are very mature, so S_b can be expressed by the first-order blind number $S_b(x)$. The residual value S_q is usually expressed as a percentage of the initial construction cost, so the blind number expression of S_q is also first-order and can be expressed as $S_q(v) = rS_b(x)$, where r is the percentage of the residual value in the initial construction investment. The operation and maintenance cost S_u is related to the service life T of the synchronous condenser and is a continuous cost that occurs every year. Therefore, it can be expressed by the cost $S_u(y)$ that occurs every year. As for the overhaul and technical transformation cost S_r and the penalty cost S_p, they do not occur every year. In order to express them more scientifically and rationally, two variables f and n_m are introduced, where f represents the frequency of occurrence of S_r in the life-cycle T of the synchronous condenser, and n_m represents the number of occurrences of S_p in the full life-cycle T. Thus, the blind number expression of each cost can be obtained.

The initial construction cost can be obtained as follows:

$$
S_b(x) = \sum_{t=0}^{k} \frac{S_{bt}(x)}{(1+t)^t}
\tag{13}
$$

The operation and maintenance cost can be obtained as follows:

$$
S_u(y) = \sum_{t=l}^{T} \frac{S_{ut}(y)}{(1+i)^t}
\tag{14}
$$

The overhaul and technical transformation cost can be obtained as follows:

$$S_r(z) = \sum_{t=1}^{T/f} \frac{C_{rt}(z)}{\left(1+i_f\right)^t} \tag{15}$$

Where $i_f = (1+i)^{1/f} - 1$, which represents the actual discount rate of the overhaul and technical transformation cost.

The penalty cost can be obtained as follows:

$$S_p(w) = \sum_{t=1}^{n_m} \frac{S_{pt}(w)}{(1+i_m)^t} \tag{16}$$

where $i_m = (1+i)^{T/n_m}$, which represents the actual discount rate of penalty cost.

The residual value can be obtained as follows:

$$S_q(v) = \frac{rS_q(x)}{(1+i)^T} \tag{17}$$

Thus, the blind number expressions of the SCTTU and the NESC are obtained as follows:

$$S_{(LCC)} = S_b(x)\left[1 - \frac{r}{(1+i)^T}\right] + S_u(y) + S_r(z) + S_p(w) \tag{18}$$

5. Reactive Power Pricing Mechanism Based on Cost of Synchronous Condenser

5.1. Reactive Power Quotation Mechanism of Synchronous Condenser

To establish the reactive power market, all reactive power participants shall provide their quotation curves to the independent system dispatcher. The synchronous condenser can obtain the corresponding economic compensation when its quotation curve is basically consistent with the comprehensive cost curve. Therefore, the life-cycle cost of the synchronous condenser needs to be subdivided to obtain the reactive power quotation. China's reactive power market is not complete. By analogy with the cost curve of generator reactive power generation participating in market auxiliary services [20], the reactive power cost curve of the synchronous condenser can be obtained, as shown in the Figure 2.

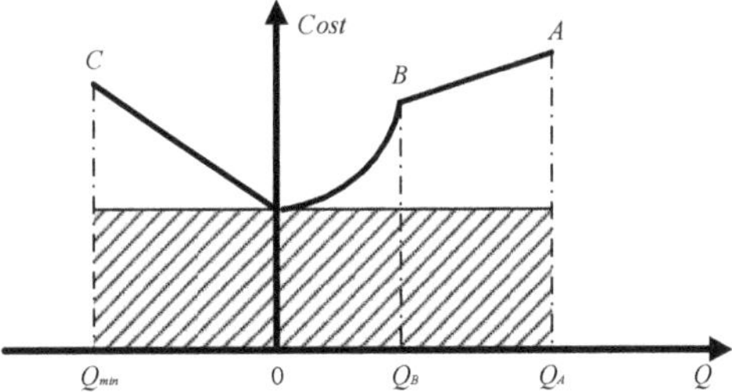

Figure 2. Comprehensive cost curve of the synchronous condenser.

The comprehensive cost curve of the synchronous condenser can be divided into three sections, as follows:

(1) In order to maintain the system voltage, the generator in this area operates in the leading phase to absorb reactive power. Similar to the generator, a certain proportion of the reactive power and quantity of the synchronous condenser can be compensated. Since the leading phase operation will cause great damage to the generator and affect the service life of the generator, the absolute value of the slope in this section is larger than that of sections Q_A to Q_B;

(2) Different from the traditional generators, the reactive power output of the section from 0 to Q_B is relative to the installed capacity of the synchronous condenser. Most of its capacity will be used as cold and hot standby, and there is no actual output. Therefore, the cost slope is larger than that of the Q_A to Q_B sections with normal output, and the rising speed is faster. Therefore, appropriate economic compensation is required;

(3) The section from Q_B to Q_A belongs to the normal output section of the synchronous condenser. The slope of the cost curve is small and the cost increases linearly. A certain proportion of the generator can be paid for.

According to the above analysis of the comprehensive cost curve of the synchronous condenser, the reactive power quotation curve of the reactive power participant is shown in Figure 3.

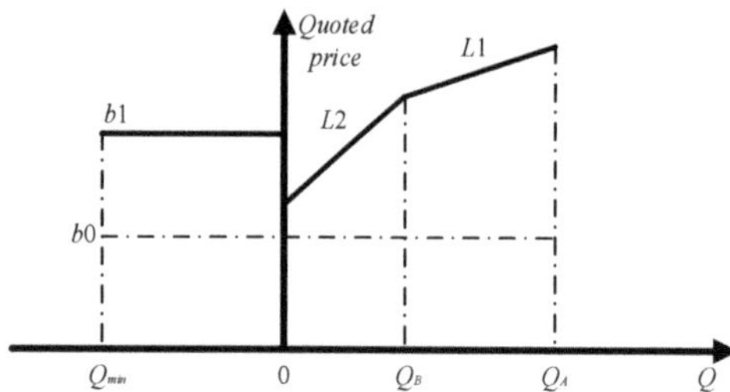

Figure 3. Reactive power quotation curve of the synchronous condenser.

It can be seen from Figure 3 that $b0$ compensates the reactive power investment cost of the synchronous condenser to encourage it to invest in reactive power and ensure that the system has sufficient reactive power sources. Here, $(b1 - b0)$ is the compensation for the leading phase operation of the synchronous condenser, and $L2$ and $L1$ are the reactive power quotations under two conditions when the synchronous condenser normally generates reactive power.

5.2. Reactive Power Market Pricing Mechanism

At present, there are the following two kinds of electricity price models: two-part electricity price and single electricity ladder electricity price. The two-part electricity price consists of the basic electricity price (capacity electricity price) and the electricity price. The basic electricity price is calculated based on the customer's electricity capacity or maximum demand, and the electricity price is calculated based on the customer's actual monthly electricity consumption. The electricity charges calculated by the two kinds of electricity prices are added together, and the electricity charges adjusted by the power factor are all the fees payable by the customer. The single step electricity price divides the monthly electricity consumption of urban and rural residents into several levels, and the electricity price is increased by levels [21]. Aiming at the cost recovery method of reactive power compensation device, combined with the electricity price recovery mode of different power grid equipment, it is proposed that reactive power compensation device can recover the cost in the following three ways [22].

(1) Unified operation of power grid. The dynamic reactive power compensation device is not an independent entity, and its asset ownership and operation right belong to the power grid company. The power grid uniformly bears the costs, principal and interest repayment, profits and taxes of the reactive power compensation device, and performs unified dispatching and unified operation. Under this management mode, the cost of the synchronous condenser is recovered by incorporating it into the transmission and distribution electricity price.

(2) Independent operation. As an independent entity, the asset ownership and operation right of the reactive power compensation device belong to the dynamic reactive power compensation device company, and the company will uniformly bear the cost, interest

payment, profit, tax and other expenses of the reactive power compensation power station, and conduct unified dispatching and unified operation. The power grid company pays for the reactive service by purchasing it, and the operation and maintenance expenses can be recovered by the reactive service expenses paid by the power grid company. Under this management mode, the dynamic reactive power compensation device recovers the cost through two modes of single electricity price and two-part electricity price [23].

Under the single electricity price mode, the government competent department verifies the feed-in tariffs of the dynamic reactive power compensation device, and the power grid company uniformly pays its costs and profits, and is responsible for the repayment of principal and interest. The power station is only responsible for operation according to the power grid dispatching requirements, and the operating income of the power station is realized through the electricity price during the operation period [24]. The capacity price is:

$$P_{ca} = \frac{P_a \times \alpha + C_f}{(1 - \eta_z \times (\eta_c + \eta_e)) \times E_o} \tag{19}$$

where P_a is the total investment, α is the capital utilization rate, C_f is the fixed cost, η_z, η_c and η_e are value-added tax rate, urban construction maintenance tax rate and education additional tax rate, E_o is the total online capacity.

Under the two-part electricity price mode, reactive power compensation price is the sum of average cost price (capacity price) and marginal cost price (electricity price) [25]. As shown in Equation (20).

$$P_e = \frac{C_k}{(1 - \eta_z \times \eta_f)) \times F_o} \tag{20}$$

where P_e is the electricity price, C_k is the variable cost, η_f is the additional tax rate, and F_o is the on grid electricity.

(3) Lease operation. The established dynamic reactive power compensation device operating company leases the equipment to the power grid company or other operating entities for operation, so as to collect the lease fee to ensure its own principal and interest repayment and appropriate profit [22], and its calculation method is similar to the calculation method of electricity price during the operation period.

6. Case Study

6.1. Case Data

This paper takes 2×300 MVar SCTTU and NESC as examples. The service time of the generator after the transformation is calculated according to the transformation level and the unit capacity in three cases of 20, 15, and 10 years. According to the engineering experience, the ending residual value is 5% of the initial construction cost. The overhaul and technical transformation period are generally 4–6 years according to the generator maintenance regulations, and the period for the main transformer is generally 10 years. Since the generator is to be reconstructed, the interval of 4 years is taken. The fault rate is taken as the engineering experience value of 0.643 times/year, from which the number of occurrences of penalty cost are 13, 10, and 7. According to the provisions on the social discount rate in the 'Economic evaluation methods and parameters of construction projects (Third Edition)' issued by the National Development and Reform Commission and the Ministry of Construction in 2006, the benchmark discount rate is set as $i = 8\%$ in this paper.

The service time of the 2×300 MVar NESC can be up to 40 years or even 60 years, and the general service time is 30 years. From the engineering experience, the ending residual value is 5% of the initial construction cost. The overhaul and technical transformation period are based on the generator maintenance regulations, and the interval of 5 years is taken as the intermediate value. The fault rate takes the engineering experience value of 0.433 times/year, from which the number of occurrences of penalty cost is 13. The benchmark discount rate is $i = 8\%$.

According to the engineering calculation, the cost corresponding to the NESC and the SCTTU with different service life after transformation is shown in Table 1.

Table 1. Life-cycle cost of the NESC and the SCTTU with different service life.

Construction Type	Construction Cost/10^4 Yuan	Operation and Maintenance Cost/10^4 Yuan		Overhaul and Transformation Cost/10^4 Yuan		Penalty Cost/10^4 Yuan	
		Reliability	Cost Range	Reliability	Cost Range	Reliability	Cost Range
Newly established	26,002	0.15	2058.5–2166.8	0.25	736.7–758.3	0.23	541.7–563.4
		0.65	2166.8–2383.5	0.65	758.3–780	0.55	563.4–585
		0.2	2383.5–2708.5	0.1	780–801.7	0.22	585–650
Transformed (20 year service life)	4802	0.15	1874.4–2003.7	0.3	884.1–909.9	0.13	650–676.1
		0.75	2003.7–2040.6	0.55	909.9–936.2	0.68	676.1–702
		0.1	2040.6–2273.5	0.15	936.2–962.1	0.19	702–780
Transformed (15 year service life)	4523	0.2	1780.68–1903.5	0.13	928.3–955.3	0.1	656.5–682.8
		0.6	1903.5–1938.5	0.76	955.3–983	0.85	682.8–709
		0.2	1938.5–2159.8	0.11	983–1,010.2	0.05	709–787.8
Transformed (10 year service life)	4039	0.05	1593.2–1703.2	0.18	946–973.6	0.25	663.1–689.7
		0.8	1703.2–1734.5	0.7	973.6–1,001.7	0.57	689.7–716.1
		0.15	1734.5–1932.5	0.12	1001.7–1029.5	0.18	716.1–795.7

Taking a city in China as example, the electricity consumption of the whole society is 98.385 billion kWh, the electricity price level is 0.4 yuan/kWh, the value-added tax rate is 13%, the urban construction and maintenance tax rate is 7%, the education additional tax rate is 3%, the on-line capacity of the synchronous condenser is subject to the rated capacity, and the on-line capacity is calculated as 80% of the occupied capacity per hour.

6.2. Result Analysis

The costs of NESC and the SCTTU with different service life are brought into the calculation formula of blind number cost in the life-cycle, and the 27th order blind number expression (reliability band edge product matrix) of the four life-cycle costs of NESC and the SCTTU with different service life can be obtained from Equation (18). Take the SCTTU with a service life of 20 years as an example, and the results are shown in Equation (19).

$$
\begin{array}{l}
0.0242 \\
0.1268 \\
0.0354 \\
0.0964 \\
0.5045 \\
0.1410 \\
1.0093 \\
0.0488 \\
0.0136
\end{array}
\left[
\begin{array}{lll}
28,083.0 - 29,501.9, & 28,107.8 - 29,526.6, & 28,132.5 - 29,600.8 \\
28,206.7 - 29,628.9, & 28,232.5 - 29,653.6, & 28,257.2 - 29,727.8 \\
28,333.7 - 29,754.1, & 28,359.5 - 29,778.8, & 28,384.2 - 29,853.0 \\
29,351.4 - 29,864.2, & 29,377.2 - 29,888.9, & 29,401.9 - 29,963.1 \\
29,476.1 - 29,991.2, & 29,501.9 - 30,015.9, & 29,526.6 - 30,090.1 \\
29,603.1 - 30,116.4, & 29,628.9 - 30,141.1, & 29,653.6 - 30,215.3 \\
29,713.7 - 32,150.8, & 29,739.5 - 32,175.5, & 29,764.2 - 32,249.7 \\
29,838.4 - 32,277.8, & 29,864.2 - 32,302.5, & 29,888.9 - 32,376.7 \\
29,965.4 - 32,402.8, & 29,991.2 - 32,427.5, & 30,015.9 - 32,501.7
\end{array}
\right]
\quad (21)
$$

$$
\qquad\qquad\quad 0.13 \qquad\qquad\qquad 0.68 \qquad\qquad\qquad 0.19
$$

The 27th order blind number expressions of the NESC and the SCTTU with different service life are optimized and calculated, and the results are shown in Figure 4.

Furthermore, the blind number mean value of annual average cost of NESC and the SCTTU with different service life can be obtained, as shown in Figure 5.

From the above results, it can be seen that the reference average value of the project cost of the NESC and the SCTTU is obtained based on the life-cycle cost analysis of the dimmer based on blind number theory, and the reliability of each interval is given. Compared with the deterministic single value obtained by the deterministic algorithm, it gives the project decision-makers greater reference value.

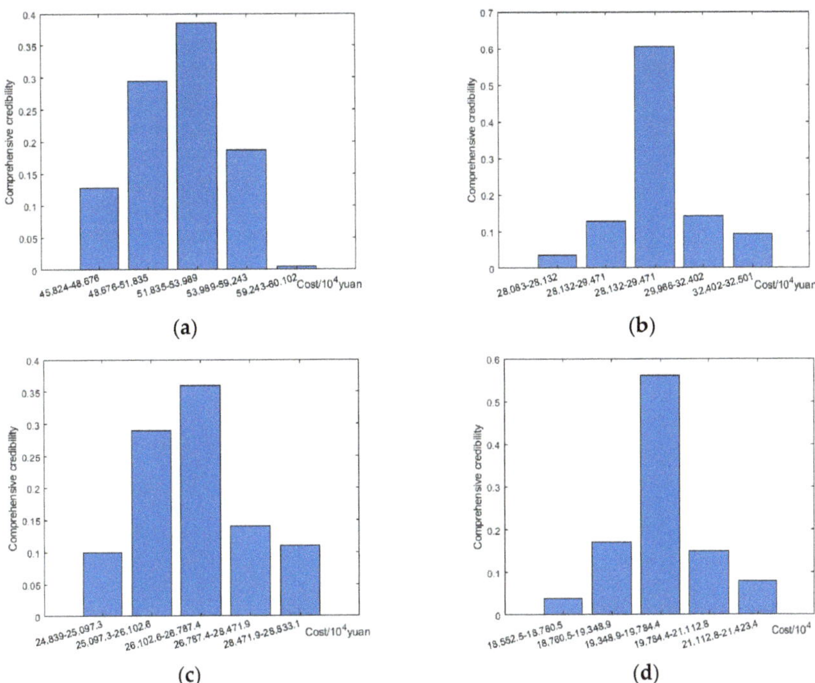

Figure 4. Life-cycle cost of NESC and the SCTTU with different service life: (**a**) NESC (30 year service life); (**b**) SCTTU (20 year service life); (**c**) SCTTU (15 year service life); (**d**) SCTTU (10 year service life).

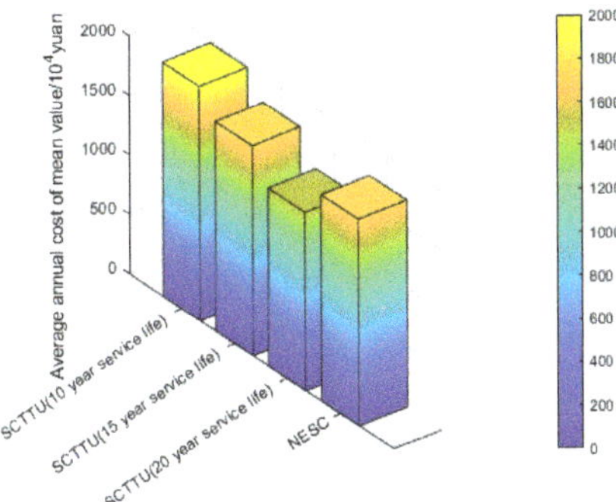

Figure 5. Blind number mean value of annual average cost of NESC and SCTTU.

The average 30 year total cost of the NESC with the same capacity is 5213.86×10^4 yuan, with an annual average of 1737.95×10^4 yuan. The average total cost of the SCTTU (20 year service life) is 2998.78×10^4 yuan, with an annual average of 1499.39×10^4 yuan; the average total cost of the SCTTU (15 year service life) is 2646×10^4 yuan, with an average annual average of 1764.7×10^4 yuan; the average total cost of the SCTTU (10 year service life) is $19,711.6 \times 10^4$ yuan, with an annual average of 1971.2×10^4 yuan. Although the service life of the four units is different, it can be seen from the average annual cost and the cost of each stage that the average annual cost of the SCTTU (20 year service life) is significantly lower than that of the NESC, meaning that the SCTTU has better economic efficiency. However, it is not certain that the entirety of the SCTTU is economical. As can be seen from Figure 5, the average annual cost of the SCTTU increases with the reduction in the service life. The average annual cost of the SCTTU with a 15 year service life is basically the same as that of the NESC. The average annual cost of the SCTTU with a 10 year service life is significantly higher than that of the NESC.

In order to analyze the reasons for the cost difference, this paper analyzes the following four aspects: investment cost (excluding residual value cost), operation cost, overhaul and technical transformation cost, and penalty cost.

We start with the annual average cost. It can be seen from Table 1 that the investment cost of the NESC is 2602×10^4 yuan, which is significantly higher than that of the SCTTU with the following three service lives: 4802×10^4 yuan, 4523×10^4 yuan, and 4039×10^4 yuan. However, the investment cost of the three types of SCTTU will gradually increase with the increase in the longevity of the service life, but the difference is in the order of one million yuan, which has advantages over the new established one. The average value of operation cost, overhaul and technical transformation cost, and penalty cost is shown in Figure 6.

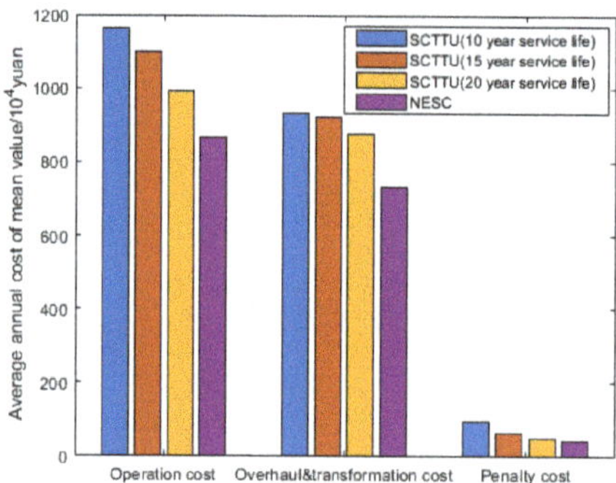

Figure 6. Average annual cost comparison of three costs between NESC and SCTTU.

It can be seen from Figure 6 that the three costs of the NESC are lower than those of the SCTTU. However, due to the high investment cost, the annual average of the life-cycle cost is higher than that of the SCTTU (20 year service life). With the increase in the service life of the transformed one, the operation cost, overhaul and technical transformation cost, and penalty cost decreases. In view of this result, this paper compares the annual change curves of the three costs, as shown in Figures 7–9.

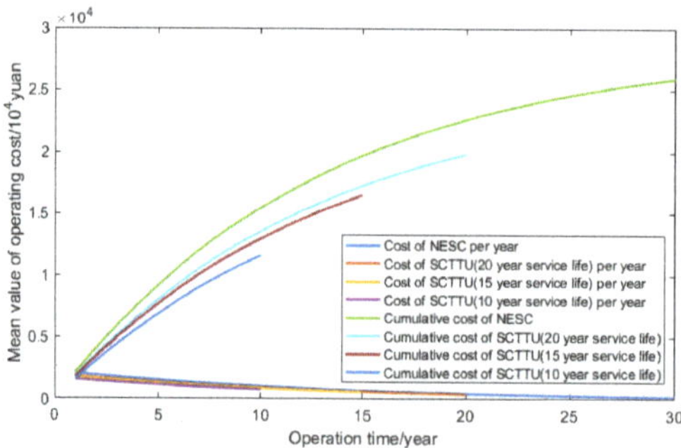

Figure 7. Comparison of operation cost curve between NESC and SCTTU.

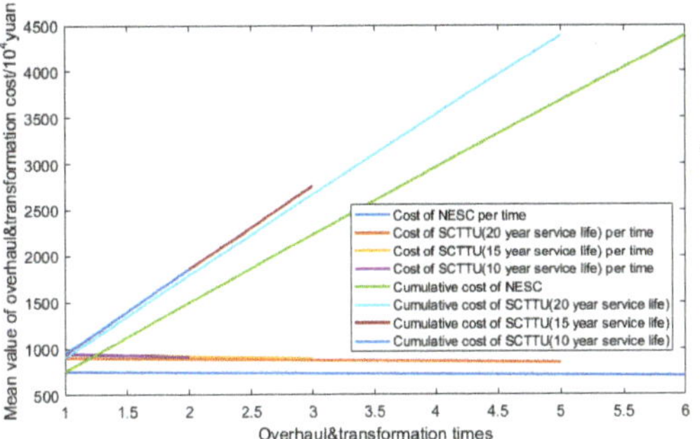

Figure 8. Comparison of overhaul and technical transformation cost curve between NESC and SCTTU.

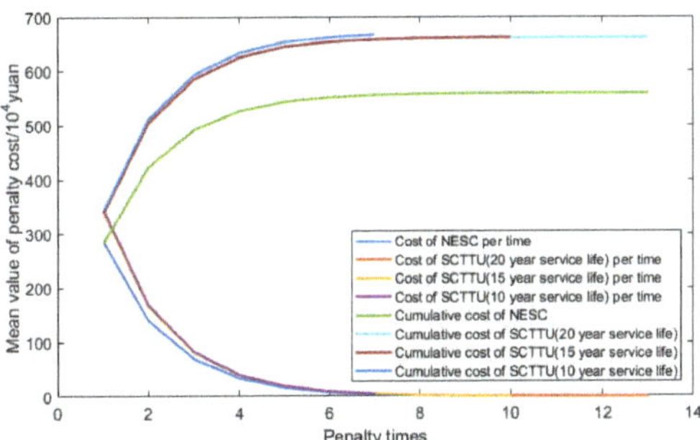

Figure 9. Comparison of penalty cost curve between NESC and SCTTU.

It can be seen that with the increase in the service year after the transformation, the average annual operation cost, and the average penalty cost decrease at the speed of the negative exponential function, resulting in the gradual increase in the accumulated cost, which largely averages the life-cycle cost. However, the overhaul and technical transformation cost has little impact on the average annual operation cost due to the low frequency of occurrence. From this, it can be concluded that, the longer the operation life-cycle of the transformed one, the better its economy. In this paper, the 20-year-old SCTTU has the best economy. It can be inferred that after 15 years of service life, the SCTTU has better economy than the NESC.

Combining the quotation curve and the market pricing mechanism, the pricing results can be obtained, as shown in Table 2.

Table 2. Price summary for unified operation, independent operation, and lease operation (yuan/(kVarh)).

Management Model	Unified Operation of Power Grid	Independent Operation		Lease Operation	
Price form	Increment price	Electricity price	Capacity price	Electricity price	Capacity price
NESC	0.000530	0.0416	55.871	0.0161	92.34
SCTTU (20)	0.000459	0.0341	47.167	0.0129	77.06
SCTTU (15)	0.000547	0.0425	58.495	0.0131	93.85
SCTTU (10)	0.000672	0.0521	67.323	0.0143	97.46

From the quotation of the three management modes, the quotation change trend and cost change trend of the four kinds of synchronous condensers are similar. Due to its cost advantage, the SCTTU with a 20 year service life has the lowest quotation in the vertical comparison of the three management modes, which once again proves the economic advantage of the SCTTU with a 20 year service life.

7. Conclusions

In this paper, a cost calculation method based on the life-cycle cost of the synchronous condenser is proposed, and the blind number theory is introduced into the cost calculation model to solve the influence of various uncertain information on the cost of the synchronous condenser project. The distribution range of the life-cycle cost of the synchronous condenser is also obtained. At the same time, the life-cycle cost of the newly established synchronous condenser is compared with that of the synchronous condenser transformed from thermal units with different service life, the reactive power pricing based on the market mechanism is calculated for the established synchronous condenser and the synchronous condenser transformed from thermal unit, and the following conclusions are obtained:

(1) Under the "dual carbon" background, for the large-scale grid connection of new energy and HVDC transmission with power electronic converter equipment, and with the reuse of a large number of retired thermal units, the synchronous condenser transformed from a thermal unit can provide sufficient dynamic reactive power support for the power grid. According to the analysis of the life-cycle cost, it can be concluded that a synchronous condenser transformed from a thermal unit can reduce investment and operation costs, and improve the utilization rate of the retired thermal units, which has significant practical value and social and economic benefits.

(2) Compared with the traditional single value cost estimation method, the life-cycle cost calculation method based on the blind number theory proposed in this paper gives the cost interval range and its confidence of the synchronous condenser. Combined with engineering cases, the calculation results are reasonable and reliable. This method can help investors summarize the expected costs, help investors identify the probability risks of different costs, and more clearly estimate the cost–return cycle. The clear cost interval range and corresponding reliability provide more valuable reference for the decision-making and construction of the project.

(3) By comparing the life-cycle cost of the newly established synchronous condenser with that of the synchronous condensers transformed from thermal units with different service life, it can be concluded that the life-cycle cost of the synchronous condenser transformed from a thermal unit with a long service life of more than 15 years is lower than that of the newly established synchronous condenser, so its economy is better. In this paper, the 20-year-old transformed thermal unit has the best economy, which proves the economic feasibility of the construction of the synchronous condenser transformed from a thermal unit.

(4) Three types of management pricing are obtained in four different operation modes for the cost recovery of synchronous condenser. Through the comparison of management modes, the impact of management modes on reactive power price can give investors a clearer reference and help investors grasp the lowest price of cost recovery under the corresponding mode. Through the comparison of pricing of a synchronous condenser with four operation modes, it is confirmed that SCTTU with a long service life of more than 15 years has a lower reactive power pricing, and better economy and competitiveness.

Author Contributions: Conceptualization, M.L. and X.Y.; methodology, Y.G. and M.L.; software, Y.G. and H.M.; validation, M.L., and X.Y.; formal analysis, Y.G. and H.M.; investigation, M.L. and Y.G.; resources, Y.G. and M.L.; data curation, M.L. and H.W.; writing—original draft preparation, Y.G.; writing—review and editing, X.Y. and M.L.; visualization, H.M.; supervision, M.L. and X.Y.; project administration, C.L.; funding acquisition, C.L. All authors have read and agreed to the published version of the manuscript.

Funding: This work was supported by the Science and Technology Project of SGCC in 2022 (Key technology and demonstration of retrofitting thermal power unit to synchronous condenser under new-type power system voltage stability demand, no. 5100-202224023A-1-1-ZN).

Institutional Review Board Statement: Not applicable.

Informed Consent Statement: Not applicable.

Data Availability Statement: Not applicable.

Conflicts of Interest: The authors declare no conflict of interest.

References

1. Qiu, W.; He, J.; Fan, X.; Xu, T.; Yu, Z.; Zhang, J.; He, F.; Lan, H.; Ye, J.; Zhang, Y. Overview on stability measures for large disturbances of UHVDC. *Power Syst. Technol.* **2022**, *46*, 3049–3067.
2. Zhou, Y.; Sun, H.; Xu, S.; Wang, X.; Zhao, B.; Zhu, Y. Synchronous condenser optimized configuration scheme for power grid voltage support strength improvement. *Power Syst. Technol.* **2021**.
3. Hadavi, S.; Mansour, M.Z.; Bahrani, B. Optimal allocation and sizing of synchronous condensers in weak grids with increased penetration of wind and solar farms. *IEEE J. Emerg. Sel. Top. Circuits Syst.* **2021**, *11*, 199–209. [CrossRef]
4. Zhilin, G.; Liangliang, H.; Junyong, W. Application of the new generation large capacity synchronous condenser in HVDC system. *IOP Conf. Ser. Earth Environ. Sci.* **2019**, *342*, 12007.
5. Shu, Y.; Chen, G.; He, J.; Zhang, F. Building a new electric power system based on new energy sources. *Strateg. Study CAE* **2021**, *23*, 61–69. [CrossRef]
6. Jiang, Z.; Wang, A.; Tian, H.; Li, S.; Fang, Q.; Tian, H.; Wang, M. Research on synchronous condenser reconstructed from retired thermal power unit in the new power system. *Shandong Electr. Power* **2022**, *49*, 17–22.
7. Chaturvedi, D.K. Use of retired hydrogen cooled generators as synchronous condenser. *CIGRE India J.* **2019**, *8*, 12–14.
8. Karan Sareen, B.S.; Bairwa, R.N.; Pardeep, J. Utilizing retiring thermal units as synchronous condenser for reactive power management in renewable energy rich scenario. *Water Energy Int.* **2019**, *62*, 10–13.
9. An, J.; Zhang, J.; Du, X.; Li, C.; Liu, M. Enhance transient voltage stability by retrofitting thermal power unit to synchronous condenser. In Proceedings of the 2022 7th Asia Conference on Power and Electrical Engineering (ACPEE), Hangzhou, China, 15–17 April 2022.
10. Kaur, J.; Chaudhuri, N.R. Conversion of retired coal-fired plant to synchronous condenser to support weak AC grid. In Proceedings of the 2018 IEEE Power & Energy Society General Meeting (PESGM), Portland, OR, USA, 5–9 August 2018.
11. Project Management and Implementation Planning of Qaidam Converter Station 2×300 MVar Synchronous Condenser Project. Available online: https://wenku.baidu.com/view/616c733dcd7931b765ce0508763231126edb7710.html (accessed on 1 December 2017).
12. Environmental Impact Report of Jiangsu Power Grid Taizhou 2×300 MVar Synchronous Condenser Project. Available online: https://jz.docin.com/p-1466295097.html (accessed on 1 January 2016).
13. Huang, Z.; Smolenova, I.; Chattopadhyay, D.; Govindarajalu, C.; De Wit, J.; Remy, T.; Deluque Curiel, I. ACT on RE+FLEX: Accelerating coal transition through repurposing coal plants into renewable and flexibility centers. *IEEE Access* **2021**, *9*, 84811–84827. [CrossRef]
14. Zhang, S. *Theory and Method of Life Cycle Cost Control of Construction Project*; China Electric Power Press: Beijing, China, 2007; pp. 79–95.
15. Yuan, J.; Wang, Y.; Zhang, Z.; Luo, X.; Zhao, D. Research on transformation equipment technical transformation project application based on life cycle cost. *Northeast Electr. Power Technol.* **2021**, *42*, 23.
16. Liu, Y.; Su, H. Life cycle cost estimation of smart substation based on blind number theory. *Electr. Power* **2022**, *49*, 83–87.
17. Yan, R.; Luo, J.; Xu, Y. Life cycle cost analysis of power transformer based on blind number theory. *Proc. CSU-EPSA* **2019**, *31*, 15–20.
18. Lin, J. Blind Number Model of Evaluating Commercial Banks Position Value and Its Application. *Math. Pract. Theory* **2018**, *48*, 1–11.
19. Xiong, Y.; Liao, X.; Ke, F.; Zhou, Q.; Tang, x.; Ming, Y.; Li, Z.; Zhou, R. Life cycle cost analysis of main transformer based on the multi-system data fusion. *J. Electr. Power Sci. Technol.* **2020**, *35*, 3–11.
20. Xiang, Z.; Ni, Q.; Wu, C.; Ren, X.; Li, S. Research on step-wise quotation rules of reactive power ancillary services in electricity market based on cost analysis. *Zhejiang Electr. Power* **2021**, *40*, 70–75.
21. Zhao, W.; Yan, Z.; He, C.; Wang, K. Comparative study on the impact of different electricity price models on the investment income of wind power projects in China. *Price Theory Pract.* **2021**, *10*, 138–142.
22. Lin, L.; Yao, C.; Sun, Y.; Wei, M. Analysis and enlightenment of operation modes of pumped storage power stations at home and abroad. *Power Syst. Clean Energy* **2021**, *37*, 107–114.
23. Yao, J.; Wu, Y.; Wang, Y.; He, J.; Dai, S. The effect of implementing method of two-part tariff policy on market resource allocation efficiency. *Electr. Power*, **2022**; in press.
24. Pan, H.; Gao, H.; Yang, Y.; Ma, W.; Zhao, Y.; Liu, J. Multi-type retail packages design and multi-level market power purchase strategy for electricity retailers based on master-slave game. *Proc. CSEE* **2022**, *42*, 4785–4800.
25. Wang, Y.; Tian, Y.; Wu, M.; Geng, J. Two-part electricity price model for peak load regulation of natural gas power based on fuzzy clustering. *Proc. CSEE* **2017**, *37*, 1610–1618.

 *processes*

Article

Research on Low-Carbon Capability Evaluation Model of City Regional Integrated Energy System under Energy Market Environment

Zhangbin Yang and Xiaojing Wang *

School of Electrical Engineering, Xinjiang University, Urumqi 830017, China
* Correspondence: wangxiaojing345@163.com; Tel.: +86-180-9961-3605

Abstract: In the context of the "carbon peaking and carbon neutrality" goal and energy marketization, the City Regional Integrated Energy System (CRIES), as an important participant in the energy market, pursues low-carbon development as its most important goal. Without a reasonable market participation structure and a comprehensive low-carbon evaluation system, it will be difficult to meet the needs of CRIES for low-carbon development in the energy market. Therefore, this paper first proposes a framework suitable for CRIES to participate in the energy market, and under the influence of the operating characteristics of the energy market, proposes an evaluation index system suitable for CRIES' low-carbon capabilities in the energy market. The analytic network process–criteria importance through intercriteria correlation (ANP-CRITIC) method is used to determine the subjective and objective weights of each indicator, and the comprehensive weight of each indicator is calculated by the principle of moment estimation to achieve a quantitative evaluation of the low-carbon capability of CRIES in the energy market. Finally, taking a CRIES as an example, the analysis verifies that the proposed evaluation model and method can scientifically and comprehensively evaluate the low-carbon capability of CRIES in the energy market. The results show that the CRIES low-carbon capability evaluation results of different market schemes can be improved by up to 24.9%, and a fairer market transaction mechanism can promote the low-carbon development of CRIES.

Keywords: low-carbon capacity; city regional integrated energy system; energy markets; ANP-CRITIC; evaluation

Citation: Yang, Z.; Wang, X. Research on Low-Carbon Capability Evaluation Model of City Regional Integrated Energy System under Energy Market Environment. *Processes* **2022**, *10*, 1906. https://doi.org/10.3390/pr10101906

Academic Editors: Haoming Liu, Jingrui Zhang, Jian Wang and Anna Trubetskaya

Received: 2 August 2022
Accepted: 16 September 2022
Published: 20 September 2022

Publisher's Note: MDPI stays neutral with regard to jurisdictional claims in published maps and institutional affiliations.

1. Introduction

In recent years, how to improve energy utilization and reduce carbon emissions has become the focus of energy development in various countries. The single traditional energy supply system has the defects of low energy efficiency and high emission, which cannot meet the current needs of low-carbon energy development [1,2]. The regional integrated energy system (RIES) can couple different energy types and promote the consumption of renewable energy, which has become a key technology for low-carbon energy development in recent years [3]. From small industrial parks to large cities, they all belong to the category of RIES. The City Regional Integrated Energy System (CRIES) is an important form of RIES [4]. It has numerous distributed energy systems and multienergy complementary systems, which are the bridge connecting the upper energy main network and the energy load side.

With the advancement of energy marketization [5], CRIES has become an important participant in the energy market due to its advantages of high economic benefits, strong low-carbon capabilities, high system reliability, and high energy utilization rates [6,7]. Among them, the low-carbon capability is an important factor that CRIES must consider when participating in the energy market. Evaluating the low-carbon capability of CRIES after participating in the energy market is an important theoretical support for promoting the consumption of renewable energy in CRIES, improving the comprehensive energy utilization rate of CRIES, scientifically planning the operation plan of CRIES participating in the energy market, and improving the low-carbon development of CRIES. Therefore, it is necessary to conduct scientific and comprehensive research on the low-carbon capability evaluation model of CRIES in the energy market.

Some scholars have carried out related research on this and obtained rich research results. Reference [8] introduced the carbon trading mechanism into the energy market clearing of integrated energy and studied the impact of the carbon trading mechanism on the RIES auction clearing strategy. References [9,10] study the low-carbon clearing strategy of IES participating in the market with uncertain demand response and new energy output. References [11,12] based their studies on the carbon emission flow (CEF) theory for low-carbon and the economical optimal scheduling of IES. References [13,14] studied the low-carbon optimization of RIES through carbon capture and carbon trading. However, most of the existing research focuses on considering RIES as a distributed energy system for the market clearing, optimal scheduling, and design planning of the system. However, in the face of RIES, such as CRIES, which covers a wide area, has a wide variety of energy sources, and has many distributed energy sources, existing research methods will not be able to satisfy CRIES' reasonable participation in the energy market.

In terms of the RIES evaluation, Reference [15] proposed a comprehensive evaluation index with universal applicability to RIES from the links of energy, installation, the distribution network and users, and thereby proposed a scientific method for evaluating the development level of RIES. Reference [16] established a comprehensive evaluation of integrated energy systems through six characteristics of multidimensional, multivector, systematic, future, systematic, and applicability. Reference [17] proposed a decision-making method for integrated energy participation in energy market transactions, and evaluated the system considering four aspects of economy, fairness, environmental protection, and safety. Reference [18] evaluated the integrated energy system from different aspects of the integrated energy system, such as reliability under the consideration of user thermal comfort, power transaction performance, and system energy efficiency analysis. Reference [19] proposed an alternative model-assisted IES quantitative evaluation method to evaluate the operation of the IES. However, the existing evaluation system only takes low-carbon capability as a part of the evaluation system, and lacks a comprehensive evaluation model for CRIES' low-carbon capability. If there is no comprehensive evaluation system, it will not be able to meet the development process of CRIES, which will bring great challenges to the low-carbon development of CRIES after participating in the energy market.

So, for the above two aspects, this paper proposes a CRIES low-carbon capability evaluation model under the energy market. First, by fully considering the difficulties faced by CRIES' participation in the energy market, and then establishing a reasonable structure for CRIES to participate in the energy market; secondly, based on the operating characteristics of the energy market, an evaluation index system for CRIES' low-carbon capability in the energy market is proposed. The ANP-CRITIC method is used to assign the objective and subjective weights of the indicators, and the moment estimation principle is used to obtain the comprehensive weight so as to realize the quantitative evaluation of the low-carbon capability of CRIES in the energy market, and provide a reference for promoting the low-carbon development of CRIES in the energy market in the future.

2. CRIES Structure under the Energy Market

CRIES is different from other RIES in that it has a vast area, a wide variety of energy sources, and the locations of distributed energy sources are scattered, which cannot be traded with the energy market according to the traditional system architecture [20]. Therefore, this paper proposes a three-tiered structure and a multisubject CRIES to participate in the energy market structure. The three-layer structure is divided into the market layer, the CRIES layer, and the load layer. As shown in Figure 1, it involves energy transactions such as electricity, natural gas, and heat.

3.2. Quantification and Overview of Evaluation Indicators

3.2.1. Low-Carbon Transition

1. (S1-1): The energy exergy efficiency of system energy is based on the second law of thermodynamics [21,22], which focuses on the conversion efficiency corresponding to the quality of energy. Compared with the comprehensive utilization rate of energy, which focuses on the quantity of energy, its exergy efficiency can better reflect the loss of energy and the level of stepped utilization, which is defined as the output of the system. The ratio of the revenue exergy to the input cost exergy.

$$\eta_{ex} = \frac{E_{os}^e + E_{os}^h + E_{os}^c + \sum\limits_{n=1}^{N} \zeta E_s^n}{E_{is}^e + E_{is}^{gas} + E_{is}^w + E_{is}^s + \sum\limits_{n=1}^{N} (1-\zeta)E_s^n} \tag{1}$$

where $E_{os}^E, E_{os}^h, E_{os}^c$ is the electrical, hot, and cold exergy of the total output of the system, respectively; E_s^n is the exergy of energy type; n is for the energy storage; $E_{is}^e, E_{is}^{gas}, E_{is}^{ws}$ is the exergy of electricity, natural gas, and new energy input to the system; ζ is the 0–1 state of the energy storage device.

2. (S1-2): The value-added rate of energy conversion is the profitability of the CRIES in the process of coupling different energy sources through its own energy coupling equipment and then selling it.

$$V = \left(\frac{C_{rl}}{\eta_{ex}} \right) \Big/ C_s \tag{2}$$

where V is the value-added rate of energy conversion, C_s is the selling price for the system, and C_{rl} is the total profit converted for the system.

3. (S1-3): The energy conversion boundary is restricted by the CRIES hardware conditions, resource conditions, and external conditions. For example, the capacity of system equipment, the ability to absorb new energy such as wind and solar, and equipment operation constraints. In addition, it also includes factors such as the selection of system operation electric–heat ratio, demand restrictions, safety, and environmental protection restrictions. Therefore, this indicator uses relevant experts to score the qualitative evaluation.

3.2.2. Low-Carbon Technology

1. (S2-1): In the CRIES, the energy storage device is the link between different energy sources, mainly used in the energy storage inside power systems and thermal systems. It has good spatiotemporal coupling and balancing capabilities of different energy sources. In addition, it can reduce the energy waste of the system and enhance the system regulation. The energy storage configuration ratio η_{se} is the proportion of the energy storage capacity connected to the system to the installed capacity of the system.

$$\eta_{se} = \frac{\sum\limits_{n=1}^{N} W_{ac.n}}{\sum\limits_{s=1}^{S} W_{ec.s}} \tag{3}$$

where $W_{ac.n}$ is the energy storage equipment capacity corresponding to the energy storage energy type n. There are a total of S devices in the system, and $W_{ec.s}$ is S-th device capacity.

2. (S2-2): The proportion of new energy installed capacity is the proportion of the installed capacity of new energy units to the installed capacity of the entire system.

$$\eta_{ni} = \frac{\sum\limits_{i=1}^{I} W_{rn.i}}{\sum\limits_{s=1}^{S} W_{ec.s}} \tag{4}$$

3.2.3. Low-Carbon Benefits

1. (S3-1): The return-on-investment in carbon emission reduction can better judge the emission reduction intensity of the CRIES, and can intuitively reflect the income generated by the investment in carbon emission reduction of the system. It is expressed as the ratio of value to the sum of investment in system projects.

$$\eta_{eb} = \frac{\sum\limits_{k=1}^{K} \Delta CE_j \cdot C_j}{V} \tag{5}$$

where ΔCE_j is the emission of the k-th pollutant in the system, C_j is the price charged for the emission of type k pollutants, and V is the total investment of the system.

2. (S3-2): Participating in the carbon trading market is to put the excess carbon emission credits of the system into the carbon trading market, and then obtain carbon rights benefits. It is assumed that in the carbon trading market, the carbon emission allowances are obtained for each capacity unit through the baseline method, and the carbon trading market income is calculated through the stepped carbon price.

$$CE_{co_2}^{qu} = \sum\limits_{s}^{S} P_s \zeta_s \tag{6}$$

where $CE_{co_2}^{qu}$ is the total amount of carbon emission quota of the system; P_s, ζ_s is the carbon emission quota for the output power and unit active power output of the equipment.

$$CE_{co_2} = \sum\limits_{s=1}^{S} \lambda_s P_s \tag{7}$$

where λ_s is the carbon emission factor of the s-th emitting device and P_s is the output power of the s-th device.

$$f_{co_2} = \begin{cases} -30 - 3a, -(1+0.1a)CE_{co_2}^{qu} \le CE_{co_2} < 0 \\ 30 + 3a, 0 \le CE_{co_2} < (1+0.1a)CE_{co_2}^{qu} \end{cases} (a = 0,1,2\ldots) \tag{8}$$

In the carbon trading market, the carbon trading price is set through the stepped carbon price, and the stepped carbon price is shown in Equation (8).

$$C_{co_2} = f_{co_2}(CE_{co_2} - CE_{co_2}^{qu}) \tag{9}$$

where C_{co_2} is the benefit of the carbon trading market; when the total amount of carbon dioxide emitted by the system is greater than the total amount of carbon emission allowances, the system needs to purchase carbon emission allowances from the carbon market. On the contrary, get the benefit.

3.2.4. Market Subject

1. (S4-1): The market concentration index can reflect the energy concentration of the CRIES in the energy market environment and the overall competition level of the multienergy market. This paper adopts the Herfindahl–Hirschman Index (HHI), which is commonly used in industrial economics, as an indicator of market concentration.

$$M_{HHI} = \sum_{d=1}^{D} R_d^2 \cdot 10{,}000 \tag{10}$$

where R_d is the CRIES d's share in the multienergy market; for HHI, the value is between [0–10,000]. The larger the value, the more concentrated the market. According to the differentiation rule, when the HHI value is [500, 1800], the market is a competitive market.

2. (S4-2): The market fairness index is to evaluate the fairness of the energy trading results of the CRIES under different market mechanisms, and to characterize whether the CRIES has the same trading status as other market entities in the participating market.

$$S = \frac{1}{L} \frac{\left(\sum\limits_{g=1}^{G} C_g\right)^2}{\sum\limits_{g=1}^{G} \left(C_g\right)^2} \tag{11}$$

where S is the market fairness index. The larger the value is between [0, 1], the fairer the market; C_g is the operational benefit of the CRIES; G is the total number of CRIES; L is the number of evaluation systems.

3.2.5. Market Operation

1. (S5-1): The new energy clearing ratio is the proportion of the system's renewable energy clearing energy to the total system clearing energy. By default, the electricity purchased from the upper-level power grid is generated by thermal power units.

$$\eta_{ne} = \frac{\sum\limits_{i=1}^{I} E_{rn}^i}{E_{load}^e + \alpha_h Q_{load}^h + \alpha_c Q_{load}^c} \tag{12}$$

where E_{load}^e, Q_{load}^h, Q_{load}^c is the demand side electricity, heating, and cooling loads corresponding to the system; α_h, α_c is the energy conversion factor for heat and cold; E_{rn}^i is the power generation of new energy equipment i.

2. (S5-2): Clearing price and new energy indicators can analyze the relationship between market energy prices and renewable energy clearing capacity. This paper uses the Spearman correlation coefficient in statistics for characterization.

$$\vartheta_{n,p} = 1 - \frac{6\sum X^2}{n(n^2 - 1)} \tag{13}$$

where $\vartheta_{n,p}$ is the Spearman correlation coefficient; X is the difference between the data of clearing price and renewable energy output arranged from small to large; n is the total number of both data; for $\vartheta_{n,p}$, its value exists between [−1, 1]. The larger the absolute value of the value, the stronger the correlation, and the closer it is to 0, the weaker the correlation.

3. (S5-3): The equivalent utilization rate of the system is the ratio of the operating power of each device in the system to the total operating power that can be dispatched by each device in the system when the system participates in the energy market. It can effectively reflect the utilization degree of the CRIES resources under the participation in the energy market.

$$\eta_{ur} = \frac{\sum\limits_{s=1}^{S} P_s^{ao}}{\sum\limits_{s=1}^{S} P_s^{to}} \tag{14}$$

where P_s^{ao}, P_s^{to} is the actual operating power and the maximum operating power of the device S.

3.2.6. Market Benefit

1. (S6-1): The market price volatility is the fluctuation of the market clearing price in the energy market. When the value is large, the market price may fluctuate violently, causing the energy market risk to increase. When it is too small, it is not conducive to a reasonable response to the market supply and demand relationship.

$$\eta_{pf} = \frac{P_{std,T}}{P_{av,T}} \tag{15}$$

where $P_{std,T}$, $P_{av,T}$ is the standard deviation of the market price and the average price of the market during the evaluation period.

2. (S6-2): The social net income index refers to the economic benefits brought by CEPs satisfying the load-side demand under market rules and deciding on energy purchase and clearing decisions according to their own and market clearing rules. It can be reflected in the social welfare level under the energy market.

4. Evaluation and Calculation of CRIES Low-Carbon Capability under Energy Market Environment Based on ANP-CRITIC

4.1. Subjective Weight Calculation

The Analytic Network Process (ANP) [23,24] is a subjective empowerment method. It consists of two layers: a control layer and a network layer. It can analyze and calculate the network structure that is mutually influenced and dependent on the principle of the super matrix, so that it can obtain more scientific index weights.

Since the first-level indicators are summative indicators, no weight assignment is performed. The secondary and tertiary indicators are the main body of the evaluation indicators, and the ANP is used for weighting. Therefore, the second layer is subordinate to the control layer, and the 14 interdependent indicators constitute the network layer. The set of control layers for the evaluation of the low-carbon capability of the CRIES under the energy market environment is $S = \{S_1, S_2, \cdots S_6\}$. The network layer factor group should be $S_i = \{S_{i1}, S_{i2}, \cdots S_{ij}\}(i = 1, 2, \ldots 6)$. The subjective weight calculation steps are as follows:

Firstly, the control layer S_i is used as the criterion, the element $S_{il}(l = 1, 2 \cdots n_i)$ in S_i is the secondary criterion, and the elements in the control layer S_j are used to compare the dominance of S_{il} by the 1–9 scale method according to their influence. Moreover, through the consistency test, the influence judgment matrix of the network layer element corresponding to the control layer S_i on the network layer element corresponding to S_j is obtained. If the two indicators are not affected by each other, then $w_{ij} = 0$ will be used, and the initial supermatrix W will be finally constructed, as shown in Formula (16) shown.

$$W = \begin{bmatrix} w_{11} & w_{12} & \cdots & w_{1n} \\ w_{21} & w_{22} & \cdots & w_{2n} \\ \vdots & \vdots & \cdots & \vdots \\ w_{n1} & w_{n2} & \cdots & w_{nn} \end{bmatrix} \tag{16}$$

Since W is not a column normalization matrix, it needs to be weighted and normalized, and the weighted matrix A is obtained by comparing each column pairwise, as shown in Formula (17).

$$A = \begin{bmatrix} a_{11} & a_{12} & \cdots & a_{1n} \\ a_{21} & a_{22} & \cdots & a_{2n} \\ \vdots & \vdots & \cdots & \vdots \\ a_{n1} & a_{n2} & \cdots & a_{nn} \end{bmatrix} \tag{17}$$

The weighted supermatrix $\overline{W}$ is obtained by processing $A \cdot W$. In order to calculate the relationship between the factors, it is necessary to stabilize $\overline{W}$ by Formula (18), calculate the limit relative sorting vector, and if it converges and is unique, the result obtained is the limit matrix, and finally calculate the normalized eigenvectors to obtain the indicators of each network layer weight w'.

$$\lim_{k \to \infty} \left(\frac{1}{n}\right) \sum_{k=1}^{n} w^k \tag{18}$$

4.2. Objective Weight Calculation

The CRITIC method determines the objective weight of each indicator by quantifying the dispersion of each indicator value [25]. Compared with the traditional entropy method, it not only considers the contrast strength between the indicators, but also considers the contradiction between the indicators to deal with the mutual influence between the indicators. Make the weight results more scientific and reasonable. The steps are as follows:

First, it is necessary to perform dimensionless processing on the original data matrix $S = [s_{ij}]_{m \times n}$, assuming that the number of evaluation samples is m and the number of evaluation indicators is n. The benefit-type index and the cost-type index are, respectively, processed by Formulas (19) and (20) to obtain a dimensionless evaluation matrix S'.

$$s'_{ij} = \frac{s_{ij} - \min(s_j)}{\max(s_j) - \min(s_j)} \tag{19}$$

$$s'_{ij} = \frac{\max(s_j) - s_{ij}}{\max(s_j) - \min(s_j)} \tag{20}$$

CRITIC determines the amount of information by calculating the variability and conflict so as to determine the objective weight of each indicator.

The index variability is expressed by the standard deviation, as shown in Formula (21), where ζ_j is the standard deviation of the j-th index.

$$\zeta_j = \sqrt{[\sum_{i=1}^{n} s_{ij} - \bar{s}_j']/(n-1)} \tag{21}$$

The index conflict is expressed by the conflict coefficient C_j as in Equation (22), where c_{ij} is the correlation coefficient between the i-th index and the j-th index, expressed by Equation (23).

$$C_j = \sum_{i=1}^{n} (1 - c_{ij}) \tag{22}$$

$$c_{ij} = \frac{\sum_{u=1}^{m} (s'_{ui} - \bar{s}'_{ui})(s'_{ui} - \bar{s}'_{ui})}{\sqrt{\sum_{u=1}^{m} (s'_{ui} - \bar{s}_i')^2 \sum_{i=1}^{n} (s_{uj} - \bar{s}_j')^2}} \tag{23}$$

The amount of information of the j-th index is I_j, and its calculation is shown in Formula (24). I_j is the fusion of index variability and index conflict. The greater the amount of information, the greater the weight it occupies.

$$I_j = \zeta_j C_j \tag{24}$$

Finally, the objective weight w'' of the j-th index is obtained by Formula (25).

$$w'' = I_j \left[\sum_{j=1}^{n} I_j\right]^{-1} \tag{25}$$

4.3. Comprehensive Weight and Scoring Mechanism

After the subjective and objective weights are obtained by ANP-CRITIC, the moment estimation principle is used to calculate the comprehensive weight. According to the principle, the coupling coefficient corresponding to the subjective weight and the objective weight is first calculated by Formula (26).

$$\begin{cases} \varsigma_j = w'_j / (w'_j + w''_j) \\ \ell_j = w''_j / (w'_j + w''_j) \end{cases} \tag{26}$$

where ς_j is the subjective coupling coefficient; ℓ_j is the objective coupling coefficient; the comprehensive weight χ_j can be obtained from Formula (27).

$$\chi_j = (\varsigma_j w'_j + \ell_j w'') / \sum_j^n (\varsigma_j w'_j + \ell_j w'') \tag{27}$$

In order to display the evaluation results more intuitively in the actual project, this paper multiplies and sums the actual data of each indicator and the comprehensive weight of each indicator, and finally obtains the final score of the low-carbon capability evaluation of the CRIES under the energy market environment.

5. Case Study

This paper takes a CRIES in a certain area as an example. The basic structure is shown in Figure 3, which includes a nine-node power network and a seven-node thermal network. G1, WP1, GB1, ESS1, and G2, as well as WP2, GB2, and ESS2 are the CHP units, wind and solar units, gas boilers, and energy storage equipment belonging to CEP1 and CEP2, respectively. CHP operates in the way of constant heat and electricity, and the system heat load is supplied by the CHP unit and the GB unit. The cooling load of the system is supplied by an absorption chiller, so the heat load node can be replaced with a cooling load node. Select the typical daily operation data in this area, and use MATLAB to fit the electricity, heating, and cooling loads. The fitting curve is shown in Figure 4.

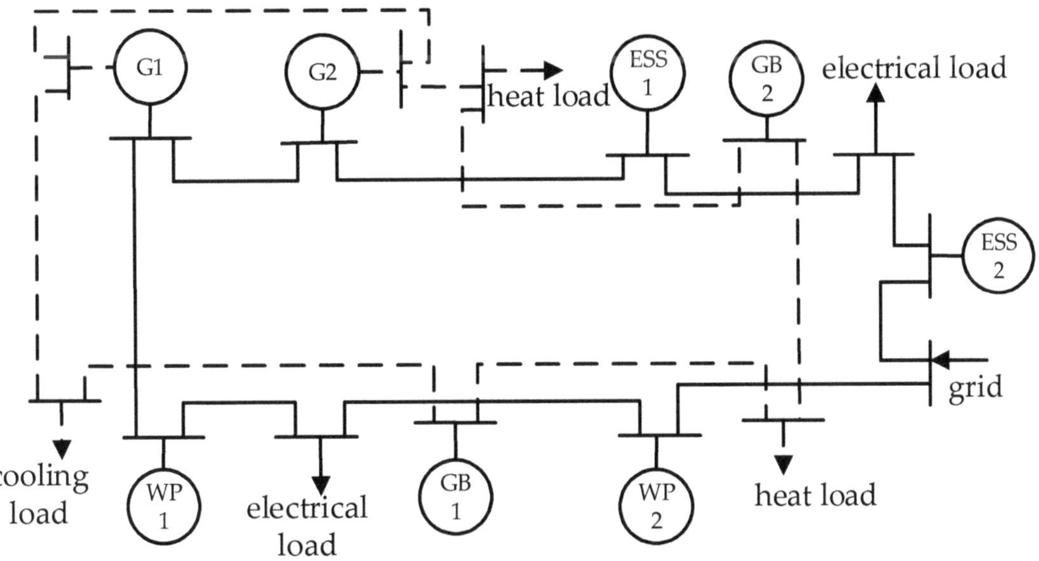

Figure 3. CRIES architecture.

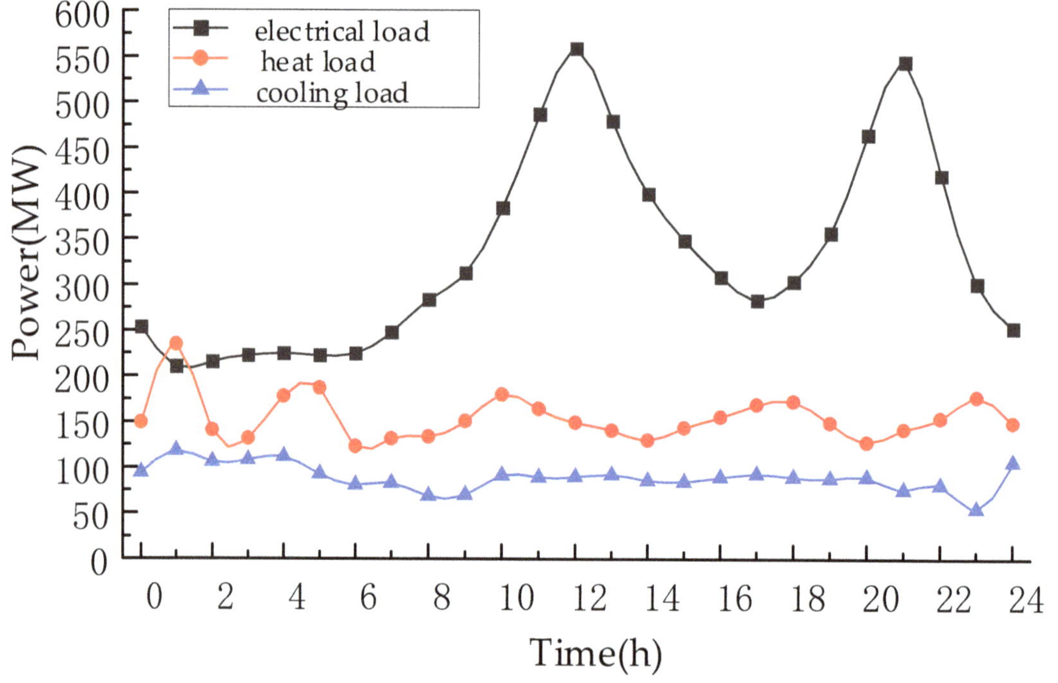

Figure 4. Typical daily load information.

The region is currently in the transitional stage of the CRIES participating in the market and has all the hardware conditions and policy support for participating in the market. Combined with the actual situation in the region, the operation plan of reference [26] and equipment constraints [7] are used to calculate the index data of each system. Tables 1 and 2 show the specific schemes.

Table 1. CRIES specific scheme.

Scheme	CEPs	Participate in the Market Scheme
1	CEP1	A
2	CEP2	A
3	CEP1	B
4	CEP2	B
5	CEP1	C
6	CEP2	C

Table 2. The parameters of each device in CEPs.

CEPs	WT	ESS	CHP	GB
CEP1	70 MW	40 MW	240 MW	200 MW
CEP2	180 MW	80 MW	240 MW	200 MW

For the design of the market participation scheme: the clearing price of Scheme A is calculated using the peak–valley electricity price, and the reverse power sales to the power grid is not considered; the clearing price of Scheme B is calculated using the real-time electricity price, and the reverse power selling to the power grid is not considered; Scheme C adopts the market real-time electricity price and sells excess electricity to the grid. From Scheme A to Scheme C, the market opening degree has gradually deepened, gradually transitioning from not participating in the market to fully participating in the market. The unit price of natural gas is 2.70 CYN/m^3.

5.1. Calculation Results of Each Indicator under Different Market Participation Schemes

Due to the limited space, only a brief analysis of the power system output is made here. Figures 5–7 is the power system output diagram under different market participation schemes. In the vertical comparison, with the deepening of the market openness, the power purchased from outside the system gradually decreases. The output of wind turbines continues to increase, and the system gradually changes from a single load mode to a power mode to sell electricity to the upper-level power grid.

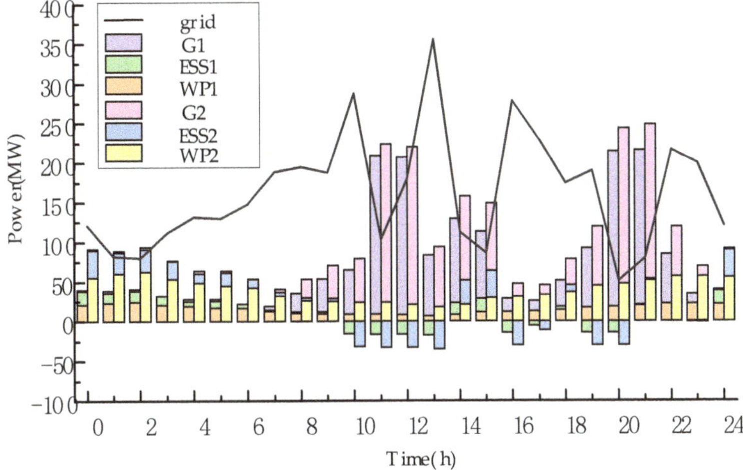

Figure 5. Power operation diagram under Scheme A.

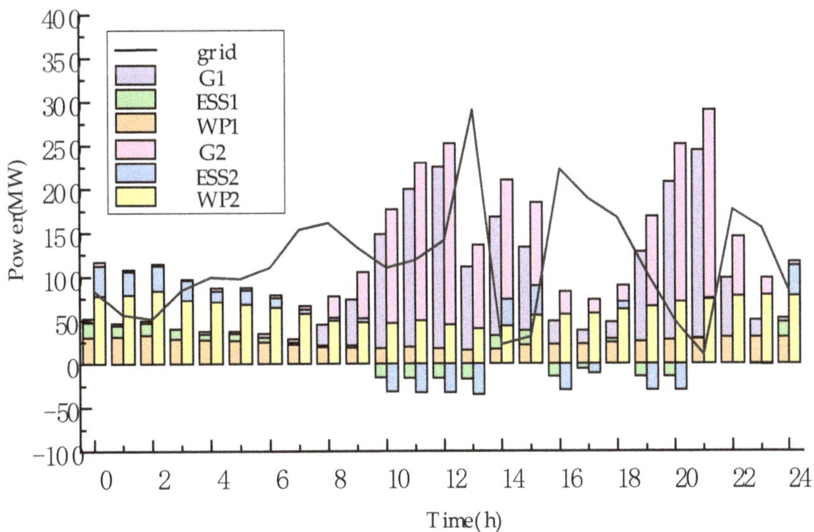

Figure 6. Power operation diagram under Scheme B.

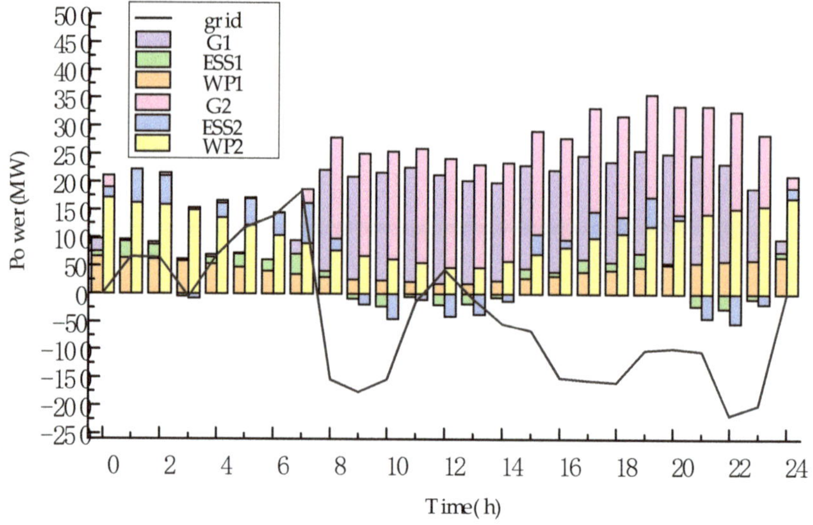

Figure 7. Power operation diagram under Scheme C.

According to the calculation method of each indicator in the low-carbon capability evaluation system (Formulas (1)–(15)), the indicators of the low-carbon situation and market structure are calculated, respectively, and the calculation results are shown in Tables 3 and 4.

Table 3. Calculation results of low-carbon situation indicators under each scheme.

Scheme	1	2	3	4	5	6
S1-1/%	56.113	57.95	58.454	64.184	62.928	66.803
S1-2/%	36.231	42.265	45.848	48.367	48.213	54.268
S1-3	74.16	84.23	74.12	84.22	74.11	84.24
S2-1/%	7.1	11.5	7.1	11.5	7.1	11.5
S2-2/%	12.7	25.7	12.7	25.7	12.7	25.7
S3-1/%	7.2	13.42	7.41	13.65	7.64	13.71
S3-2/KCNY	-72	12.1	-34	62	17.5	71

Table 4. Calculation results of market structure indicators under each scheme.

Scheme	1	2	3	4	5	6
S4-1	4800	4800	3500	3500	1600	1600
S4-2	0.667	0.671	0.756	0.761	0.891	0.868
S5-1/%	18.94	24.07	22.82	30.46	23.62	34.98
S5-2/%	21.46	25.1	28.17	32.64	46.18	48.82
S5-3/%	75.21	78.16	78.64	80.12	83.41	85.13
S6-1/%	23.6	24.1	42.5	41.9	56.4	56.7
S6-1/KCYN	243.0	267.8	275.4	305.9	321.8	349.6

5.2. Analysis of Indicator Results

Through the comparative analysis of the data in Table 3 in Section 5.1, under the same market participation scheme, the energy exergy efficiency of CEP2 can be improved by up to 9.8% compared with CEP1 because CEP2 has larger capacity new energy units, and new energy units rely on renewable energy such as wind energy instead of consuming fossil energy. Therefore, the input exergy of the new energy unit is considered to be zero [27], so the CEP2 with a large installed capacity of the new energy units has a higher exergy efficiency. Under different market participation schemes, from Scheme A to Scheme C, the system can sell more electric energy generated by clean energy in the region to the energy market, reducing the occurrence of energy waste and improving the exergy efficiency of the system. CEP2, with larger energy storage capacity, can sell electricity when the load is high and generate electricity when the load is low. It has a stronger energy translation ability and reduces the cost of electricity, so it can obtain higher income and improve the value-added rate of energy conversion. In the face of different CEPs, the qualitative evaluation results of relevant experts on the energy conversion boundary are consistent with the actual situation, and CEPs with more new energy and energy storage equipment capacity obtain higher qualitative evaluation results. For CEP2, from Scheme A to Scheme C, the profit obtained in the carbon market increased from 12.1 KCYN to 71 KCYN. This is because, with the deepening of the market opening, it can promote the consumption of new energy, thus increasing the emission rights sold in the carbon market. However, for CEP1 under Scheme A, due to the low capacity of new energy and energy storage equipment, and the inability to sell electricity to the upper power grid, the wind and solar energy are seriously abandoned, not only unable to make profits in the carbon market, but also needing to purchase carbon emission rights in the carbon market. With the deepening of market openness, it can improve the consumption of new energy, so that it has more carbon emission rights, and carbon benefits can be obtained under the final Scheme C.

From the data in Table 4 in Section 5.1, the Herfindahl–Hirschman Index for Scheme A is higher. This is because, at this time, the energy market only takes the role of CEPs as energy receivers, and they do not have the ability to compete in the energy market. With the deepening of the market openness, the HHI index value gradually decreases, and the HHI index value is within the range of the competitive market in the case of Scheme C. With the deepening of the market openness, its market fairness also becomes relatively fair. The proportion of new energy clearing has a positive correlation with the capacity of new energy equipment, and a more open market is more conducive to the increase in the proportion of new energy clearing. For CEP2, the proportion of new energy clearing in Scheme C has increased by 45.3% compared to Scheme A. The increase in the proportion of energy storage equipment can promote the space–time coupling and balancing ability of different energy sources, thereby improving the equivalent utilization rate of the system. The degree of market openness affects the relationship between supply and demand in the market, and the relationship between supply and demand guides the fluctuation of market prices. Therefore, the price volatility of a market with a high degree of openness maintains a higher level than other market solutions. Under Scheme C, CEPs can participate in the market competition as the main body of the energy market and obtain more social benefits in the energy market, while under Scheme A, CEPs can only passively act as energy receivers to obtain lower social benefits.

5.3. Calculation Results of Index Weights Based on ANP-CRITIC

Through the ANP-CRITIC indicator weight calculation method proposed in Section 4, and the actual indicator data of each scheme, the subjective and objective weights and comprehensive weights of the secondary and tertiary indicators are obtained, as shown in Table 5 and Figure 8.

From the weight distribution of secondary indicators, it can be seen that market benefit accounts for the highest proportion. This is because the CRIES first pursues the maximization of social welfare in the process of participating in the market, so that the main body of integrated energy can be motivated to improve the energy service level, optimize the system operation plan and upgrade, and invest in lower-carbon and efficient equipment good positive cycle. The high proportion of low-carbon transition and low-carbon technical indicators reflects higher energy coupling efficiency, stronger energy space–time translation capability, and lower-carbon and efficient equipment, which can minimize primary energy consumption and build a green energy consumption model to improve the system low-carbon capacity. The market operation indicator is the embodiment of the system's low-carbon capability in the market transaction mechanism, which can reflect the relationship between the system's new energy output, supply and demand, and energy prices in the energy market. A reasonable market transaction mechanism can promote the system's low-carbon capability improve. Each weight is consistent with the actual low-carbon performance of the system, which also verifies the scientificity and rationality of the indicators proposed in this paper.

Table 5. Calculation results of three-level indicator weights.

Indicator Number	Type	Subjective Weight	Objective Weight	Comprehensive Weight
S1-1	benefit-type	0.0892	0.0397	0.0737
S1-2	benefit-type	0.0230	0.0369	0.0769
S1-3	benefit-type	0.0067	0.0386	0.0289
S2-1	benefit-type	0.0173	0.0781	0.0872
S2-2	benefit-type	0.0552	0.0826	0.0704
S3-1	benefit-type	0.0222	0.1054	0.0775
S3-2	benefit-type	0.1081	0.1181	0.1104
S4-1	cost-type	0.0196	0.0830	0.0605
S4-2	benefit-type	0.0589	0.0786	0.0399
S5-1	benefit-type	0.1172	0.0384	0.0835
S5-2	intermediate-type	0.0245	0.0599	0.0423
S5-3	benefit-type	0.0710	0.0439	0.0518
S6-1	benefit-type	0.0556	0.0794	0.0592
S6-2	benefit-type	0.3339	0.1185	0.1571

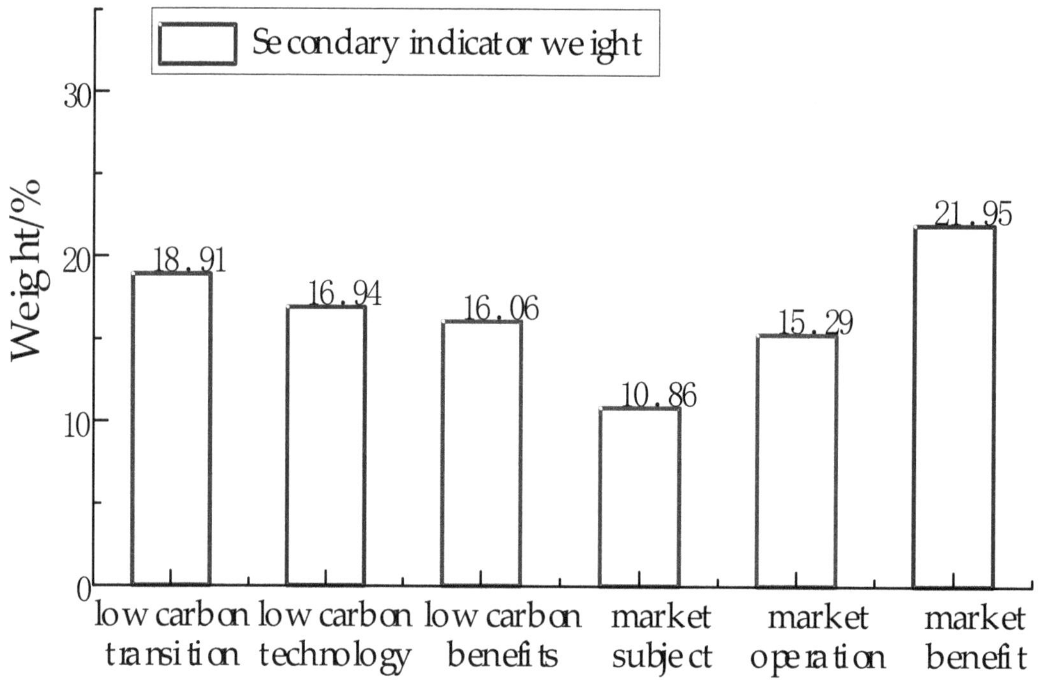

Figure 8. Weight distribution of secondary indicators.

5.4. Analysis of Evaluation Results

Based on the index calculation results calculated above, the comprehensive evaluation results of the six schemes and the second-level index evaluation results are shown in Table 6 and Figure 9.

Table 6. Valuation results and ranking.

Scheme	1	2	3	4	5	6
Fraction	2.186	2.4305	2.3268	2.601	2.516	3.037
sort	6	4	5	2	3	1

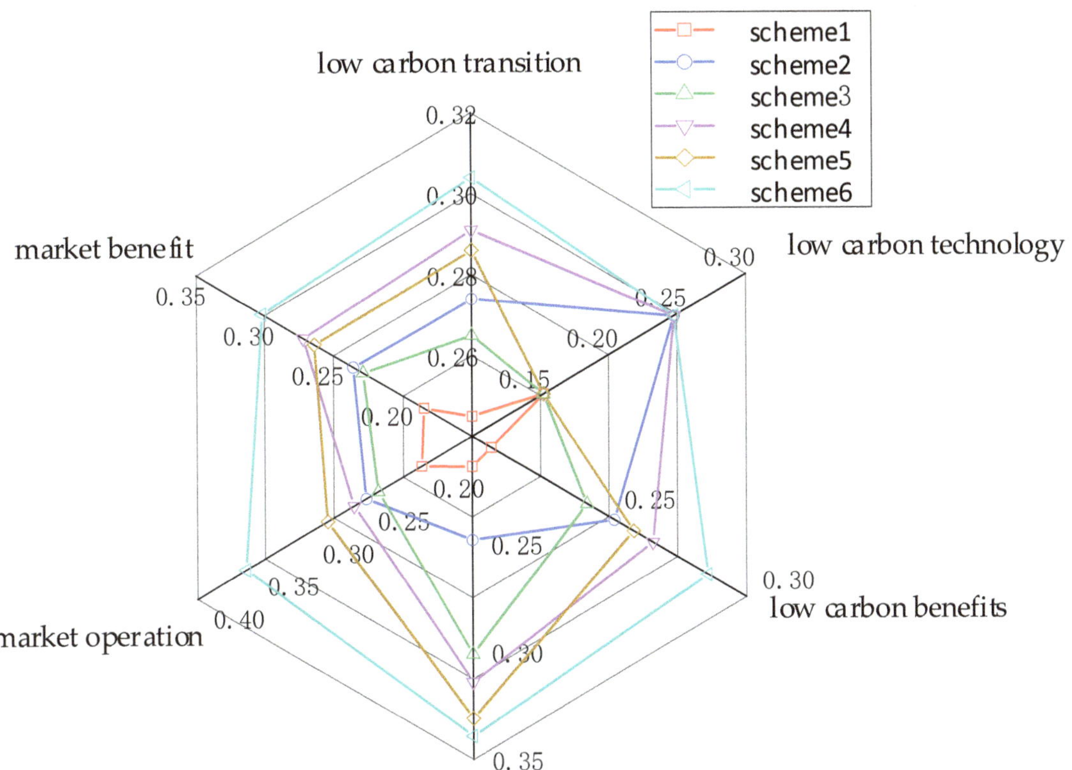

Figure 9. Radar chart of secondary index evaluation results.

It can be seen from Table 6 and Figure 9 that the comprehensive evaluation results under different market schemes are, from high to low, Scheme 6, Scheme 4, Scheme 5, Scheme 2, Scheme 3, Scheme 1. Among the six schemes, Scheme 1 cannot conduct energy interaction with the energy market, and due to the small energy storage capacity and insufficient energy translation capability, the phenomenon of energy abandonment is relatively serious. Therefore, the low-carbon capacity evaluation result is the lowest. Scheme 5 has a higher degree of market openness, so the market subject indicator it has a higher score, but due to the small capacity of new energy and energy storage equipment and serious energy abandonment, the evaluation results of low-carbon technology and market operation are low. So, the final evaluation result is in the third place. In contrast, Scheme 4 has a high proportion of new energy and energy storage, which can promote the coupling efficiency and space–time translation capability of different energy sources in the system, so that low-carbon transition, low-carbon technology, and market operations have high scores. So, it comes in second. For Scheme 6, it can participate in the competition in the energy market and sell the new energy that cannot be absorbed in the region in the energy market to reduce the occurrence of energy waste in the system. Therefore, the evaluation result is the best. For CEP1 and CEP2 from Scheme A to Scheme C, the low-carbon capacity assessment results increased by 15.08% and 24.9%, respectively.

5.5. Comparative Analysis of Evaluation Methods

In order to verify the effectiveness and superiority of the ANP-CRITIC method, the original data were compared with three traditional methods of the fuzzy analytic hierarchy process (Fuzzy-AHP) [28], entropy weight method (EWM) [29], and AHP–antientropy weight method (AHP-AEWM) [30]. The final comprehensive evaluation results are shown in Table 7.

Table 7. Results and rankings under different methods.

Evaluation Methodology	Fraction/Sort											
	1		2		3		4		5		6	
Fuzzy-AHP	5.424	6	6.361	4	6.184	5	6.764	2	6.657	3	7.154	1
EWM	0.473	6	0.697	4	0.618	5	0.748	3	0.794	2	0.869	1
AHP-AEWM	0.846	6	1.268	4	1.195	5	1.593	2	1.386	3	1.895	1
ANP-CRITIC	2.1862	6	2.4305	4	2.3268	5	2.601	2	2.516	3	3.037	1

It can be seen from Table 7 that the results obtained by the other methods, except EWM, are the same, which also verifies the effectiveness of the evaluation model proposed in this paper. The EWM relies too much on objective indicator data. Although it can reflect the correlation between indicators, it ignores the guiding role of decision-makers in the low-carbon development of CRIES, which leads to deviations in the evaluation results. Although the Fuzzy-AHP and AHP-AEWM are the same as the comprehensive evaluation results of the method proposed in this paper, the Fuzzy-AHP is too much affected by the subjective factors of decision-makers and cannot reflect the objective impact of system data on the evaluation of CRIES low carbon capacity in the energy market, which will adversely affect the final evaluation result. Although the AHP-AEWM considers both subjective and objective factors, it lacks the consideration of the correlation between the indicators. The method proposed in this paper makes up for the shortcomings of traditional methods, and can obtain more detailed and comprehensive scientific evaluation results for the CRIES low-carbon capability evaluation model in the energy market.

6. Conclusions

This paper fully considers the impact of the energy market on the low-carbon capability of CRIES and takes into account the characteristics of low-carbon development and energy market operation to construct a low-carbon capability evaluation system for CRIES under the energy market. The ANP-CRITIC comprehensive empowerment method is used to evaluate the low-carbon capacity of CRIES under six different schemes. The following conclusions are drawn from the analysis:

(1) The low-carbon capacity evaluation system and comprehensive empowerment method constructed in this paper can quantitatively analyze and compare the low-carbon capacity of the CRIES, and then provide the system construction and policies for the CRIES to participate in the energy market under the low-carbon target. It provides a useful reference for the formulation and improvement of market rules.

(2) Improving the installed capacity and consumption level of new energy, promoting the coupling efficiency and translation ability of different energy sources in the system, and a more open energy market are the key factors for improving the low-carbon capability of the CRIES.

(3) Establishing a fairer market transaction mechanism and taking CRIES as a participant in the energy market to participate in the energy market competition rather than just the role of energy receiver. It can enable CRIES to obtain more market benefits in the energy market. Promoting the low-carbon upgrade of CRIES equipment has a positive effect on the low-carbon development of CRIES.

This paper establishes the CRIES low-carbon capability index under the current energy market. However, with the continuous development of the CRIES and the advancement of energy marketization, it is necessary to continuously improve and refine the index system for the evaluation of the CRIES low-carbon capability in the energy market environment in the future.

Author Contributions: Conceptualization, Z.Y. and X.W.; methodology, X.W.; software, Z.Y.; validation, X.W.; investigation, Z.Y.; resources, X.W.; writing—original draft preparation, Z.Y.; writing—review and editing, X.W.; supervision, X.W. All authors have read and agreed to the published version of the manuscript.

Funding: This research was supported by Natural Science Foundation of Xinjiang Uygur Autonomous Region under Grant 2020D01C031.

Institutional Review Board Statement: Not applicable.

Informed Consent Statement: Not applicable.

Data Availability Statement: The data presented in this study are available on request from the corresponding author. The data are not publicly available due to data confidentiality requirements.

Conflicts of Interest: The authors declare no conflict of interest.

Nomenclature

City Regional Integrated Energy System (CRIES); Regional Integrated Energy System (RIES); analytic network process–criteria importance through intercriteria correlation (ANP-CRITIC); comprehensive energy producers (CEPs); Power Trading Center (PTC); Power Generator (PG); City Regional Integrated Energy Trading Centre (CRIETC); Electricity Retailer (ER); Natural Gas Trading Centers (NGTC); Natural Gas Producers (NGP); Natural Gas Retailers (NGR); gas boiler (GB); combined heat and power (CHP); wind turbine (WT); Energy Storage Systems (ESS); photovoltaic (PV); vapor-driven absorption refrigerating machine (VAR); Herfindahl–Hirschman Index (HHI); fuzzy analytic hierarchy process (Fuzzy-AHP); entropy weight method (EWM); analytic hierarchy process–antientropy weight method (AHP-AEWM).

References

1. Yang, Y.; Luo, Z.; Yuan, X.; Lv, X.; Liu, H.; Zhen, Y.; Yang, J.; Wang, J. Bi-Level Multi-Objective Optimal Design of Integrated Energy System Under Low-Carbon Background. *IEEE Access* **2021**, *9*, 53401–53407. [CrossRef]
2. Yang, L.; Huang, W.; Guo, C.; Zhang, D.; Xiang, C.; Yang, L.; Wang, Q. Multi-Objective Optimal Scheduling for Multi-Renewable Energy Power System Considering Flexibility Constraints. *Processes* **2022**, *10*, 1401. [CrossRef]
3. Miao, B.; Lin, J.; Li, H.; Liu, C.; Li, B.; Zhu, X.; Yang, J. Day-Ahead Energy Trading Strategy of Regional Integrated Energy System Considering Energy Cascade Utilization. *IEEE Access* **2020**, *8*, 138021–138035. [CrossRef]
4. Cheng, H.; Hu, X.; Wang, L.; Liu, Y.; Yu, Q. Summary of Research on Regional Integrated Energy System Planning. *Autom. Electr. Power Syst.* **2019**, *43*, 2–13.
5. Xiang, E.; Gao, H.; Liu, C.; Liu, Y.; Liu, J. Optimal Decision of Energy Trading for Community Multi-energy Operator Based on Game Interaction With Supply and Demand Sides. *Proc. CSEE* **2021**, *41*, 2744–2757.
6. Shubhangi, M.; Gaurav, S.; Saikat, S.; Anurag, C.; Anuj, K.; Santanu, M. A survey on multi-criterion decision parameters, integration layout, storage technologies, sizing methodologies and control strategies for integrated renewable energy system. *Sustain. Energy Technol. Assess.* **2022**, *52*, 102246.
7. Lv, J.; Zhang, C.; Cheng, H.; Han, F.; Yuan, K.; Song, Y.; Fang, S. Review on District-level Integrated Energy System Planning Considering Interconnection and Interaction. *Proc. CSEE* **2021**, *41*, 4001–4021.
8. Peng, C.; Zhang, H.; Sun, H.; Xu, R. Equilibrium bidding strategy for multiple energy supply entities in the comprehensive energy market under the carbon trading mechanism. *Power Syst. Technol.* **2022**, *46*, 463–471.
9. Wu, B.; Zhang, S.; Wang, Y.; Liu, S. Research on the Demand Response Strategy of Integrated Energy Service Providers in a Multi-Energy Market Environment. *Power Syst. Technol.* **2022**, *46*, 1800–1811.
10. Han, Z.; Zhou, L.; Liu, S.; Li, G.; Fu, X. A day-ahead market declaration strategy for integrated energy systems considering uncertainty. *Electr. Power Constr.* **2021**, *42*, 121–131.
11. Xiong, Z.; Luo, S.; Wang, L.; Jiang, C.; Zhou, S.; Gong, K. Bi-level optimal low-carbon economic operation of regional integrated energy system in electricity and natural gas markets. *Front. Energy Res.* **2022**, *10*, 959201. [CrossRef]
12. Feng, J.; Nan, J.; Wang, C.; Sun, K.; Deng, X.; Zhou, H. Source-Load Coordinated Low-Carbon Economic Dispatch of Electric-Gas Integrated Energy System Based on Carbon Emission Flow Theory. *Energies* **2022**, *15*, 3641. [CrossRef]
13. Luo, Z.; Wang, J.; Xiao, N.; Yang, L.; Zhao, W.; Geng, J.; Lu, T.; Luo, M.; Dong, C. Low Carbon Economic Dispatch Optimization of Regional Integrated Energy Systems Considering Heating Network and P2G. *Energies* **2022**, *15*, 5494. [CrossRef]

14. Wang, R.; Wen, X.; Wang, X.; Fu, Y.; Zhang, Y. Low carbon optimal operation of integrated energy system based on carbon capture technology, LCA carbon emissions and ladder-type carbon trading. *Appl. Energy* **2022**, *311*, 118664. [CrossRef]
15. Chen, B.; Liao, Q.; Liu, D.; Wang, W. Comprehensive evaluation indicators and methods of regional integrated energy systems. *Autom. Electr. Power Syst.* **2018**, *42*, 174–182.
16. Berjawi, A.; Walker, S.; Patsios, C.; Hosseini, S. An evaluation framework for future integrated energy systems: A whole energy systems approach. *Renew. Sustain. Energy Rev.* **2021**, *145*, 111163. [CrossRef]
17. Hu, M.; Gao, H.; Wang, C.; He, S.; Cai, W.; Yang, J. Multi-party co-governance decision-making method for multi-microgrid coordination transactions. *Autom. Electr. Power Syst.* **2022**, *46*, 30–42.
18. Wang, S.; Zhang, S.; Cheng, H.; Yuan, K.; Song, Y.; Han, F. Reliability Index and Evaluation Method of Integrated Energy System Considering User Thermal Comfort. *Autom. Electr. Power Syst.* **2022**. Available online: https://kns.cnki.net/kcms/detail/32.1180.TP.20220211.1725.004.html (accessed on 20 May 2022).
19. Jiang, J.; Yu, H.; Song, G.; Zhao, J.; Zhao, K.; Ji, H.; Li, P. Surrogate model assisted multi-criteria operation evaluation of community integrated energy systems. *Sustain. Energy Technol. Assess.* **2022**, *53*, 102656. [CrossRef]
20. Huang, D.; Wang, D.; Jia, H.; Hu, Q.; Chen, J.; Li, J. Energy retail market double auction clearing strategy for interconnected region. *Power System Protection and Control* **2022**, *5*, 1–12.
21. Tahir, M.; Chen, H.; Han, G. Evaluating individual heating alternatives in integrated energy system by employing energy and exergy analysis. *Energy* **2022**, *249*, 123753. [CrossRef]
22. Wang, Y.; Huang, F.; Tao, S.; Ma, Y.; Ma, Y.; Liu, L.; Dong, F. Multi-objective planning of regional integrated energy system aiming at exergy efficiency and economy. *Appl. Energy* **2022**, *306*, 118120. [CrossRef]
23. Mastura, M.; Nadlene, R.; Jumaidin, R.; Kudus, S.; Firdaus, H. Concurrent Material Selection of Natural Fibre Filament for Fused Deposition Modeling Using Integration of Analytic Hierarchy Process/Analytic Network Process. *J. Renew. Mater.* **2022**, *10*, 1221–1238. [CrossRef]
24. Paweł, Z. Inter-Criteria Dependencies-Based Decision Support in the Sustainable wind Energy Management. *Energies* **2019**, *12*, 749.
25. Shi, H.; Li, Y.; Jiang, Z.; Zhang, J. Comprehensive power quality evaluation method of microgrid with dynamic weighting based on CRITIC. *Meas. Control* **2021**, *54*, 5–6. [CrossRef]
26. Chen, B.; Wang, C.; Wu, C.; Liu, N.; He, S. Equilibrium analysis of electricity-heat integrated energy market based on joint clearing mechanism. *Power Syst. Technol.* **2020**, *44*, 2848–2858.
27. Hu, X.; Shang, C.; Chen, D.; Wang, L.; Zhang, F.; Cheng, H. A multi-objective planning method for regional integrated energy systems considering energy quality. *Autom. Electr. Power Syst.* **2019**, *43*, 19.
28. Wang, Z.; Li, B.; Jiang, T.; Chen, H.; Li, X.; Liu, F. Screening and sorting of AC faults in AC-DC receiving-end power grid based on improved fuzzy -AHP. *Power Syst. Technol.* **2021**, *45*, 4047–4056.
29. Lin, H.; Du, L.; Liu, Y. Soft Decision Cooperative Spectrum Sensing with Entropy Weight Method for Cognitive Radio Sensor Networks. *IEEE Access* **2020**, *8*, 1. [CrossRef]
30. Xiang, S.; Cai, Z.; Liu, P.; Li, L. Fuzzy comprehensive evaluation of low-carbon operation of distribution network based on AHP-anti-entropy weight method. *J. Electr. Power Sci. Technol.* **2019**, *34*, 69–76.

processes

Article

Virtual Voltage Vector-Based Model Predictive Current Control for Five-Phase Induction Motor

Qingfei Zhang, Jinghong Zhao, Sinian Yan *, Yiyong Xiong, Yuanzheng Ma and Hansi Chen

School of Electrical Engineering, Naval University of Engineering, Wuhan 430033, China
* Correspondence: ysnian0504@126.com

Abstract: The high-performance control technology of multi-phase motors is a key technology for the application of multi-phase motors in many fields, such as electric transportation. The model predictive current control (MPCC) strategy has been extended to multi-phase systems due to its high dynamic performance. Model-predictive current control faces the problem that it cannot effectively regulate harmonic plane currents, and thus cannot obtain high-quality current waveforms because only one switching state is applied in a sampling period. To solve this problem, this paper uses the virtual vector-based MPCC to select the optimal virtual vector and apply it under the premise that the average value of the harmonic plane voltage in a single switching cycle is zero. Taking a five-phase induction motor as an example, the steady-state and dynamic performance of the proposed virtual vector MPCC and the traditional model predictive current control were simulated, respectively. Simulation results demonstrated the effectiveness of the proposed method in improving waveform quality while maintaining excellent dynamic performance.

Keywords: five-phase induction motor; model predictive current control; virtual voltage vectors

Citation: Zhang, Q.; Zhao, J.; Yan, S.; Xiong, Y.; Ma, Y.; Chen, H. Virtual Voltage Vector-Based Model Predictive Current Control for Five-Phase Induction Motor. *Processes* **2022**, *10*, 1925. https://doi.org/10.3390/pr10101925

Academic Editors: Haoming Liu, Jingrui Zhang and Jian Wang

Received: 29 July 2022
Accepted: 19 September 2022
Published: 23 September 2022

Publisher's Note: MDPI stays neutral with regard to jurisdictional claims in published maps and institutional affiliations.

1. Introduction

Due to the massive consumption of fossil energy and the increasingly serious air pollution problems caused by the exhaust emissions of fuel vehicles, new energy systems focusing on green and low-carbon development have achieved great progress. Vehicles utilizing new forms of energy with low pollution emissions are an important application of new energy systems in transportation; as an example, pure electric vehicles and hybrid vehicles are gradually gaining favor in the market [1,2]. The motor is a key component of new energy vehicles; excellent motor control technology directly affects the operation and safety performance of electric vehicles, so a high-performance motor and its control technology are very important for the development of electric vehicles [3,4]. Usually, variable-speed AC drives are powered by power electronic converters, but with the development of power electronic technology, the phase number of the motor is no longer restricted by the traditional three-phase power supply. The multi-phase motor drive system has received a lot of attention and research in the fields of ship power propulsion, aerospace and so on [5,6]. Compared with the traditional three-phase system, the multi-phase motor drive system has the following advantages: (1) high reliability and strong fault tolerance [7,8]; (2) low voltage and high power drive can be realized by the reduction of per-phase current [9]; and (3) low torque ripple reduces vibration and noise [10]. Among the various multi-phase drive system solutions, the five-phase system is a typical representative, which has the typical advantages of the multi-phase system and a relatively simple structure [11].

Model predictive control (MPC) first appeared in the 1960s and was applied in the chemical industry. With the rapid development of microprocessors, MPC schemes have been extended to the fields of power electronics and have been widely researched and applied [12,13]. As one of the most commonly used schemes, finite-control-set model predictive control (FCS-MPC) has the advantages of simple structure, fast dynamic response, simple concept, and easy handling of nonlinear multivariable constraints [14,15]. FCS-MPC uses the discrete characteristics of the inverter output to enumerate all the alternative voltage vectors and select the optimal voltage vector as the output by minimizing the cost function, eliminating the need for pulse width modulation [16,17]. For motor systems, FCS-MPC can be further divided into two categories, namely model-predictive current control (MPCC) and model-predictive torque control (MPTC). Since the cost function in the model-predicted torque includes two variables with different physical units, torque and flux linkage, it is difficult to select the weight coefficients [18]. Therefore, the MPCC scheme is more widely used.

Due to its high dynamic performance and flexibility to handle constraints, MPCC is replacing the most commonly used field-oriented control (FOC) in the field of motor control [19]. MPCC has been successfully extended from the three-phase motor drive system to the multiphase motor drive system, which proves the feasibility of model predictive current control in the multiphase system [20,21]. However, in the MPCC scheme of multiphase motors, it is often faced with the problem that too many alternative vectors lead to the complexity of the algorithm and an increase of the calculation amount. At the same time, due to the use of a single inverter-switching state, generating voltage in the harmonic plane of the multiphase motor is inevitable, so as to bring harmonic current, resulting in excess harmonic loss and affecting the performance of the system [22,23].

In order to solve the problem of the large computational load of model predictive control in multiphase systems, a simplified algorithm using adjacent voltage vectors was proposed in the literature [24], which reduced the number of alternative voltage vectors and was applied in multilevel cascade inverters. The literature [25] limited the number of switches in adjacent sampling periods in advance, so as to exclude some voltage vectors with a large number of switches and narrow the range of voltage vector candidates. In order to reduce the harmonic current of the multi-phase motor, the traditional MPC uses a cost function including two plane current tracking errors, but this method introduces a weighting factor, and the selection of the weighting factor is very troublesome and the harmonic current suppression effect is relatively general [15,22].

In order to eliminate the harmonic current of the multiphase motor from the source, this paper uses virtual voltage vectors (VVs) to solve this problem. The basic voltage vector is synthesized according to the fixed ratio of VVs, and the VVs are used as the set of selected vectors to carry out the model predictive current control of five-phase induction motor. In the literature [26], large vectors and medium vectors were used to synthesize virtual vectors and construct switching tables, and voltage vectors were selected for direct torque control according to torque, flux error and speed range. The literature [27] redefined a new virtual vector to realize fault-tolerant operation of the five-phase induction motor on the premise of open-circuit induction motor. The literature also [28,29] introduced the concept of virtual vectors into the model predictive current control of the six-phase induction motor, and proved that the use of virtual vectors can effectively reduce the loss and improve the quality of current waveform.

The remaining chapters are organized as follows. In the second part, the model of the five-phase induction motor and the two-level five-phase voltage source inverter is introduced, and the distribution characteristics of the basic voltage vector are given. The third part introduces the traditional model predictive current control scheme. The fourth part explains the process of virtual vector synthesis and implementation, and introduces the virtual vector MPC scheme proposed in this paper in detail. In the fifth part, the simulation results and analysis are presented. Finally, the sixth part gives the conclusion.

2. Five-Phase Induction Motor Drive

2.1. Five-Phase Induction Motor Model

Before using the model predictive current control strategy for the five-phase induction motor, it is necessary to establish the mathematical model of the five-phase inverter and the five-phase induction motor. The topology of the five-phase voltage source inverter and the five-phase induction motor connection system is shown in Figure 1. The motor windings are star-connected and the DC side voltage V_{dc} is considered constant.

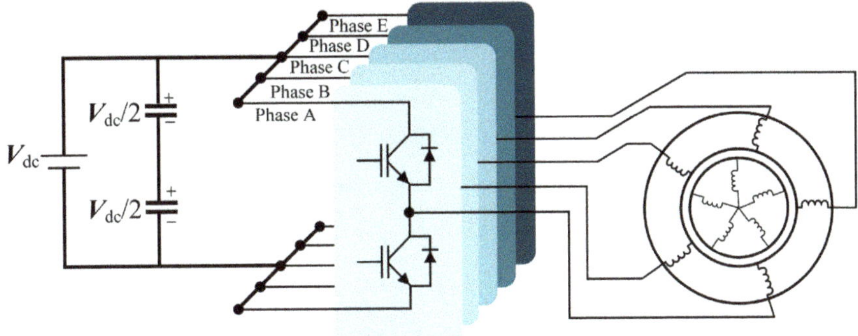

Figure 1. Diagram of five-phase induction motor system.

Applying the decoupling transformation matrix of the five-phase system defined in (1), variables in the stationary coordinates of five-phase can be transformed into three normal subspaces in (2). They are the α-β subspace (fundamental subspace), the x-y subspace (third harmonic subspace), and the o1-o2 subspace (zero-sequence harmonic subspace), respectively:

$$T = \sqrt{\frac{2}{5}} \begin{bmatrix} 1 & \cos\alpha & \cos 2\alpha & \cos 3\alpha & \cos 4\alpha \\ 0 & \sin\alpha & \sin 2\alpha & \sin 3\alpha & \sin 4\alpha \\ 1 & \cos 3\alpha & \cos 6\alpha & \cos 9\alpha & \cos 12\alpha \\ 0 & \sin 3\alpha & \sin 6\alpha & \sin 9\alpha & \sin 12\alpha \\ 1 & 1 & 1 & 1 & 1 \end{bmatrix} \tag{1}$$

$$\begin{bmatrix} x_\alpha & x_\beta & x_x & x_y & x_o \end{bmatrix}^T = [T] \cdot \begin{bmatrix} x_1 & x_2 & x_3 & x_4 & x_5 \end{bmatrix}^T \tag{2}$$

where $\alpha = 2\pi/5$.

Therefore, the control of the five-phase motor can be realized through the control of these two two-dimensional spaces. The five-phase system includes four degrees of freedom and zero-sequence components. For a five-phase induction motor, the α-β subspace corresponds to the first and second rows of the decoupling matrix, x-y subspace corresponds to the third and fourth rows of the decoupling matrix, and the o1-o2 subspace corresponds to the last row of the decoupling matrix, respectively. These subspaces have the following characteristics:

(1) The three subspaces are orthogonal subspaces to each other.

(2) The α-β subspace generates rotating magnetomotive force, and the fundamental wave and $(10n \pm 1)$ harmonics in the magnetic flux of the motor winding are projected into this space, which is the electromechanical energy conversion subspace.

(3) The motor variable in the x-y subspace is the $(10n \pm 3)$ harmonic in the motor winding flux. This harmonic component does not generate rotating magnetic motive force and does not participate in electromechanical energy conversion, which will generate harmonic current and cause harmonic loss.

(4) The $5k$ ($k = 1, 3, 5\ldots$) harmonics in the motor variables are projected into the zero-sequence harmonic subspace, which is usually negligible for a five-phase induction motor.

After decoupling transformation, the stator voltage equation of the five-phase induction motor is:

$$v_{s\alpha} = (R_s + L_s \tfrac{d}{dt})i_{s\alpha} + L_m \tfrac{di_{r\alpha}}{dt}$$
$$v_{s\beta} = (R_s + L_s \tfrac{d}{dt})i_{s\beta} + L_m \tfrac{di_{r\beta}}{dt}$$
$$v_{sx} = (R_s + L_{ls} \tfrac{d}{dt})i_{sx}$$
$$v_{sy} = (R_s + L_{ls} \tfrac{d}{dt})i_{sy}$$
(3)

The rotor voltage equation is:

$$0 = (R_r + L_r \tfrac{d}{dt})i_{r\alpha} + L_m \tfrac{di_{s\alpha}}{dt} + \omega_r L_r i_{r\beta} + \omega_r L_m i_{s\beta}$$
$$0 = (R_r + L_r \tfrac{d}{dt})i_{r\beta} + L_m \tfrac{di_{s\beta}}{dt} - \omega_r L_r i_{r\alpha} - \omega_r L_m i_{s\alpha}$$
(4)

The torque equation is:

$$T_e = n_p L_m (i_{s\beta} i_{r\alpha} - i_{s\alpha} i_{r\beta})$$
(5)

where L_m is the equivalent mutual inductance between stator and rotor windings, and L_s and L_r are the equivalent self-inductance of the stator winding and stator winding, respectively. L_{ls} is the stator leakage inductance. n_p is the number of pole pairs. ω_r is the rotor electrical speed. R_s, R_r are the stator resistor and the rotor resistor. The subscripts "s" and "r" represent the stator-side and rotor-side variables, respectively.

2.2. Voltage Vectors Distribution

The switch function can be defined as S_x, $x \in \{A,B,C,D,E\}$, where x = A,B,C,D,E represents the switching state of the inverter bridge arm, $S_x = 1$ represents that the upper arm of the bridge opens and the lower arm closes, while $S_x = 0$ represents that the lower arm opens and the upper arm closes. A total of $2^5 = 32$ switching states can be generated by the inverter phase number. Therefore, the stator phase voltage represented by switching state (S_x) and DC bus voltage (V_{dc}) is:

$$
\begin{bmatrix} U_{ao} \\ U_{bo} \\ U_{co} \\ U_{do} \\ U_{eo} \end{bmatrix}
= \frac{1}{5} V_{dc}
\begin{bmatrix}
4 & -1 & -1 & -1 & -1 \\
-1 & 4 & -1 & -1 & -1 \\
-1 & -1 & 4 & -1 & -1 \\
-1 & -1 & -1 & 4 & -1 \\
-1 & -1 & -1 & -1 & 4
\end{bmatrix}
\cdot
\begin{bmatrix} S_A \\ S_B \\ S_C \\ S_D \\ S_E \end{bmatrix}
$$
(6)

By the decoupling transformation matrix (1), the stator phase voltage is mapped to two orthogonal subspaces and an additional zero-order subspace, and the voltage vector can be written as a switching state in two separate spaces:

$$V_{\alpha\text{-}\beta} = V_\alpha + jV_\beta = \tfrac{2}{5}V_{dc}(S_a + aS_b + a^2 S_c + a^3 S_d + a^4 S_e)$$
$$V_{x\text{-}y} = V_x + jV_y = \tfrac{2}{5}V_{dc}(S_a + aS_c + a^2 S_e + a^3 S_b + a^4 S_d)$$
(7)

where $a = \exp(j\pi 2/5)$.

The distribution of 32 voltage vectors generated by the five-phase voltage source inverters in these two subspaces is shown in Figure 2.

These 32 voltage vectors consist of 2 zero vectors and 30 non-zero vectors, which can be grouped into three groups based on the magnitude of the voltage vector in the fundamental wave space: large vectors, medium vectors and small vectors. The specific groupings are shown in Table 1. The amplitudes of each group of voltage vectors in the fundamental wave space are: $0.6472\,V_{dc}$, $0.4\,V_{dc}$, $0.2472\,V_{dc}$. Therefore, the voltage vector in α-β subspace can be divided into three decagons. From the outside to the inside, the ratio of the radius amplitudes of two adjacent decagons is 1.618.

Table 1. Space vector and vorresponding switching states.

Group	Switching States
Large	V_{25} (11001) V_{24}(11000) V_{28} (11100) V_{12} (01100) V_{14} (01110) V_6 (00110) V_7 (00111) V_3 (00011) V_{19} (10011) V_{17} (10001)
Medium	V_{16} (10000) V_{29} (11101) V_8 (01000) V_{30} (11110) V_4 (00100) V_{15} (01111) V_2 (00010) V_{23} (10111) V_1 (00001) V_{27} (11011)
Small	V_9 (01001) V_{26} (11010) V_{20} (10100) V_{13} (01101) V_{10} (01010) V_{22} (10110) V_5 (00101) V_{11} (01011) V_{18} (10010) V_{21} (10101)

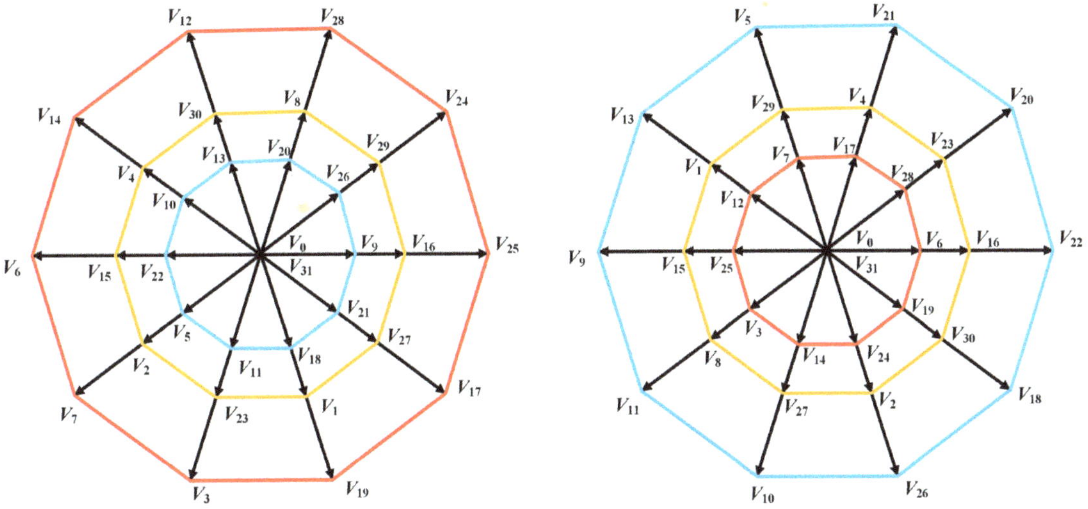

(a)

(b)

Figure 2. Voltage space vectors in two subspaces. (**a**) α-β subspace. (**b**) x-y subspace.

3. Traditional FCS-MPCC Scheme

3.1. Prediction Model of Induction Motor

In general, MPCC controls the stator current in the d-q coordinate system [22,30,31], but considering that there are 32 alternative voltage vectors for the two-level five-phase inverter, if the control is still carried out in the d-q coordinate system, there are 32 candidate voltage vectors needed to carry out the rotation transformation into the d-q coordinate system. In order to reduce the number of rotation transformations, the control is carried out in the α-β two-phase stationary coordinate system [14,17]. It is necessary to transform the generated given current from the d-q coordinate system to the α-β two-phase stationary coordinate system, and the number of rotation transformations is reduced to only once. The structure of traditional FCS-MPCC scheme is shown in the Figure 3.

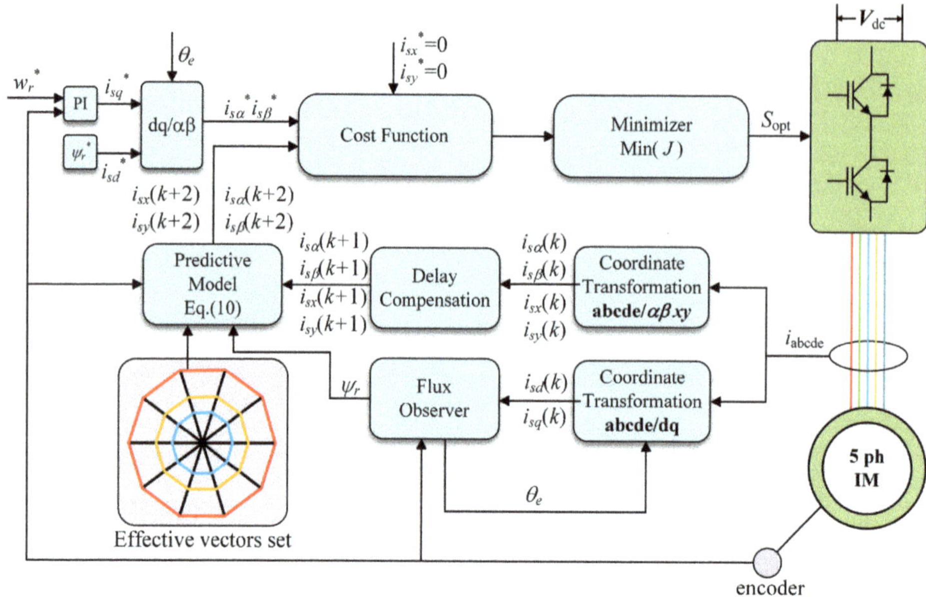

Figure 3. Traditional FCS-MPCC scheme for a five-phase IM (Induction Motor) drive.

In the α-β two-phase coordinate system, the stator current and rotor flux linkage of the motor are selected as state variables:

$$X_{\alpha\beta xy} = \begin{bmatrix} i_{s\alpha} & i_{s\beta} & i_{sx} & i_{sy} & \psi_{r\alpha} & \psi_{r\beta} \end{bmatrix}^T \tag{8}$$

where $i_{s\alpha}, i_{s\beta}$ are the α-β axis components of the stator phase current in the fundamental wave plane, i_{sx}, i_{sy} are the α-β axis components of the stator phase current in the harmonic plane, $\psi_{r\alpha}, \psi_{r\beta}$ are the α-β axis components of the rotor flux linkage in the stationary frame.

The input variables are the α-β axis components of the stator voltage in the fundamental wave plane and the α-β axis components in the harmonic plane:

$$U_{\alpha\beta xy} = \begin{bmatrix} u_{s\alpha} & u_{s\beta} & u_{sx} & u_{sy} \end{bmatrix}^T \tag{9}$$

The state equation of the motor can be obtained as follows:

$$p \cdot X_{\alpha\beta xy} = A \cdot X_{\alpha\beta xy} + B \cdot U_{\alpha\beta xy} \tag{10}$$

where

$$A = \begin{bmatrix} A_1 & 0 & 0 & 0 & A_2 & -A_3 \\ 0 & A_1 & 0 & 0 & A_3 & A_2 \\ 0 & 0 & -\frac{R_s}{L_{ls}} & 0 & 0 & 0 \\ 0 & 0 & 0 & -\frac{R_s}{L_{ls}} & 0 & 0 \\ \frac{L_m}{T_r} & 0 & 0 & 0 & -\frac{1}{T_r} & -\omega_r \\ 0 & \frac{L_m}{T_r} & 0 & 0 & \omega_r & -\frac{1}{T_r} \end{bmatrix} ; \quad B = \begin{bmatrix} \frac{1}{\sigma L_s} & 0 & 0 & 0 \\ 0 & \frac{1}{\sigma L_s} & 0 & 0 \\ 0 & 0 & \frac{1}{L_{ls}} & 0 \\ 0 & 0 & 0 & \frac{1}{L_{ls}} \\ 0 & 0 & 0 & 0 \\ 0 & 0 & 0 & 0 \end{bmatrix} ;$$

$$\sigma = 1 - \frac{L_m{}^2}{L_s L_r}; \quad A_1 = \frac{\sigma-1}{\sigma T_r} - \frac{1}{\sigma T_s}; \quad A_2 = \frac{1-\sigma}{\sigma L_m T_r}; \quad A_3 = \frac{(\sigma-1)\omega_r}{\sigma L_m}$$

In the formula: σ is the flux leakage factor, and T_r is the rotor time constant, $T_r = L_r/R_r$. The sampling period is selected as T_s, and the first-order Euler method is used to discretize the Equation (10). Since the electrical time constant is usually much smaller than the mechanical time constant, the motor speed is assumed to remain approximately constant during a single sampling period. The motor model is described by discrete state space equations:

$$
\begin{bmatrix} i_{s\alpha}(k+1) \\ i_{s\beta}(k+1) \\ i_{sx}(k+1) \\ i_{sy}(k+1) \\ \psi_{r\alpha}(k+1) \\ \psi_{r\beta}(k+1) \end{bmatrix} = (AT_s + I) \begin{bmatrix} i_{s\alpha}(k) \\ i_{s\beta}(k) \\ i_{sx}(k) \\ i_{sy}(k) \\ \psi_{r\alpha}(k) \\ \psi_{r\beta}(k) \end{bmatrix} + T_s B \cdot \begin{bmatrix} u_{s\alpha} \\ u_{s\beta} \\ u_{sx} \\ u_{sy} \end{bmatrix} \tag{11}
$$

where I is the sixth-order identity matrix.

3.2. Cost Function

MPCC needs to establish a cost function after obtaining the prediction model. In the cost function, not only the error between the α-β plane current reference value, but also the predicted value needs to be considered. At the same time, the current of the harmonic plane will cause unnecessary loss, so it needs to be suppressed, and therefore the reference value of the harmonic plane current is set to 0. Since the cost function includes the currents in both the fundamental plane and the harmonic plane, a weighting factor is introduced to represent the importance of the harmonic plane current tracking error relative to the fundamental plane current tracking error. However, there is no perfect theory for the selection of weight factors at present. In practice, it can be selected through multiple experiments, so as to choose appropriate weight factors to better balance the influence of fundamental wave plane current and harmonic plane current on the quality of current waveform. In the model predictive control, the switching state of the inverter finally selected has the smallest cost function; that is to say, the value of the cost function corresponding to different switching states directly determines what switching state will be selected. The established cost function is:

$$
J = \left[(i_{s\alpha}{}^* - i_{s\alpha})^2 + (i_{s\beta}{}^* - i_{s\beta})^2 \right] + W_{xy} * \left[(i_{sx}{}^* - i_{sx})^2 + (i_{sy}{}^* - i_{sy})^2 \right] \tag{12}
$$

where the superscript * of i represents the current reference value.

3.3. Delay Compensation

Ideally, the sampling and computation time is negligible, but in the actual system [32], since the current and rotational speed sampling and current prediction process take a certain time, the calculated optimal voltage vector cannot be used immediately at the current k time, but will be applied at the $k + 1$ time. Therefore, the predictive control needs to consider the delay effect; that is, after the sampling at the current time k is completed, based on the switch state used at the time k, calculate the current I $(k + 1)$ at the time $k + 1$ firstly. This current is used as the starting point for all switching states to be predicted, then the current at time $k + 2$ is predicted, and the optimal vector is selected to be used at time $k + 1$.

Therefore, in the actual system, the following discrete state equation is used as:

$$
X(k+2) = FX(k+1) + GU(k+1) \tag{13}
$$

where, the $F = AT_s + I$, $G = BT_s$.

The cost function at this time is:

$$
J = \left[(i_{s\alpha}{}^*(k+2) - i_{s\alpha}(k+2))^2 + (i_{s\beta}{}^*(k+2) - i_{s\beta}(k+2))^2 \right] + W_{xy} * \left[(0 - i_{sx}(k+2))^2 + (0 - i_{sy}(k+2))^2 \right] \tag{14}
$$

4. Proposed VV-MPCC Schemes

4.1. Prediction Model of Induction Motor

The prediction model can be obtained from the distribution of vectors in Figure 2.

The outermost large vector in the α-β subspace is mapped as the innermost small vector in the x-y subspace, and the median vector in the α-β subspace is still a median vector in the x-y subspace after mapping. However, the direction of small vectors and middle vectors mapped to x-y subspace are opposite to each other.

Since the third harmonic plane current is generated by the third harmonic plane voltage, if the third harmonic plane voltage can be eliminated, the third harmonic plane current can also be eliminated. Therefore, as long as the large vector and the medium vector in the same direction in the fundamental wave plane are combined by a certain ratio, the virtual voltage vector with an average harmonic voltage of 0 in the third harmonic plane can be obtained to achieve the effect of eliminating harmonic current. Figure 4 shows the synthesis process of virtual voltage vector VV$_1$ by taking medium vector V_{16} and large vector V_{25} as examples.

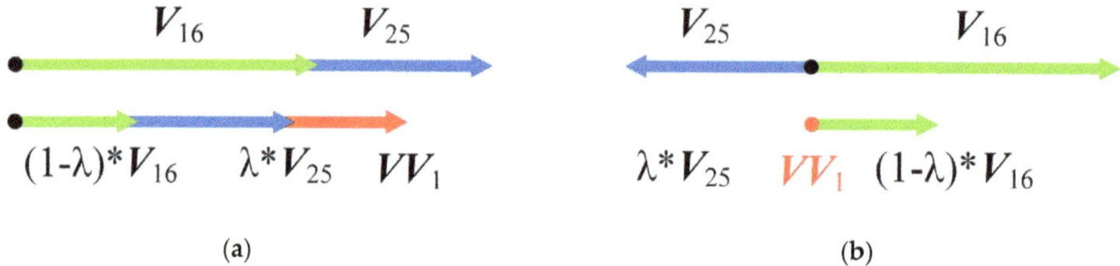

Figure 4. Schematic of virtual voltage vector synthesis (**a**) α-β subspace; (**b**) x-y subspace.

Select the large vector and the medium vector of the fundamental wave space, according to the principle of volt-second balance, the relationship satisfies

$$\begin{cases} 0.6472V_{dc}\lambda T_s + 0.4V_{dc}(1-\lambda)T_s = |V_{VL}|T_s \\ 0.6472V_{dc}(1-\lambda)T_s - 0.4V_{dc}\lambda T_s = 0 \end{cases} \tag{15}$$

In Equation (15), $|V_{VL}|$ represents the amplitude of the synthesized virtual voltage vector, and λT_s represents the action time of large vectors in a single period. From Equation (15), we could achieve:

$$\begin{cases} |V_{VL}| = 0.5527V_{dc} \\ \lambda = 0.618 \end{cases} \tag{16}$$

According to the above calculation time, the large vector and the medium vector are applied respectively in one cycle. In the fundamental wave space, the new voltage vector synthesized by the large vector and the medium vector in the same direction is called the virtual vector, whose direction is in the same direction with the large and medium vector, and the amplitude is $0.5527V_{dc}$ ($0.618 \times 0.6472V_{dc} + 0.382 \times 0.4V_{dc}$). However, compared with the traditional MPCC, the bus voltage utilization rate is reduced, and the speed regulation range is reduced. Figure 5 shows the distribution of virtual vectors in α-β subspace.

For any virtual vector, the ratio of action time between a large vector and medium vector is constant. No matter whether the virtual vector is synthesized in the order of first large vector followed by a medium vector, or first medium vector followed by a large vector, the waveform symmetry of the virtual vector cannot be guaranteed, so it is difficult to implement in hardware, and it will bring certain high-order harmonics. Therefore, under the condition that the average value of the output voltage remains unchanged, the effective pulse sequence is rearranged according to the principle of centrosymmetry. Figure 6 shows the pulse sequence of VV$_1$ before and after optimization according to the central symmetry principle, where $t_{25} = \lambda T_s$, $t_{16} = (1-\lambda)T_s$. That is, in a single cycle, the inverter first outputs the switching state of 10,000 with a duration of $t_{16}/2$, then outputs the switching state of 11,001 with a duration of t_{25}, and then outputs the switching state of 10,000 with a duration of $t_{16}/2$. The optimized pulse sequence of the remaining nine virtual vectors is similar to this.

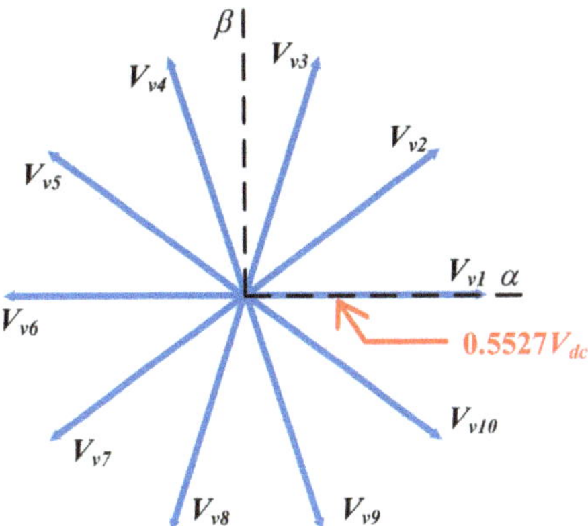

Figure 5. Virtual voltage vectors in α-β subspace.

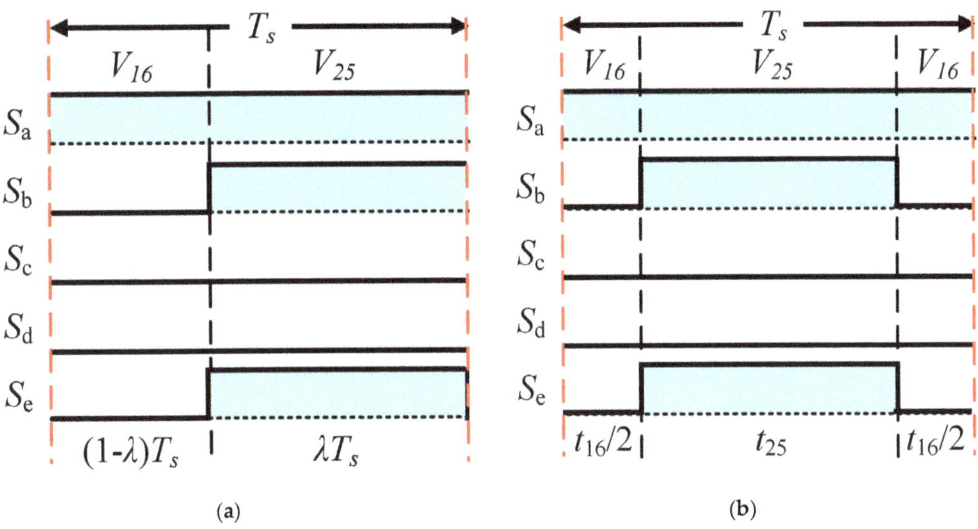

(a)   (b)

Figure 6. Schematic diagram of the virtual vector implementation method (VV_1). (**a**) Initial mode; (**b**) Symmetrical mode.

4.2. Cost Function Optimization

When applying virtual vector MPCC, because the third harmonic voltage has been eliminated theoretically, there is no need to predict the harmonic current in the prediction process, and there is no need to include the harmonic current term in the cost function. Therefore, the prediction model and the cost function are simplified. The simplified prediction block diagram is shown in Figure 7. The part in the red dotted line box in the figure shows the difference from the traditional MPCC. The simplified prediction model is:

$$p \cdot X_{\alpha\beta} = A \cdot X_{\alpha\beta} + B \cdot U_{\alpha\beta} \tag{17}$$

where

$$X_{\alpha\beta} = \begin{bmatrix} i_{s\alpha} & i_{s\beta} & i_{sx} & i_{sy} & \psi_{r\alpha} & \psi_{r\beta} \end{bmatrix}^T$$

$$U_{\alpha\beta} = \begin{bmatrix} u_{s\alpha} & u_{s\beta} & 0 & 0 \end{bmatrix}^T$$

Because the virtual vector is used as the alternative vector set, the number of voltage vectors is also reduced to 10, which reduces the computational burden. The simplified cost function is:

$$J = \left[(i_{s\alpha}{}^* - i_{s\alpha})^2 + (i_{s\beta}{}^* - i_{s\beta})^2 \right] \tag{18}$$

Because the average harmonic plane voltage value is 0 only in a single period, and the dead zone effect exists and the motor winding is impossible to make perfectly symmetric, the harmonic plane current will still exist, but the harmonic plane current value will be very small.

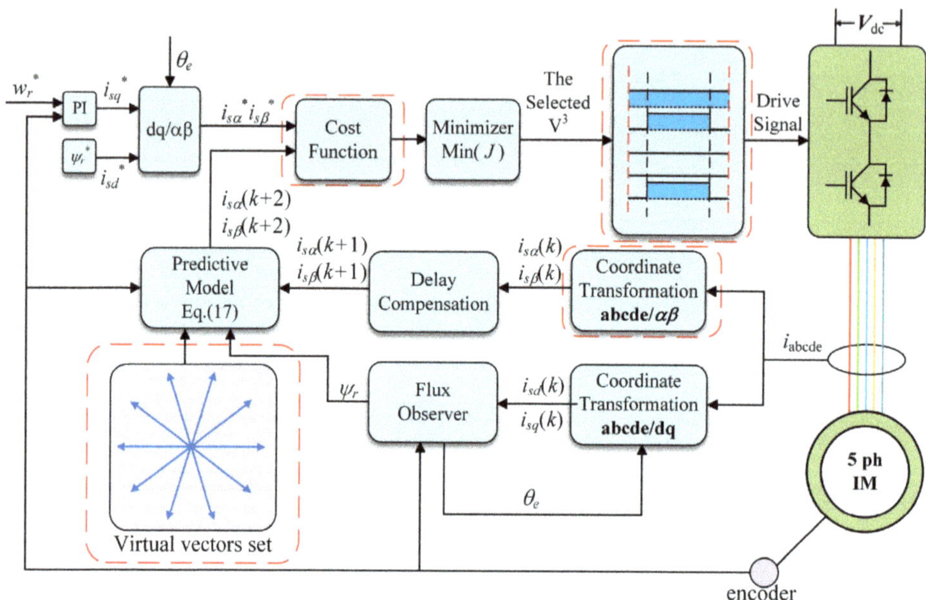

Figure 7. Proposed FCS-MPCC with VVs scheme for a five-phase IM drive.

5. Simulation Results

In order to prove the effectiveness of the proposed virtual vector MPCC, the motor model and its control system were built in MATLAB/SIMULINK, and the simulation comparison between the proposed method and the traditional MPCC was carried out. The traditional MPCC was written as T-MPC, and the proposed method was denoted as VV-MPC. The parameters of the five-phase induction motor used are shown in Table 2.

Table 2. Parameters of five-phase induction motor.

Parameter	Symbol/Unit	Value
Stator resistance	$R_s[\Omega]$	1.9
Rotor resistance	$R_r[\Omega]$	3.4
Stator leakage inductance	$L_{ls}[\text{mH}]$	35
Rotor leakage inductance	$L_{lr}[\text{mH}]$	20
Mutual inductance	$L_m[\text{mH}]$	530
Rotational inertia	$J[\text{kg·m}^2]$	0.04
Pole pairs	n_p	2
Rated speed	$[\text{r/min}]$	1500
Rated power	$[\text{kW}]$	2.2

5.1. Performance of T-MPC with Different Weighting Factors

The first set of simulations verified the effect of weighting factors on motor dynamic performance and harmonic current suppression. The preset speed of the motor was 500 r/min at 0 s, and the speed command suddenly changed to 1000 r/min at 0.5 s, running with no load. The left side of Figures 8–11 represent the operation when the weighting factor was 0.5, and the right side the operation when the weighting factor was 0.1.

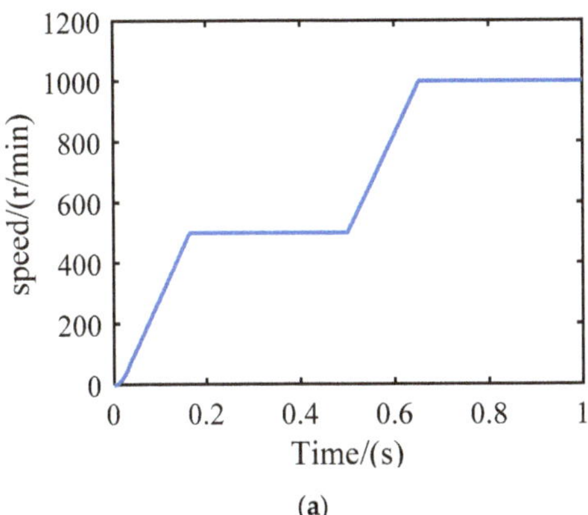

(a)

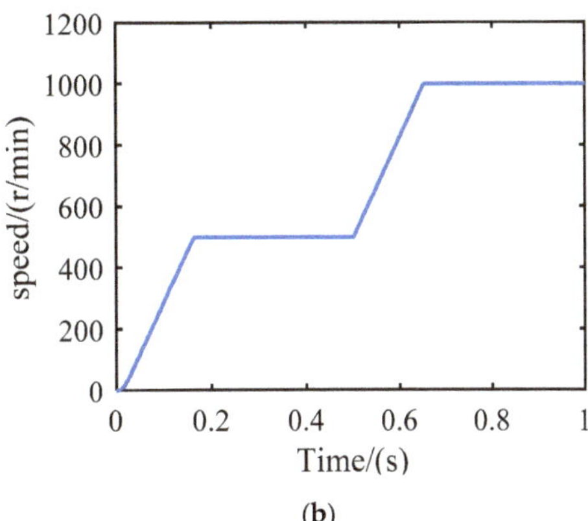

(b)

Figure 8. Speed performance of T-MPC scheme with different weighing factors at same sampling frequency (10 kHz). (**a**) $W_{xy} = 0.1$; (**b**) $W_{xy} = 0.5$.

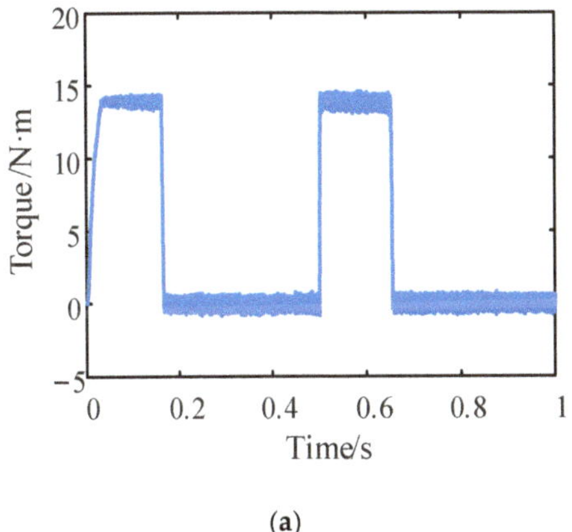

(a)

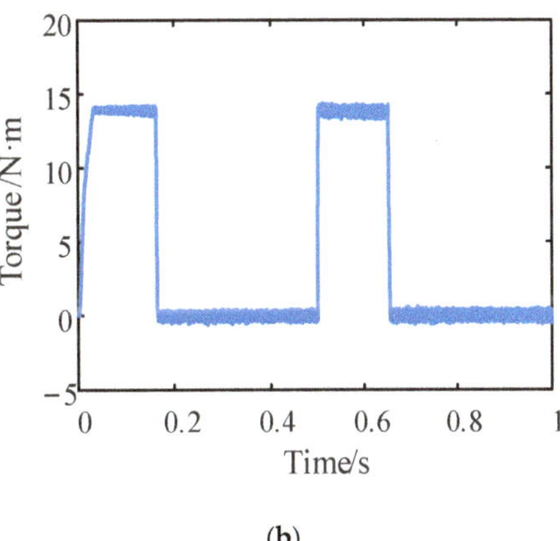

(b)

Figure 9. Torque performance of T-MPC scheme with different weighing factors at same sampling frequency (10 kHz). (**a**) $W_{xy} = 0.1$; (**b**) $W_{xy} = 0.5$.

Figure 8 shows that the dynamic performance of the motor is similar with different weighting factors. Figure 8 shows that the motor speed is accelerated to 500 r/min within 0.2 s, and then it only requires 0.15 s (0.65 s–0.5 s) to accelerate from 500 r/min to 1000 r/min when it accelerates again. Figure 9 shows that both weighting factors demonstrate fast torque responses, and the torque ripple is smaller when the weighting factor is larger. Figure 10 shows that when the weighting factor is 0.5, the harmonic current is concentrated within ±0.5A, while when the weighting factor is 0.1, the harmonic current is concentrated between ±1A. This indicates that when the weight factor is larger, more attention is paid to the influence of the harmonic plane current and its suppression effect is enhanced, so the amplitude of harmonic current decreases. Figure 11 demonstrates harmonic analysis of phase currents with different weight factors during 1000 r/min operation, and total harmonic distortion (THD) decreases from 23.42% to 18.76%, which further illustrates the improvement effect of larger weighting factors on waveform quality.

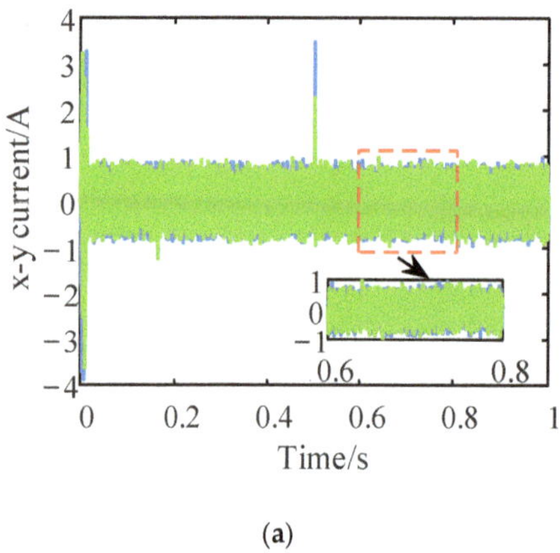

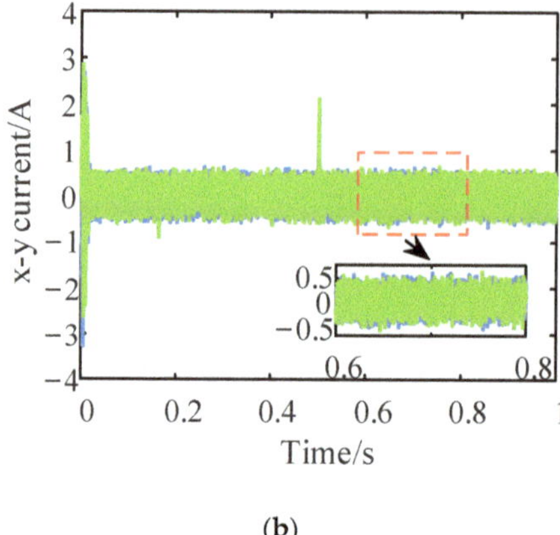

(a)   **(b)**

Figure 10. Harmonic currents of T-MPC scheme with different weighing factors at same sampling frequency (10 kHz). (**a**) $W_{xy} = 0.1$; (**b**) $W_{xy} = 0.5$.

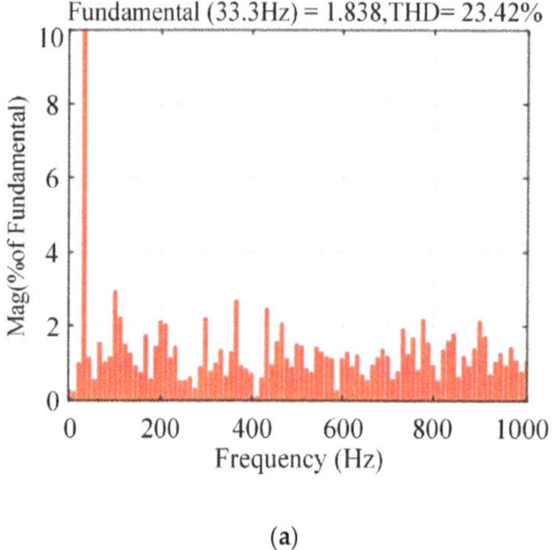

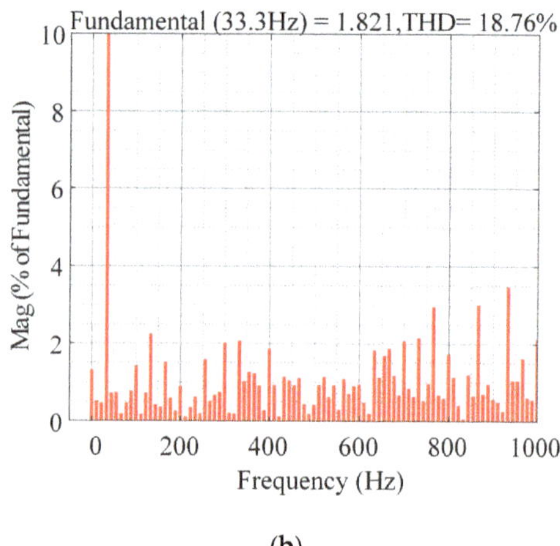

(a)   **(b)**

Figure 11. Total harmonic distortion (THD) performance of T-MPC scheme with different weighing factors at same sampling frequency (10 kHz). (**a**) $W_{xy} = 0.1$; (**b**) $W_{xy} = 0.5$.

This set of simulations shows that although the weighting factor is effective in regulating harmonic currents, since T-MPC only uses a single basic voltage vector in a single switching period, it always generates voltage in the harmonic plane and thus generates large harmonic currents. T-MPC has inherent defects in suppressing harmonic current and improving waveform quality, and it is always unable to obtain better current sinusoids.

5.2. Performance of Different Control Strategies

The T-MPC and VV-MPC strategies were used, respectively, to make the motors run at 1200 r/min, and a load of 5 N·m was suddenly applied at 0.5 s.

Figure 12 shows that the motor can run stably at the set speed under different control strategies. When the load is suddenly added, the speed drop is small and the motor can recover previous speed quickly. The superscript * of "Torque" in Figure 13 represents the reference torque. Figure 13 shows that when the load is suddenly added, the output torque of the motor under the two strategies rises rapidly, in which the torque overshoot of VV-MPC is larger and the response is more rapid. Figure 14 shows that the phase current waveform of VV-MPC is less glitchy and the current is more sinusoidal. Figure 15 shows the harmonic currents under different strategies. It can be seen that during the whole loading process, the harmonic currents of VV-MPC are smaller than those of T-MPC under this working condition.

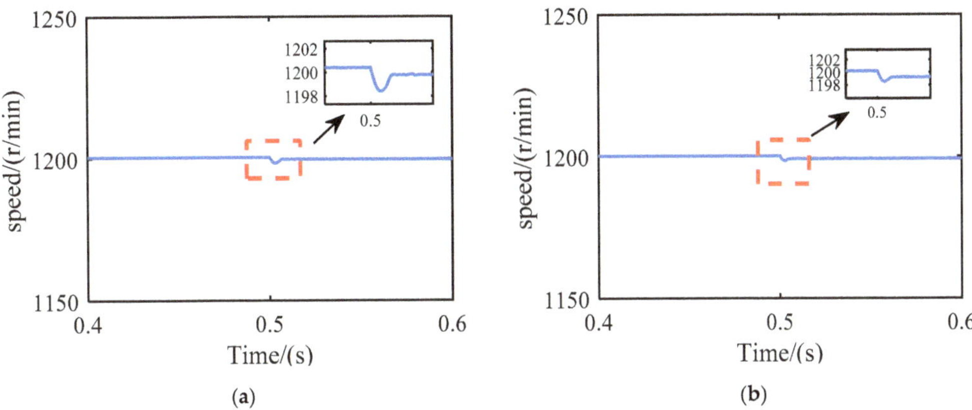

Figure 12. Performance of speed when loaded for two FCS-MPCC schemes at same sampling frequency (10 kHz). (**a**) T-MPC; (**b**) VV-MPC.

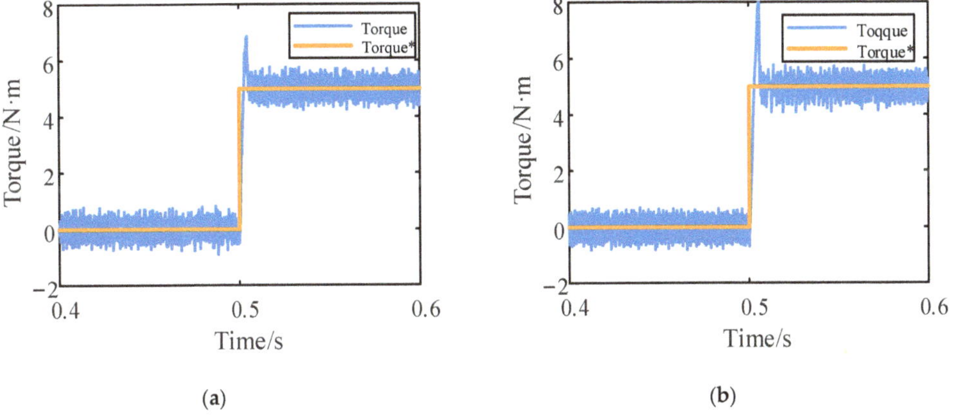

Figure 13. Performance of torque when loaded for two FCS-MPCC schemes at same sampling frequency (10 kHz). (**a**) T-MPC; (**b**) VV-MPC.

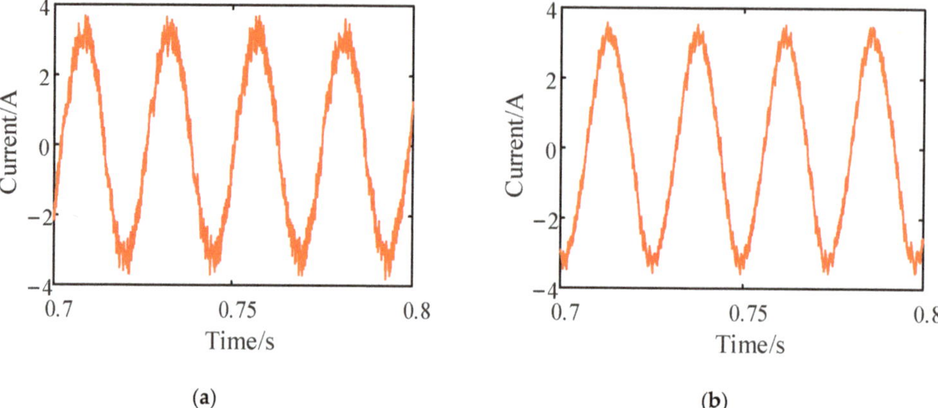

Figure 14. Performance of phase current in steady-state for two FCS-MPCC schemes at same sampling frequency (10 kHz). (**a**) T-MPC; (**b**) VV-MPC.

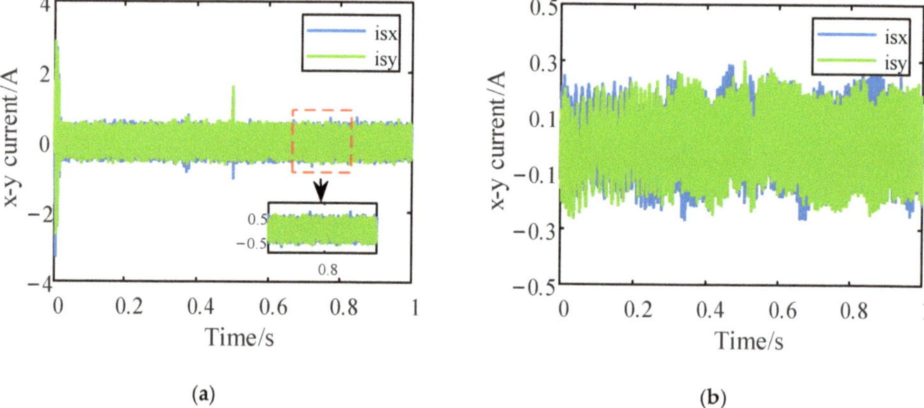

Figure 15. Performance of harmonic current for two FCS-MPCC schemes at same sampling frequency (10 kHz). (**a**) T-MPC; (**b**) VV-MPC.

Figure 16 shows the trajectories of fundamental and harmonic plane currents, where the outer ring is the fundamental current trajectory and the inner ring is the harmonic current trajectory. The fundamental current trajectory of VV-MPC is smoother, and the circle formed by the harmonic current has a smaller radius than that of T-MPC, indicating that VV-MPC has an obvious suppressing effect on harmonic currents during steady operation.

This set of simulation shows that after the virtual voltage vectors were adopted, the voltage average value of the harmonic plane was 0, so the current value of the harmonic plane in the VV-MPC method is smaller than that of the T-MPC method. Compared with T-MPC, VV-MPC also has good torque response characteristics when the load is suddenly added, and has a good harmonic current suppression effect when running with load at a steady state.

In order to compare the steady-state performance of the two methods at different speeds, the THD analysis of the phase current of VV-MPC and T-MPC in steady-state operation with load was carried out, respectively. Table 3 shows the THD analysis of phase current of the two methods at different speeds. It can be seen from Table 3 that, at the same speed, the THD of VV-MPC is about half smaller than that of T-MPC, indicating that VV-MPC has a better suppression effect on harmonic currents and thus obtains a better sinusoidal current. The THD of T-MPC does not change much at different speeds, and the minimum THD is achieved in the medium speed range (750 r/min). VV-MPC can effectively suppress the harmonic current in the whole rate range, and the current quality is better at medium and low speed (300–750 r/min).

Table 3. THD analysis of T-MPC and VV-MPC.

Scheme	1200 r/min	750 r/min	300 r/min
T-MPC	10.67%	9.91%	10.77%
VV-MPC	6.65%	5.68%	5.82%

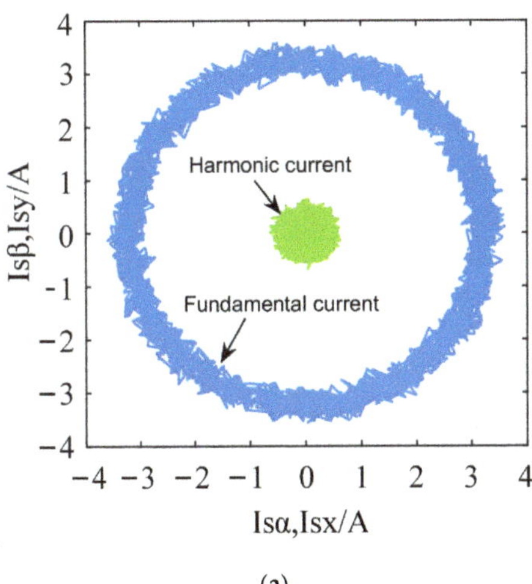

(a)

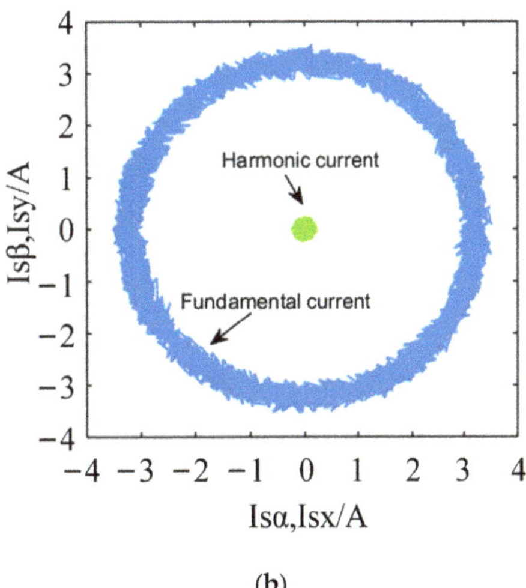

(b)

Figure 16. Performance of current trajectories of fundamental and harmonic currents for two FCS-MPCC schemes at same sampling frequency (10 kHz). (**a**) T-MPC; (**b**) VV-MPC.

5.3. Forward and Reverse Performance under the Proposed Strategy

In order to further test the dynamic performance of the proposed VV-MPC, the simulation of forward-to- reverse switching under no-load condition was carried out. After acceleration in advance, the motor was stable at 1000 r/min at 0.5 s, and the reference speed became −1000 r/min at 2.5 s. The superscript * of "ωm" in Figure 17 represents the reference speed. Figure 17 shows that the actual speed of the motor quickly tracks the reference speed and runs stably at −1000 r/min after 0.6 s. The d-q axis current and its reference value during the whole switching period are shown in Figure 18. The superscript * in Figure 18 represents the reference d-axis current and reference q-axis current. It can be seen that the reference current of the d-axis keeps constant, while the actual current value of the d-axis fluctuates more than during the steady-state operation only during the conversion process of 2.5 s–3.1 s, but the reference value can still be tracked. When the speed reference command changes, in order to force the motor to decelerate rapidly and reverse, the reference value of the q-axis current quickly changes to −8A, and the actual value is immediately tracked, which reflects that it still has good dynamic performance in the extreme case of forward and reverse switching. Figure 19 shows that during the entire operation process, the harmonic current is always maintained between −0.3A and 0.3A, and the harmonic suppression ability is not affected by the working conditions.

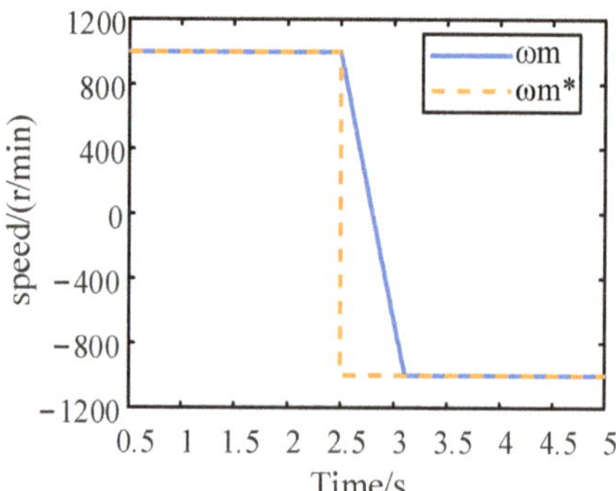

Figure 17. Performance of speed when used VV-MPC in speed reversal simulation.

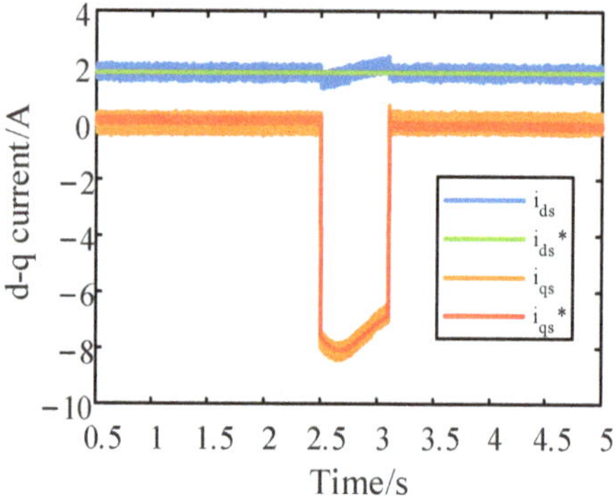

Figure 18. Performance of d-q current tracking error when used VV-MPC in speed reversal simulation.

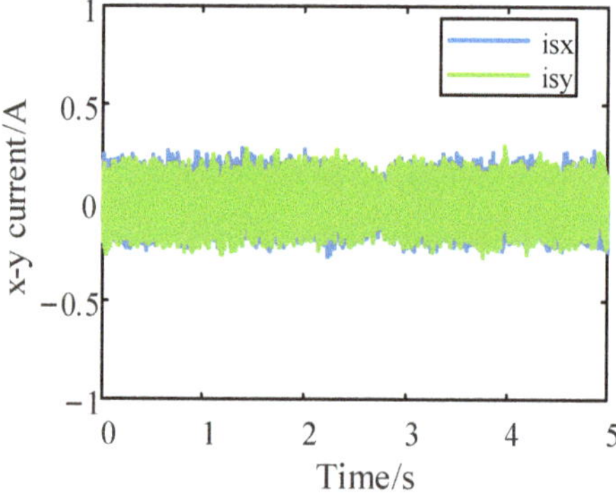

Figure 19. Performance of harmonic current when used VV-MPC in speed reversal simulation.

5.4. The Performance of the Proposed Strategy under All Operating Conditions

In order to compare the performance differences between the proposed VV-MPC and T-MPC at different speeds and different loads, the speed values used were 300 r/min, 600 r/min, 900 r/min, 1200 r/min, and the loads used were 2 N·m, 4 N·m, 6 N·m, 8 N·m, 10 N·m. Multiple sets of simulations were performed, and the results are displayed in three-dimensional diagrams as shown in Figures 20 and 21.

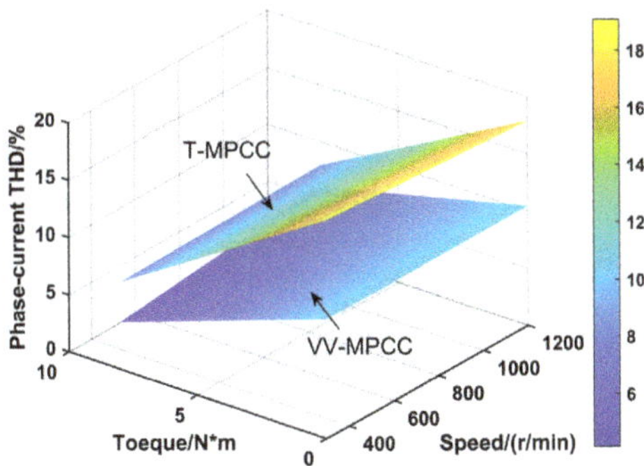

Figure 20. Effects of torque and speed on phase current.

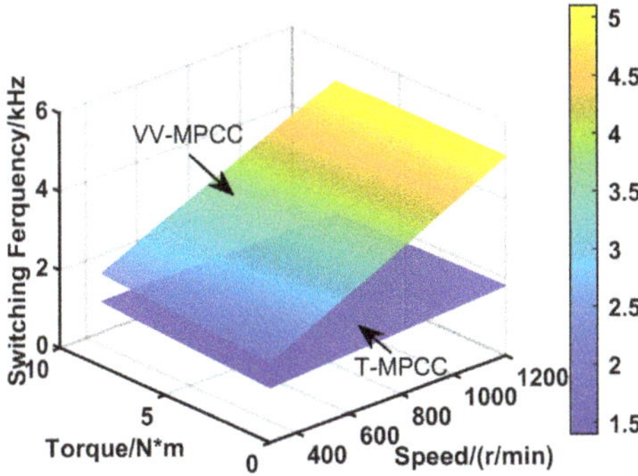

Figure 21. Effects of torque and speed on switching frequency.

Figure 20 shows that the THD-plane of VV-MPC and T-MPC does not overlap at full speed and under full working conditions, which fully shows that VV-MPC can effectively suppress harmonic current at any speed and load compared with T-MPC. At the same time, it can be seen that for these two control strategies, the THD of the motor with light load is greater than that of the motor with heavy load, indicating that the waveform quality of the two strategies under load is improved compared with that of no load. With the same load, THD does not change much with the increase of the motor speed, indicating that the speed has little influence on the phase current THD of the motor.

Figure 21 shows that the switching frequency of T-MPC changes slowly and does not change greatly with the change of load. The switching frequency at low speed is slightly lower than that at high speed. At 300 r/min and no load, the minimum switching frequency is 1.4 kHz and the maximum is 1.8 kHz at 1200 r/min and no load. This indicates that the same voltage vector will be used for a long period of time, which means that the switching device will not act. The switching frequency of VV-MPC varies greatly with the speed. The switching frequency is low at low speeds, about 2.1 kHz at 300 r/min, and high at high speeds, about 5 kHz at 1200 r/min. It can be speculated that when the speed increases, the stator current frequency also increases, so that the switching frequency increases. For these two strategies, the control frequency is set to 10 kHz, and the switching frequency of VV-MPC, which has the highest switching frequency, is only about 5 kHz, indicating that the MPC has the characteristic of low switching frequency.

6. Conclusions

Although T-MPC still has good dynamic performance when extended to multiphase systems, it lacks effective suppression of harmonic plane currents. The single switching state used in T-MPC will inevitably bring the problem of high harmonic plane currents. In this paper, ten new virtual vectors were synthesized from the large and medium vectors in the same direction in α-β space according to the fixed ratio by using the vector distribution characteristics of five-phase inverter. Taking the virtual vector set as the alternative vector set, the VV-MPC method based on the virtual vector set was proposed and applied to a five-phase induction motor. The advantages of the proposed VV-MPC can be summarized as follows:

(1) Compared with the traditional method, the proposed method has excellent dynamic performance, and the harmonic plane current value is greatly reduced, so as to achieve the purpose of suppressing harmonic current, reducing harmonic loss and improving power quality.

(2) Due to the reduction of the number of vectors in the alternative vector set, the calculation amount of MPC in each cycle is reduced.

(3) In the proposed method, the theoretical average value of the harmonic plane voltage is 0, so there is no need to predict the harmonic plane currents in the prediction process, the weighting factor is omitted, and the prediction model is simplified.

Author Contributions: Conceptualization, Q.Z. and J.Z.; methodology, Q.Z.; software, Y.M.; investigation, H.C. and Y.X.; writing—original draft preparation, Q.Z.; writing—review and editing, S.Y. All authors have read and agreed to the published version of the manuscript.

Funding: This research received no external funding.

Data Availability Statement: Not applicable.

Conflicts of Interest: The authors declare no conflict of interest.

References

1. Chan, C.C. The State of the Art of Electric, Hybrid, and Fuel Cell Vehicles. *Proc. IEEE* **2007**, *95*, 704–718. [CrossRef]
2. Sun, X.; Jin, Z.; Wang, S.; Yang, Z.; Li, K.; Fan, Y.; Chen, L. Performance Improvement of Torque and Suspension Force for a Novel Five-Phase BFSPM Machine for Flywheel Energy Storage Systems. *IEEE Trans. Appl. Supercond.* **2019**, *29*, 5–9. [CrossRef]
3. Chau, K.T.; Chan, C.C.; Liu, C. Overview of Permanent-Magnet Brushless Drives for Electric and Hybrid Electric Vehicles. *IEEE Trans. Ind. Electron.* **2008**, *55*, 2246–2257. [CrossRef]
4. Frieske, B.; Kloetzke, M.; Mauser, F. Trends in Vehicle Concept and Key Technology Development for Hybrid and Battery Electric Vehicles. *World Electr. Veh. J.* **2013**, *6*, 9–20. [CrossRef]
5. Bojoi, R.; Rubino, S.; Tenconi, A.; Vaschetto, S. Multiphase Electrical Machines and Drives: A Viable Solution for Energy Generation and Transportation Electrification. In Proceedings of the 2016 International Conference and Exposition on Electrical and Power Engineering (EPE), Iasi, Romania, 20–22 October 2016; pp. 632–639. [CrossRef]
6. Cao, W.; Mecrow, B.C.; Atkinson, G.J.; Bennett, J.W.; Atkinson, D.J. Overview of Electric Motor Technologies Used for More Electric Aircraft (MEA). *IEEE Trans. Ind. Electron.* **2012**, *59*, 3523–3531. [CrossRef]
7. Tao, T.; Zhao, W.; He, Y.; Cheng, Y.; Saeed, S.; Zhu, J. Enhanced Fault-Tolerant Model Predictive Current Control for a Five-Phase PM Motor with Continued Modulation. *IEEE Trans. Power Electron.* **2021**, *36*, 3236–3246. [CrossRef]
8. Fnaiech, M.A.; Betin, F.; Capolino, G.-A.; Fnaiech, F. Fuzzy Logic and Sliding-Mode Controls Applied to Six-Phase Induction Machine with Open Phases. *IEEE Trans. Ind. Electron.* **2010**, *57*, 354–364. [CrossRef]
9. Zheng, L.; Fletcher, J.E.; Williams, B.W.; He, X. Dual-Plane Vector Control of a Five-Phase Induction Machine for an Improved Flux Pattern. *IEEE Trans. Ind. Electron.* **2008**, *55*, 1996–2005. [CrossRef]
10. Diana, M.; Guglielmi, P.; Vagati, I.A. Very Low Torque Ripple Multi-3-Phase Machines. In Proceedings of the IECON 2016—42nd Annual Conference of the IEEE Industrial Electronics Society, Florence, Italy, 24–27 October 2016; IEEE: Piscataway, NJ, USA, 2016; pp. 1750–1755.
11. Li, G.; Hu, J.; Li, Y.; Zhu, J. An Improved Model Predictive Direct Torque Control Strategy for Reducing Harmonic Currents and Torque Ripples of Five-Phase Permanent Magnet Synchronous Motors. *IEEE Trans. Ind. Electron.* **2019**, *66*, 5820–5829. [CrossRef]
12. Cortés, P.; Kazmierkowski, M.P.; Kennel, R.M.; Quevedo, D.E.; Rodriguez, J. Predictive Control in Power Electronics and Drives. *IEEE Trans. Ind. Electron.* **2008**, *55*, 4312–4324. [CrossRef]
13. Elmorshedy, M.F.; Xu, W.; El-Sousy, F.F.M.; Islam, M.R.; Ahmed, A.A. Recent Achievements in Model Predictive Control Techniques for Industrial Motor: A Comprehensive State-of-the-Art. *IEEE Access* **2021**, *9*, 58170–58191. [CrossRef]
14. Martinez, J.C.R.; Kennel, R.M.; Geyer, T. Model Predictive Direct Current Control. In Proceedings of the 2010 IEEE International Conference on Industrial Technology, Vina del Mar, Chile, 14–17 March 2010; IEEE: Piscataway, NJ, USA, 2010; pp. 1808–1813.

15. Iqbal, A.; Alammari, R.; Mosa, M.; Abu-Rub, H. Finite Set Model Predictive Current Control with Reduced and Constant Common Mode Voltage for a Five-Phase Voltage Source Inverter. In Proceedings of the 2014 IEEE 23rd International Symposium on Industrial Electronics (ISIE), Istanbul, Turkey, 1–4 June 2014; pp. 479–484. [CrossRef]
16. Stolze, P.; Tomlinson, M.; Kennel, R.; Mouton, T. Heuristic Finite-Set Model Predictive Current Control for Induction Machines. In Proceedings of the 2013 IEEE ECCE Asia Downunder—5th IEEE Annual International Energy Convers, Melbourne, VIC, Australia, 3–6 June 2013; pp. 1221–1226. [CrossRef]
17. Xue, Y.; Meng, D.; Zhao, Z.; Zhao, L.; Diao, L. Model Predictive Current Control in the Stationary Coordinate System for a Three-Phase Induction Motor Fed by a Two-Level Inverter. In Proceedings of the 2019 IEEE International Symposium on Predictive Control of Electrical Drives and Power Electronics (PRECEDE), Quanzhou, China, 31 May–2 June 2019; IEEE: Piscataway, NJ, USA, 2019; pp. 1–5.
18. Kousalya, V.; Rai, R.; Singh, B. Predictive Torque Control of Induction Motor for Electric Vehicle. In Proceedings of the 2020 IEEE Transportation Electrification Conference & Expo (ITEC), Chicago, IL, USA, 23–26 June 2020; IEEE: Piscataway, NJ, USA, 2020; pp. 890–895.
19. Rivera, M.; Riveros, J.A.; Rodriguez, C.; Wheeler, P. Field-Oriented Control with a Predictive Current Strategy of an Induction Machine Fed by a Two-Level Voltage Source Inverter. In Proceedings of the 2021 IEEE International Conference on Automation/XXIV Congress of the Chilean Association of Automatic Control (ICA-ACCA), Valparaíso, Chile, 22–26 March 2021. [CrossRef]
20. Lim, C.S.; Levi, E.; Jones, M.; Abdul Rahim, N.; Hew, W.P. Experimental Evaluation of Model Predictive Current Control of a Five-Phase Induction Motor Using All Switching States. In Proceedings of the 2012 15th International Power Electronics and Motion Control Conference (EPE/PEMC), Novi Sad, Serbia, 4–6 September 2012; pp. 1–7. [CrossRef]
21. Xue, C.; Song, W.; Feng, X. Model Predictive Current Control Schemes for Five-Phase Permanent-Magnet Synchronous Machine Based on SVPWM. In Proceedings of the 2016 IEEE 8th International Power Electronics and Motion Control Conference (IPEMC-ECCE Asia), Hefei, China, 22–26 May 2016; pp. 648–653. [CrossRef]
22. Lim, C.S.; Levi, E.; Jones, M.; Rahim, N.A.; Hew, W.P. FCS-MPC-Based Current Control of a Five-Phase Induction Motor and Its Comparison with PI-PWM Control. *IEEE Trans. Ind. Electron.* **2014**, *61*, 149–163. [CrossRef]
23. Bermúdez, M.; Martín, C.; González-Prieto, I.; Durán, M.J.; Arahal, M.R.; Barrero, F. Predictive Current Control in Electrical Drives: An Illustrated Review with Case Examples Using a Five-phase Induction Motor Drive with Distributed Windings. *IET Electr. Power Appl.* **2020**, *14*, 1291–1310. [CrossRef]
24. Cortes, P.; Wilson, A.; Kouro, S.; Rodriguez, J.; Abu-Rub, H. Model Predictive Control of Multilevel Cascaded H-Bridge Inverters. *IEEE Trans. Ind. Electron.* **2010**, *57*, 2691–2699. [CrossRef]
25. Duran, M.J.; Prieto, J.; Barrero, F.; Toral, S. Predictive Current Control of Dual Three-Phase Drives Using Restrained Search Techniques. *IEEE Trans. Ind. Electron.* **2011**, *58*, 3253–3263. [CrossRef]
26. Zheng, L.; Fletcher, J.E.; Williams, B.W.; He, X. A Novel Direct Torque Control Scheme for a Sensorless Five-Phase Induction Motor Drive. *IEEE Trans. Ind. Electron.* **2011**, *58*, 503–513. [CrossRef]
27. Bermudez, M.; Gonzalez-Prieto, I.; Barrero, F.; Guzman, H.; Duran, M.J.; Kestelyn, X. Open-Phase Fault-Tolerant Direct Torque Control Technique for Five-Phase Induction Motor Drives. *IEEE Trans. Ind. Electron.* **2017**, *64*, 902–911. [CrossRef]
28. Romero, C.; Delorme, L.; Gonzalez, O.; Ayala, M.; Rodas, J.; Gregor, R. Algorithm for Implementation of Optimal Vector Combinations in Model Predictive Current Control of Six-Phase Induction Machines. *Energies* **2021**, *14*, 3857. [CrossRef]
29. Gonzalez-Prieto, I.; Duran, M.J.; Aciego, J.J.; Martin, C.; Barrero, F. Model Predictive Control of Six-Phase Induction Motor Drives Using Virtual Voltage Vectors. *IEEE Trans. Ind. Electron.* **2018**, *65*, 27–36. [CrossRef]
30. Ryu, H.M.; Kim, J.H.; Sul, S.K. Analysis of Multiphase Space Vector Pulse-Width Modulation Based on Multiple d-q Spaces Concept. *IEEE Trans. Power Electron.* **2005**, *20*, 1364–1371. [CrossRef]
31. Wang, F.; Li, S.; Mei, X.; Xie, W.; Rodriguez, J.; Kennel, R.M. Model-Based Predictive Direct Control Strategies for Electrical Drives: An Experimental Evaluation of PTC and PCC Methods. *IEEE Trans. Ind. Inform.* **2015**, *11*, 671–681. [CrossRef]
32. Cortes, P.; Rodriguez, J.; Silva, C.; Flores, A. Delay Compensation in Model Predictive Current Control of a Three-Phase Inverter. *IEEE Trans. Ind. Electron.* **2012**, *59*, 1323–1325. [CrossRef]

Article

Static Voltage Stability Assessment Using a Random UnderSampling Bagging BP Method

Zhujun Zhu [1], Pei Zhang [1], Zhao Liu [1,*] and Jian Wang [2]

[1] School of Electrical Engineering, Beijing Jiaotong University, Beijing 100044, China
[2] College of Energy and Electrical Engineering, Hohai University, Nanjing 211100, China
* Correspondence: liuzhao1@bjtu.edu.cn

Abstract: The increase in demand and generator reaching reactive power limits may operate the power system in stressed conditions leading to voltage instability. Thus, the voltage stability assessment is essential for estimating the loadability margin of the power system. The grid operators urgently need a voltage stability assessment (VSA) method with high accuracy, fast response speed, and good scalability. The static VSA problem is defined as a regression problem. Moreover, an artificial neural network is constructed for online assessment of the regression problem. Firstly, the training sample set is obtained through scene simulation, power flow calculation, and local voltage stability index calculation; then, the class imbalance problem of the training samples is solved by the random under-sampling bagging (RUSBagging) method. Then, the mapping relationship between each feature and voltage stability is obtained by an artificial neural network. Finally, taking the modified IEEE39 node system as an example, by setting up four groups of methods for comparison, it is verified that the proposed method has a relatively ideal modeling speed and high accuracy, and can meet the requirements of power system voltage stability assessment.

Keywords: static voltage stability; machine learning; class imbalance problem; random under-sampling; bagging; artificial neural network

Citation: Zhu, Z.; Zhang, P.; Liu, Z.; Wang, J. Static Voltage Stability Assessment Using a Random UnderSampling Bagging BP Method. *Processes* **2022**, *10*, 1938. https://doi.org/10.3390/pr10101938

Academic Editor: Mohd Azlan Hussain

Received: 22 August 2022
Accepted: 19 September 2022
Published: 26 September 2022

Publisher's Note: MDPI stays neutral with regard to jurisdictional claims in published maps and institutional affiliations.

1. Introduction

Voltage stability [1] is the main limiting factor for the safe and reliable operation of power systems. With continued load growth and the penetration of new energy sources, modern power systems have been pushed to operate closer to their voltage stability limits. Over the past few decades, great efforts have been devoted to investigating the mechanisms of voltage instability and developing effective voltage stability assessment (VSA) methods [2].

Generally, voltage profiles show no anomalies before undergoing a voltage collapse due to load changes. Voltage stability margin (VSM) is a static voltage stability index that quantifies how "close" a particular operating point is to the point of voltage collapse [3]. Therefore, the VSM can be used to estimate the steady-state voltage stability limit of a power system. Knowing voltage stability margins is critical for utilities to operate their systems safely and with reliability. The system operator must provide an accurate and fast method to predict the voltage stability margin to initiate the necessary control actions [4].

That proposed a static voltage stability prediction method based on gradient boosting, which has better prediction accuracy [5]. However, its training set data are obtained through the calculation of the cumulative probability function (CPF), which is only applicable to a fixed load power factor case. Ghiocel et al. [6] proposed a new method to directly eliminate the singularity by reformulating the power flow problem. The central idea is to introduce an AQ bus in which the bus angle and the reactive power consumption of a load bus are specified. However, the computation burden is still heavy, and the solution speed cannot meet the requirement of real-time assessment.

With the boom of wide-area measurement systems in smart grids [7–9], the availability of large amounts of data acquired by phasor measurement units (PMUs) presents a huge opportunity for data-driven stability assessments. Great efforts have been made to perform such tasks through machine learning techniques.

In [10], a static stability assessment method for a power system based on a decision tree algorithm is proposed, which improves the assessment speed. However, there is no countermeasure for the decision tree over-fitting problem. Lai et al. [11] proposed a transient voltage stability assessment model based on convolutional neural networks, which improves the assessment speed by using statistical analysis for data dimensionality reduction. However, relying only on statistical analysis for data dimensionality reduction, it is easy to ignore individual features. Liang et al. proposed a random forest model for static voltage stability assessment, which makes up for the assessment defects of a single decision tree [12]. However, the selection of features is based on subjective judgment. The voltage stability assessment problem is treated as a classification problem of machine learning, making it difficult to accurately know the degree of voltage stability.

A common feature of these machine learning-based efforts is that they assume that the learning dataset can be generated by system simulations in the desired quantity [13]. Since accurate simulation and modeling, especially load modeling, are considered a great challenge in power systems, errors are inevitably introduced into the learning dataset. Preferably, the learning dataset can be obtained from PMU records, which will significantly improve the quality and reliability of the knowledge base. However, learning machines are likely to suffer from severe class imbalance problems. The system remains stable after most disturbances and becomes unstable only in a few cases. If not handled properly, this imbalance can greatly deteriorate the performance of the learning machine, and the minority class will be ignored and thus leading to misjudging. The class imbalance problem exists not only in the field of power systems, but also widely in other academic and industrial contexts, such as credit fraud detection, biomedical diagnosis, equipment fault diagnosis, and Internet intrusion [14].

Faced with class imbalance problems [15], considerable efforts have been made by machine learning researchers to deal with them [16,17]. Synthetic sampling is the most commonly used method for rebalancing class distributions. However, it cannot be directly applied to voltage stability assessment. Because datasets created by naive replication or linear interpolation may not exist in practice. Besides sampling-related techniques, some cost-sensitive tricks are proposed to build cost-sensitive classifiers. By attaching costs to different classes, these techniques manage to enhance minority learning by drawing more attention to minority classes [18].

To meet the requirements of voltage stability assessment and solve the problem of class imbalance and poor model generalization in machine learning, an online assessment method of static voltage stability using the RUSBagging method is proposed. The method differs from other methods in that:

The problem of VSA is defined as a machine learning regression problem, which is helpful for grid operators to observe the voltage stability state of the power system.

The bagging method of the ensemble framework is used to build the model to improve the generalization ability of the model.

The random under-sampling method is added to bagging, which solves the class imbalance problem to a certain extent and improves the assessment accuracy on minority class samples.

2. Local Voltage Stability Index

Commonly used static voltage stability indexes are [19,20]: the Jacobi singular value index, voltage sensitivity index, load margin index, VCPI index, and local voltage stability index. Compared to other voltage stability indices, the local voltage stability index (L index), which can give normalized index values for different systems, and which is not limited by the randomness of the direction of load growth, are highly applicable and highly accurate.

By the KCL law (Kirchhoff's current law) there is $YV = I$, where Y stands for node admittance, V stands for node voltage, and I stands for node current. In addition, according to the value of the node injection current, the network nodes are divided into generator nodes, load nodes, and contact nodes, and the equations of the node network after the division are as follows.

$$\begin{bmatrix} I_G \\ I_L \\ 0 \end{bmatrix} = \begin{bmatrix} Y'_{GG} & Y'_{GL} & Y'_{GK} \\ Y'_{LG} & Y'_{LL} & Y'_{LK} \\ Y'_{KG} & Y'_{KL} & Y'_{KK} \end{bmatrix} \begin{bmatrix} V_G \\ V_L \\ V_K \end{bmatrix} \tag{1}$$

where V_G and I_G are the voltage and current vectors at the generator node, V_L and I_L are the voltage and current vectors at the load node, and V_K is the voltage vector at the contact node.

By eliminating the contact nodes, the remaining nodes in the network are divided into the set of generator nodes (α_G) and the set of load nodes (α_L), and Equation (1) can be transformed as:

$$\begin{bmatrix} I_G \\ I_L \end{bmatrix} = \begin{bmatrix} Y_{GG} & Y_{GL} \\ Y_{LG} & Y_{LL} \end{bmatrix} \begin{bmatrix} V_G \\ V_L \end{bmatrix} \tag{2}$$

where $Y_{GG} = Y'_{GG} - Y'_{GK}Y'^{-1}_{KK}Y'_{KG}$, $Y_{GL} = Y'_{GL} - Y'_{GK}Y'^{-1}_{KK}Y'_{KL}$, $Y_{LG} = Y'_{LG} - Y'_{LK}Y'^{-1}_{KK}Y'_{KG}$, $Y_{LL} = Y'_{LL} - Y'_{LK}Y'^{-1}_{KK}Y'_{KL}$.

Substituting $Z_{LL} = Y^{-1}_{LL}$ into Equation (2) converts to:

$$\begin{bmatrix} I_G \\ V_L \end{bmatrix} = \begin{bmatrix} Y_{GG} - Y_{GL}Z_{LL}Y_{LG} & Y_{GL}Z_{LL} \\ -Z_{LL}Y_{LG} & Z_{LL} \end{bmatrix} \begin{bmatrix} V_G \\ I_L \end{bmatrix} \tag{3}$$

Reference [15] gives the local voltage stability index L_j for load node j:

$$L_j = \frac{\left| \sum_{i \in \alpha_L} \left| \frac{Z^*_{ji} \widetilde{S}_i}{Z^*_{jj} \dot{V}_i} \right| \dot{V}_j \right|}{V_j^2 Y_{jj}} = \frac{\left| \sum_{i \in \alpha_L} \frac{Z^*_{ji} \widetilde{S}_i}{\dot{V}_i} \right|}{V_j} \tag{4}$$

where $\dot{V}_i$, $\dot{V}_j$ are the voltage phases of nodes i, j respectively. $\widetilde{S}_i$ is the equivalent load of node i. Z^*_{ji} is the mutual impedance conjugate between loads j,i of the equivalent load impedance matrix Z_{LL}. Y_{jj} is the self-conductance of the jth node of the equivalent load conductance matrix Y_{LL}.

The local voltage stability index for all load nodes in the network forms the overall system stability index vector $L = [L_1, L_2, \cdots L_n]$, $n \in \alpha_L$ and the maximum index value for the load is selected to define the voltage stability index for the system.

$$L = \|L\|_\infty \tag{5}$$

The relationship between local voltage stability index and system voltage stability is [21]:

$L < 1$, system voltage stability.

$L = 1$, system voltage critical stability.

$L > 1$, system voltage instability.

3. Random Under-Sampling Bagging BP Method for VSA

3.1. BP Neural Network for Regression of VSA

Since the VSA problem is defined as a machine learning regression problem in this article, a model is built to implement the regression. The back-propagation (BP) neural network model is easy to build, has a wide range of adaptability, and the algorithm is easy to implement. Hence, BP neural network is selected to solve this regression problem.

3.1.1. Model of BP Neural Network for Regression of VSA

Neural networks have an adaptive character, changing the weight values during the training process to suit different requirements [22,23]. The internal neural network can be divided into input, hidden and output layers according to the different functional layers.

For regression of the VSA problem, the input variables are the operating states of the power system, such as nodal voltage, branch power flow, and load demand. The output of the model is the L index, which is the result of the voltage stability assessment.

Where X is the input column vector; x_i is the element in row i. W is the weight matrix; specifically, an element of W can be represented by $w_{f,ij}$. The subscript f indicates the corresponding layer, and the subscript ij indicates the connection between node i in this layer and node j in the next layer. Y is the output column vector; y_i is the element in row i. Σ is the summation symbol, which sums multiple input signals; φ_i is the activation function of the i-th neuron in the hidden layer and ϕ_i is the i-th neuron activation function in the output layer; θ_i is the i-th neuron threshold in the hidden layer and b_i is the i-th neuron threshold in the output layer.

Neural networks use a large number of hidden layer neurons for data flow processing and network training. Without loss of generality, the neural network data stream processing process is briefly described using a single neuron in Figure 1 as an example.

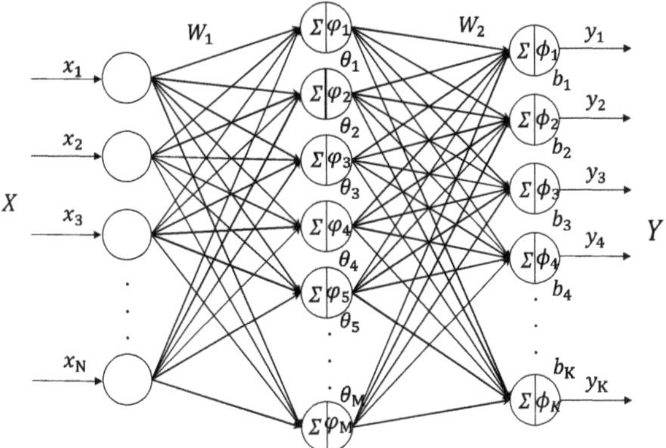

Figure 1. Structure diagram of typical BP neural network.

Assuming that the output of the i-th neuron in the hidden layer is, the output of the ith neuron can be represented by Equation (6) from Figure 1.

$$o_i = \varphi_i\left(\sum_{j=1}^{n} w_{1,ij} \cdot x_j + \theta_i\right) \tag{6}$$

Similarly, it can be deduced that the output of the ith neuron in the output layer is:

$$y_i = \phi_i\left(\sum_{j=1}^{m} w_{2,ij} \cdot o_j + b_i\right) \tag{7}$$

3.1.2. Algorithm of BP Neural Network for Regression of VSA

The learning process consists of two processes: forward propagation of the signal and backward propagation of the error. The forward propagation process is shown in Equations (6) and (7). If the actual output of the output layer is not equal to the label value, then it is transferred to the error backpropagation process. The core of the BP neural network is the error back propagation process.

The error back propagation is to backpropagate the output error in a certain form through the hidden layer, and the error is apportioned to all neurons in each layer according to certain rules. To obtain the error signal of the neurons in each layer, and use it as the basis for correcting the weights of each neuron, the specific correction method is shown in Equations (8) to (11). The process of weight correction is also the learning process of the network. In general, this process continues until the network output error is within the set range or until a predetermined learning time or iterations.

$$\Delta w_{2,ij} = \eta \sum_{p=1}^{P} \sum_{k=1}^{K} \left(T_k^p - o_k^p \right) \phi'_i o_j \tag{8}$$

$$\Delta b_i = \eta \sum_{p=1}^{P} \sum_{k=1}^{K} \left(T_k^p - o_k^p \right) \phi'_i \tag{9}$$

$$\Delta w_{1,ij} = \eta \sum_{p=1}^{P} \sum_{k=1}^{K} \left(T_k^p - o_k^p \right) \phi'_i w_{2,ij} \phi'_j x_j \tag{10}$$

$$\Delta \theta_i = \eta \sum_{p=1}^{P} \sum_{k=1}^{K} \left(T_k^p - o_k^p \right) \phi'_i w_{2,ij} \phi'_j \tag{11}$$

where $\Delta w_{2,ij}$ is the weight correction between the i-th neuron in the hidden layer to the j-th neuron in the output layer. Δb_i is the threshold correction for the i-th neuron in the output layer. $\Delta w_{1,ij}$ is the weight correction between the i-th neuron in the input layer to the j-th neuron in the hidden layer. $\Delta \theta_i$ is the threshold correction for the i-th neuron in the hidden layer. p is the sample index and P is the total number of training samples. η is the weight correction learning rate. T_k^p is the expected output value of the kth output neuron for the pth sample data.

3.2. Random Under-Sampling Bagging for Improving Model Accuracy of VSA

In the actual operation of the power system, the system is in a stable state in most cases. Stable data is far more than unstable data or critically stable data, which is a typical data imbalance problem. In fact, for the stable operation of the power system, the value of unstable or critically stable sample data is higher than that of stable sample data. Therefore, solving the data imbalance problem in the voltage stability assessment of the power system has a positive effect on improving the accuracy of the model in the critical stable operating state.

3.2.1. Random Under-Sampling Method for Solving Class Imbalance Problem in VSA

Data sampling is a type of data preprocessing method, which can solve the learner bias problem caused by data imbalance to a certain extent. Data sampling is generally divided into two categories: under-sampling and over-sampling. Over-sampling achieves the balance of original skewed data by introducing a new minority of instances, while under-sampling does the opposite. However, the over-sampling method will generate the wrong samples, which damages the learning result of minority samples [23]. Therefore, an under-sampling method is selected in this article. The schematic diagram of random under-sampling is shown as Figure 2.

To generate a balanced data set for training, the under-sampling method is used to resample the original set. Assuming that the size of the resampled data sets is S, where $S \leq N_P \times 2$, N_P is the size of the majority set P. Randomly sample the number of instances from both the majority set P and the minority set N without replacement and put them into the new training data set D [24].

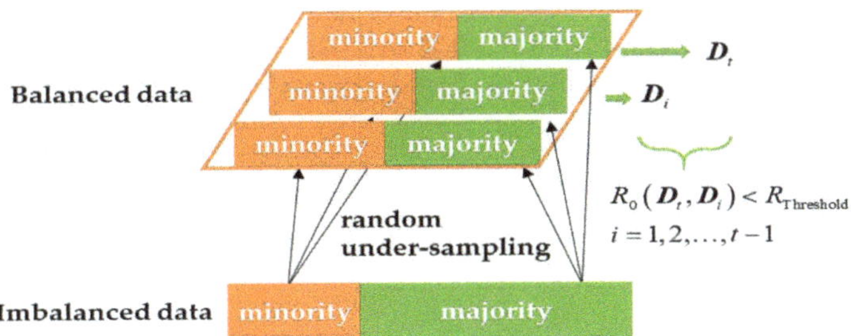

Figure 2. Schematic diagram of random under-sampling.

To make sure that every subset D_i for training is relatively independent and as many instances of the original sets as possible are covered, a concept of overlap rate was proposed in [25]. The overlap rate of two data sets is defined as follows:

Given two data sets D_1, D_2 with the size of M, M_S is the number of the same samples in D_1 and D_2, so the overlap rate R_0 of D_1 and D_2 is:

$$R_0(D_1, D_2) = \frac{M_S}{M} \tag{12}$$

The threshold value $R_{\text{Threshold}}$ is set to limit the subset D_t obtained from the t under-sampling by:

$$R_0(D_t, D_i) < R_{\text{Threshold}}, i = 1, 2, \ldots, t-1 \tag{13}$$

3.2.2. Bagging with Random Under-Sampling for Improving Model Accuracy of VSA

Bagging is the abbreviation of bootstrap AGGregatING, the representative of the parallel ensemble learning method. Multiple different training sets are constructed by the method of bootstrap sampling (re-sampling). Then the corresponding weak learners are trained in each training set. Finally, the final model after the aggregation of the weak learners is obtained [26].

For the bagging method, each weak learner uses the same model. It is necessary to distinguish the training data sets of each weak learner. If the training data set is directly divided and different weak learners are trained on each subset, the weak learner will miss the key information in the original training set, which limits the performance of the weak learner.

The proposal of bootstrap sampling solves the above problems well. This sampling method ensures the independence of different training subsets as much as possible while using more samples. Specifically, the method of repeatable sampling is adopted. There are repeated samples in these samples, so they are the true subset of the original data set. Assuming that the probability of each sample being sampled during the sampling process is equal, it is not difficult to calculate that when the number of samples n is large, about 63.2% of the original samples will be drawn in one bootstrap. Use bootstrap to sample M times to obtain M sample sets, and build a basic learner on each sample set to obtain M different learners.

Another step in bagging is model aggregation. For classification problems, the one with the largest proportion of the results of M weak learners is selected as the final classification result by voting; for regression problems, the outputs of M weak learners are averaged. Table 1 is a brief flow of the bagging method, which can be used for both classification and regression.

However, the original bagging does not take into account the imbalance problem. The data set of every training group is still imbalanced, and integration does not contribute to solving the imbalance problem. After the bootstrap sampling method, the random under-sampling method is added to improve the applicability of bagging to imbalanced data.

Table 1. Algorithm of the random under-sampling bagging method.

Input: dataset $D = \{(x_1, y_1), (x_2, y_2), \cdots, (x_n, y_n)\}$, weak learner algorithm, number of weak learner M
For $i = 1, 2, \ldots, $ M:
Using the bootstrap sampling method on the dataset D to generate a subsampled set D_i
Using the random under-sampling method on the dataset D_i to generate a balanced subsample set D'_i
Training the ith weak learner $G_i(x)$ with a balanced subsample set D'_i
End
Output: the final model

Figure 3 shows a schematic diagram of the framework of the RUSBagging method. First, the training set is randomly sampled to form multiple training subsets with differences in data characteristics. Then, using random under-sampling to obtain balanced data subsets. The weak learner is trained based on each balanced subset. Finally, the results are synthesized to obtain the final comprehensive result. After training, the test set is used to test the training effect of the model.

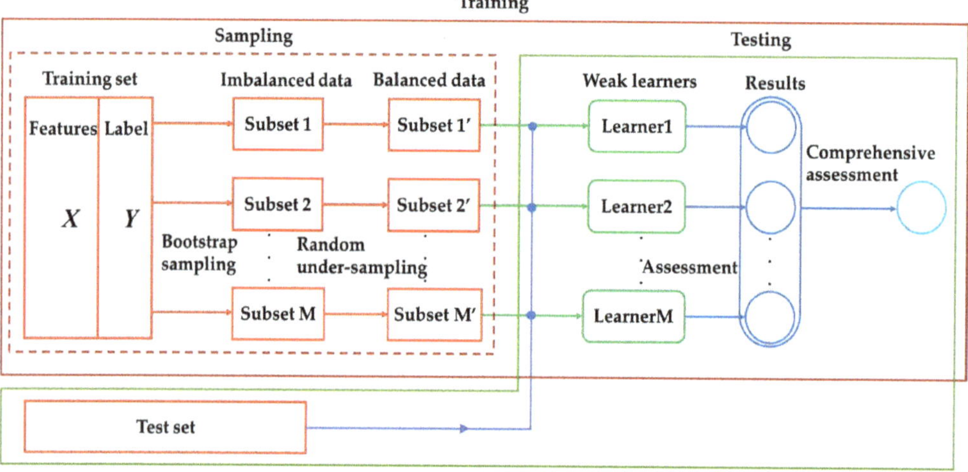

Figure 3. Schematic diagram of the random under-sampling bagging method.

4. Modeling of Static Voltage Stability Assessment Based on Machine Learning

Based on power flow calculation and local voltage stability index calculation, the static voltage stability assessment problem of the power system is treated as a supervised machine learning problem. With the help of the machine learning method, the mapping relationship between the operating state and voltage stability is mined. The idea frame diagram is shown in Figure 4.

In the framework shown in Figure 4, there are mainly four parts: scene generation, sample generation, model building, and model training. The scene simulation is carried out considering the characteristics of the actual operation scene. In addition, the power flow calculation is performed on the simulated scene. The power flow calculation result is a feature variable of the sample corresponding to the scene. Based on the power flow calculation result, the local voltage stability index is calculated too. The index value is the corresponding label value (true value) of the sample in the scene.

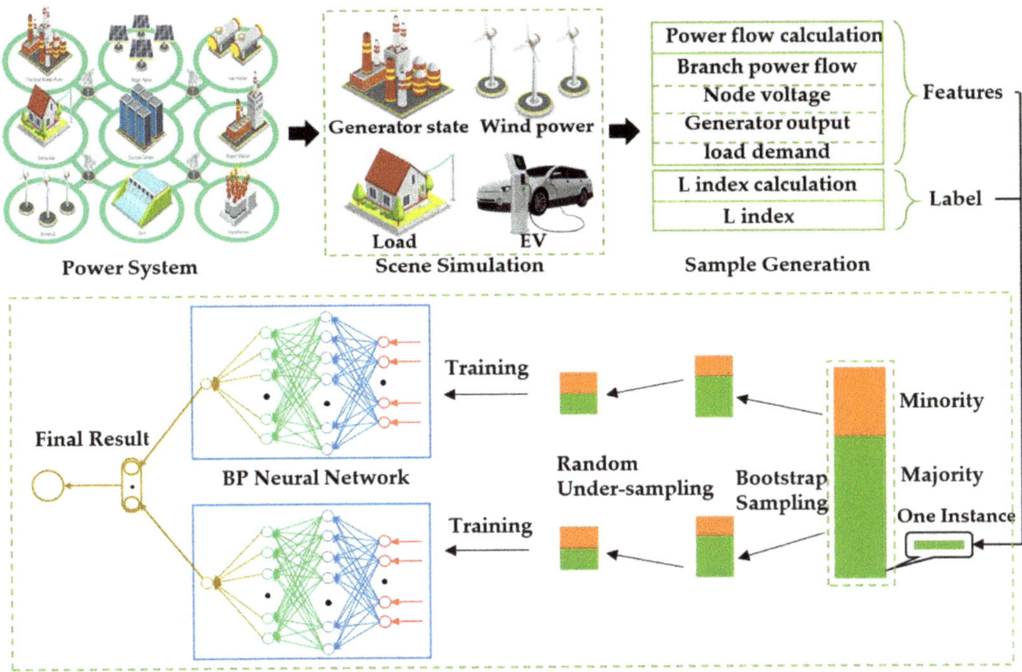

Figure 4. The framework for static VSA of power system based on RUSBagging.

The scenario simulation mainly considers the following factors: load demand, generator status, and new energy output power. Specifically, for the load demand, there are heavy load demand and light load demand. For the generator status, the generator does not reach the limit of reactive power and the reactive power of some units reaches the limit. There are two reasons for considering this factor: First, when the generator node transforms into a PQ node, the voltage stability state of the system will change abruptly. Second, the calculation of the L index needs to determine the type of system nodes in advance. When the type of node changes, the L index calculation model needs to be updated. For the output of new energy, the node where the unit is located is regarded as the PQ node. The load side also considers the batch connection to the grid and withdraws from the grid of electric vehicles.

The above factors only consider typical scenarios, so the number of scenarios is limited. Therefore, in the simulation, a mixed simulation of various factors is adopted to expand the number of scenarios. After the scenario simulation is completed, the power flow calculation is carried out for each scenario to obtain the voltage amplitude and phase angle of each node, the active and reactive power output of the generator, and the line power flow. These power flow calculation results and the load demand together constitute the features of samples. The L index corresponds to the features of samples packaged into complete training data. Since the L index can be directly calculated based on the power flow state, it does not require continuous power flow calculation like PV analysis, so the sample collection speed is very fast. The training set, validation set, and test set are randomly selected according to the ratio of 90%, 5%, and 5%.

BP network belongs to supervised learning [27]. In the process of neural network modeling, the selection of activation function, loss function, and the optimization algorithm is required. In the design of hyperparameters, such as the number of hidden layers, the number of neurons in the hidden layer, the number of parallel BP networks, etc., it needs to be set according to specific problems, and these hyperparameters rely more on empirical values.

(1)   Activation function

The activation function is the key to the nonlinear mapping function of the neural network. Common activation functions include Sigmoid, ReLU classes (ReLU, LReLU, RReLU), Tanh and Softmax. Through the nonlinearization of the input data by the above activation function, combined with the deep superposition of the neural network, the fitting of the nonlinear function is realized. In this paper, the LReLU activation function is selected as the activation function of the BP network, because the dead zone of LReLU has a small range. At the same time, LReLU can effectively avoid the problem of gradient disappearance, and also alleviate the problem of neuron death of ReLU, which is beneficial to the neural network. The curve of LReLU is shown in Figure 5. The expression of the LReLU activation function is:

$$f(x) = \max\{0.1x, x\} \tag{14}$$

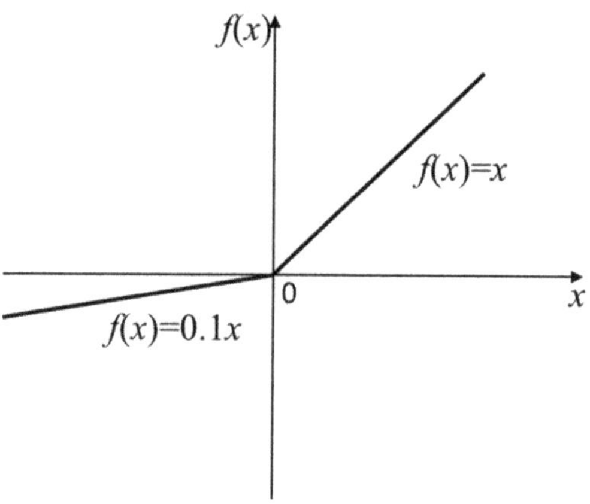

Figure 5. The curve of LReLU.

The domain of the LReLU function is negative infinity to positive infinity. LReLU alleviates the problem of ReLU neuron death and solves the problem that some neurons cannot be activated.

(2)   Loss function

The loss function is used to measure the difference between the output value of the model and the true value of the sample. Through the back-propagation process, the loss function is minimized, the weight of the network is corrected, and the gap between the output value of the model and the true value of the sample is continuously narrowed to achieve network convergence.

For different learning models, such as regression models and classification models, the type of loss function needs to be selected. For the classification model, the cross-entropy loss function is generally used. For the regression model, the mean square error loss function is generally used. In this paper, the problem of voltage stability assessment is defined as a regression problem, so the mean square error function is chosen as the loss function.

$$E = \frac{1}{N} \sum_{i=1}^{N} (y_i - \hat{y}_i)^2 \tag{15}$$

where E represents the output value of the loss function, N represents the number of samples used in a parameter update process, and y_i and $\hat{y}_i$ represent the true value and predicted value of the ith sample label, respectively.

(3)   Optimization algorithm

After completing the construction of the loss function, it is necessary to implement parameter correction through the optimization algorithm. The deep learning optimization algorithm mainly includes the basic optimization algorithm and the adaptive parameter optimization algorithm. The representative algorithm of the basic optimization algorithm is the stochastic gradient descent method, which keeps the learning rate unchanged during the training process. It cannot dynamically adapt to the training requirements. In addition, it is easy to fall into the local optimum point. The representative algorithm of the adaptive parameter optimization algorithm is Adam. The learning rate is gradually attenuated to better adapt to the training requirements as the learning progresses, shorten the training time, and improve the training effect. In this paper, the Adam algorithm [28] is used as the optimization algorithm for network training.

$$
\begin{cases}
m_t = \mu \cdot m_{t-1} + (1 - \mu) \cdot g_t \\
n_t = v \cdot n_{t-1} + (1 - v) \cdot g_t^2 \\
\hat{m}_t = m_t / (1 - \mu) \\
\hat{n}_t = n_t / (1 - v) \\
\Delta\theta_t = -\hat{m}_t \cdot \eta \cdot g_t / \sqrt{n_t + \varepsilon}
\end{cases}
\tag{16}
$$

In the formula, g_t represents the gradient, m_t represents the first-order moment estimation of the gradient, n_t represents the second-order moment estimation of the gradient, $\hat{m}_t$ and $\hat{n}_t$ represent the corrected values of m_t and n_t, respectively. μ and v represent the first-order momentum and the second-order momentum, respectively. Momentum coefficient, ε means avoiding smoothing terms with 0 denominators, η means learning rate. The standard settings for μ and v are 0.9 and 0.999, respectively, and the default value for η is 0.001. A satisfactory training effect can be obtained by applying this set of hyperparameters during the training process, and no special adjustment is generally required.

5. Results

To verify the effectiveness of the proposed model, the modified IEEE39 system case is taken as an example, as shown in Figure 6. The IEEE39 system [29] has 39 nodes, 19 load nodes, 10 thermal power units, and 46 branches (including transformers) with the following modifications: replacing the thermal power generator on bus-39 with wind turbines with a capacity of 650 MVA and removing the load of bus-39.

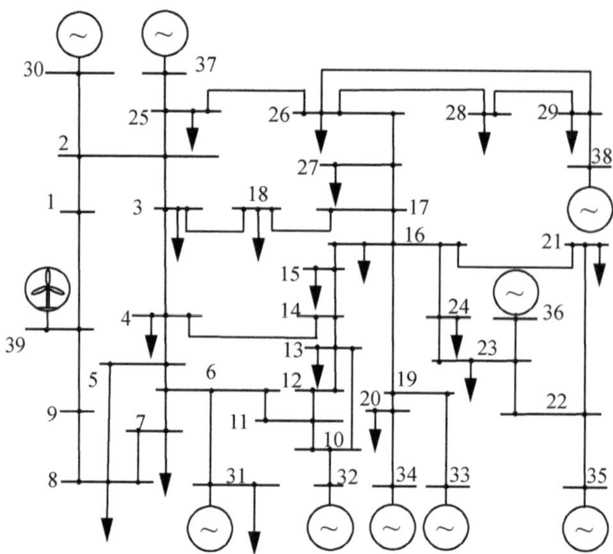

Figure 6. Modified IEEE 39 bus system.

The neural network model is built based on the Pytorch framework, and the power system simulation is performed based on the PSSE simulation platform. An analysis is developed from the perspectives of model training time, mean square error (MSE), and mean absolute percentage error (MAPE).

In machine learning, MSE is generally used as the error of model training, and it is used as the objective function to update the parameters. The expression of MSE is shown in Formula (17):

$$\text{MSE} = \frac{1}{n}\sum_{i=1}^{n}(y_i' - y_i)^2 \tag{17}$$

where y_i' is the predicted value of the ith sample, and y_i is the true value of the ith sample. The advantage of MSE is to amplify extreme errors and avoid huge deviations in the model. The disadvantage is that it is not intuitive and it is difficult to explain its meaning after squaring.

In order to intuitively reflect the difference between the actual value and the predicted value, there is MAPE, which is expressed as Formula (18):

$$\text{MAPE} = \frac{1}{n}\sum_{i=1}^{n}\frac{|y_i' - y_i|}{y_i} \tag{18}$$

The value of MAPE is intuitive and has a clear meaning, but when the actual value is very small, it is easy to produce misleading information. Therefore, MAPE is generally not used for the loss function of regression problems with small real values, but it can be used as a more intuitive method to measure the model error.

In summary, MSE is used to evaluate the overall performance of the model, and MAPE is used to evaluate the performance of the model on batch instances.

The training time of different methods and the MSE and MAPE error of each method on the test set are shown in Table 2.

Table 2. Comparison of the performance of different methods on the whole test set.

Num	Methods	Train-Time/s	MSE	MAPE
1	BP-RUSBagging	312.44	6.0252×10^{-7}	0.0011
2	BP-Bagging	290.35	4.7632×10^{-7}	0.0014
3	BP	213.54	2.2391×10^{-6}	0.0018
4	SVR	564.21	9.3797×10^{-6}	0.0022

SVR stands for support vector regression. The test set is divided into multiple batches of data, and each batch of data is calculated to obtain the batch MAPE index and plot the results, as shown in Figure 7.

As shown in Figure 7, the errors of the four methods on the test set are all small, and the advantage of method 1 is not obvious. This is because the test set is also a class-imbalanced data set, so the advantage of the under-sampling method is not prominent on the whole test set.

To further illustrate the applicability of the proposed method to the class imbalance problem, the minority class samples in the test set are screened out, and then four methods are used for comparison based on the minority test set. The results are shown in Table 3.

As shown in Figure 8, on the screened test set, the proposed method has obvious advantages over other methods. It has a lower error on minority class samples. Compared with methods 3 and 4, method 2 also shows the adaptability to the class imbalance problem to a certain extent. This should be credited to the bagging framework.

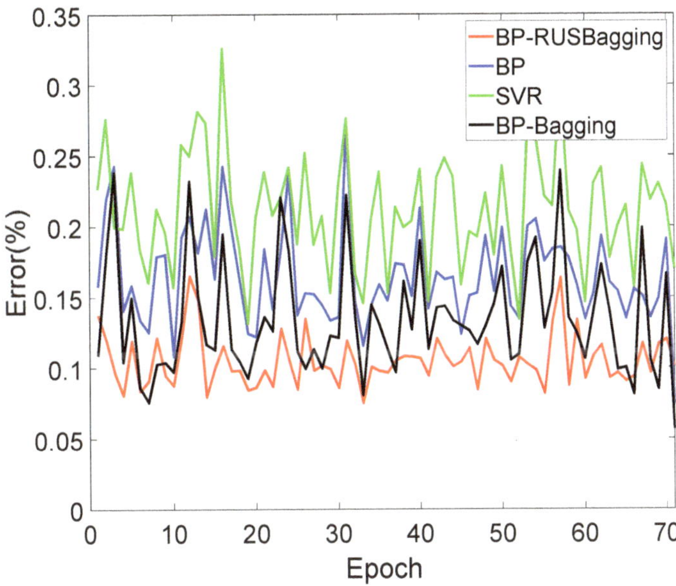

Figure 7. MAPE results of each method on the whole test set.

Table 3. Comparison of the performance of different methods on the minority test set.

Num	Methods	Train-Time/s	MSE	MAPE
1	BP-RUSBagging	312.44	1.7763×10^{-5}	0.0167
2	BP-Bagging	290.35	8.1022×10^{-5}	0.0284
3	BP	213.54	2.2391×10^{-4}	0.0411
4	SVR	564.21	1.4397×10^{-3}	0.0622

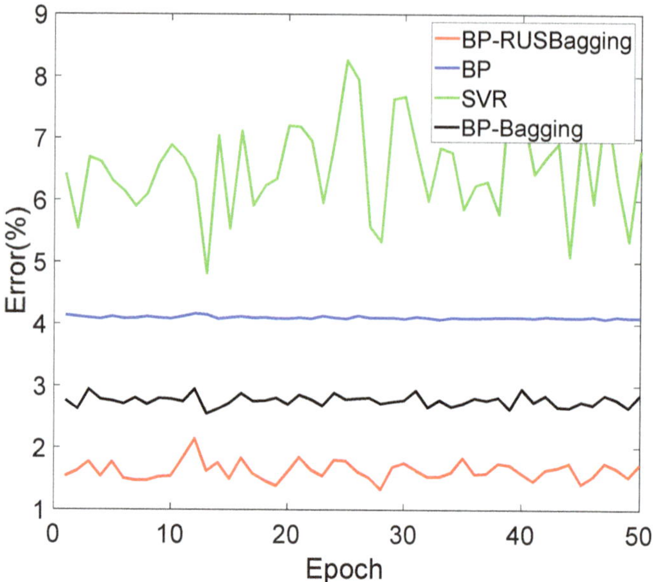

Figure 8. MAPE results of each method on the minority test set.

6. Conclusions

The comparative analysis proves that the proposed static voltage stability assessment method not only has high accuracy, strong adaptability, and short modeling time but also has the following advantages:

(1) In the sample preparation stage, various operating scenarios such as load demand, new energy output power, and EV status were considered, which greatly improved the scenario applicability of the model.

(2) Defining the static voltage stability assessment problem as a regression problem of machine learning, which essentially improves the model assessment accuracy, provides voltage stability information with a quantitative index, and helps grid operators to better observe the grid state.

(3) Using the random under-sampling bagging framework provides a method to solve the problem of imbalanced data in the field of power system operating, which comprehensively improves the accuracy of the model.

Author Contributions: Conceptualization, P.Z. and J.W.; methodology, P.Z. and Z.L.; software, Z.Z. and Z.L.; validation, Z.Z. and J.W.; formal analysis, P.Z. and Z.L.; investigation, J.W. and Z.L.; resources, P.Z.; data curation, Z.Z. and J.W.; writing—original draft preparation, Z.Z.; writing—review and editing, P.Z. and J.W.; visualization, Z.Z.; supervision, P.Z.; project administration, P.Z.; funding acquisition, Z.L. All authors have read and agreed to the published version of the manuscript.

Funding: This research was funded by National Nature Science Foundation of China, grant number 52107068.

Data Availability Statement: Not applicable.

Acknowledgments: Not applicable.

Conflicts of Interest: The authors declare no conflict of interest.

References

1. Kundur, P. *Power System Stability and Control*; McGraw-Hill: New York, NY, USA, 1993.
2. Van Cutsem, T.; Vournas, C. *Voltage Stability of Electric Power Systems*; Kluwer: Cambridge, UK, 1998.

3. Srivastava, L.; Singh, S.N.; Sharma, J. Estimation of loadability margin using parallel self-organizing hierarchical neural network. *Comput. Electr. Eng.* **2000**, *26*, 151–167. [CrossRef]
4. Suganyadevi, M.V.; Babulal, C.K. *Online Voltage Stability Assessment of Power System by Comparing Voltage Stability Indices and Extreme Learning Machine*; Springer: Berlin, Germany, 2013.
5. Qiang, W.; Hao, C.; Lian, L. Static Voltage Stability Margin Prediction Based on Natural Gradient Boosting and Its Influencing Factors Analysis. In Proceedings of the CSU-EPSA, Prague, Czech Republic, 23–25 June 2022; pp. 1–9.
6. Ghiocel, S.G.; Chow, J.H. A Power Flow Method Using a New Bus Type for Computing Steady-State Voltage Stability Margins. *IEEE Trans. Power Syst.* **2014**, *29*, 958–965. [CrossRef]
7. Luo, F.; Dong, Z.Y.; Chen, G.; Xu, Y.; Meng, K.; Chen, Y.Y. Advanced pattern discovery-based fuzzy classification method for power system dynamic security assessment. *IEEE Trans. Ind. Informat.* **2015**, *11*, 416–426. [CrossRef]
8. Gholami, M.; Sanjari, M.J.; Safari, M.; Akbari, M.; Kamali, M.R. Static security assessment of power systems: A review. *Int. Trans. Electr. Energy Syst.* **2020**, *30*, e12432. [CrossRef]
9. Liu, S.; Liu, L.; Yang, N.; Mao, D.; Shi, R. A data-driven approach for online dynamic security assessment with spatial-temporal dynamic visualization using random bits forest. *Int. J. Electr. Power Energy Syst.* **2021**, *124*, 106316. [CrossRef]
10. Ding, C.; Zhang, P.; Meng, X.; Li, W.; Wang, Y. Online Evaluation on Static Voltage Stability Margin Based on Classification and Regression Tree Algorithm. *Proc. CSU-EPSA* **2020**, *32*, 93–100.
11. Wenqing, L.; Qingwu, G.; Chunhui, G.; Liu, H.; Liu, W.; Liuchuang, W.U. Transient voltage stability evaluation based on feature and convolutional neural network. *Eng. J. Wuhan Univ.* **2019**, *52*, 815–823.
12. Xiurui, L.; Daowei, L.; Hongying, Y. Data-Driven Situation Assessment of Power System Static Voltage Stability. *Electr. Power Constr.* **2020**, *41*, 126–132.
13. Zhu, L.; Lu, C.; Dong, Z.Y. Imbalance learning machine-based power system short-term voltage stability assessment. *IEEE Trans. Ind. Inform.* **2017**, *13*, 2533–2543. [CrossRef]
14. Johnson, J.M.; Khoshgoftaar, T.M. Survey on deep learning with class imbalance. *J. Big Data* **2019**, *6*, 1–54. [CrossRef]
15. Li, Y.; Zhang, M.; Chen, C. A Deep-Learning intelligent system incorporating data augmentation for Short-Term voltage stability assessment of power systems. *Appl. Energy* **2022**, *308*, 118347. [CrossRef]
16. Haixiang, G.; Yijing, L.; Shang, J.; Mingyun, G.; Yuanyue, H.; Bing, G. Learning from class-imbalanced data: Review of methods and applications. *Expert Syst. Appl.* **2017**, *73*, 220–239. [CrossRef]
17. Malave, N.; Nimkar, A.V. A survey on effects of class imbalance in data pre-processing stage of the classification problem. *Int. J. Comput. Syst. Eng.* **2020**, *6*, 63–75. [CrossRef]
18. Khan, S.H.; Hayat, M.; Bennamoun, M.; Sohel, F.; Togneri, R. Cost-sensitive learning of deep feature representations from imbalanced data. *IEEE Trans. Neural Netw. Learn. Syst.* **2017**, *29*, 3573–3587. [PubMed]
19. Rao, N.A.; Vijaya, P.; Kowsalya, M. Voltage stability indices for stability assessment: A review. *Int. J. Ambient. Energy* **2021**, *42*, 829–845.
20. Salama, H.S.; Vokony, I. Voltage stability indices–A comparison and a review. *Comput. Electr. Eng.* **2022**, *98*, 107743. [CrossRef]
21. Kessel, P.; Glavitsch, H. Estimating the Voltage Stability of a Power System. *IEEE Trans. Power Deliv.* **2007**, *1*, 346–354. [CrossRef]
22. Wei, W.; Xu, Y. Deterministic convergence of an online gradient method for neural networks. *J. Comput. Appl. Math.* **2002**, *144*, 335–347.
23. Zhang, L.; Wang, F.; Sun, T.; Xu, B. A constrained optimization method based on BP neural network. *Neural Comput. Appl.* **2018**, *29*, 413–421. [CrossRef]
24. Shi, X.; Xu, G.; Shen, F. Solving the data imbalance problem of P300 detection via random under-sampling bagging SVMs. In Proceedings of the 2015 International Joint Conference on Neural Networks, Killarney, Ireland, 12–17 July 2015; pp. 1–5.
25. Błaszczyński, J.; Stefanowski, J. Actively balanced bagging for imbalanced data. In Proceedings of the International Symposium on Methodologies for Intelligent Systems, Graz, Austria, 23–25 September 2017; pp. 271–281.
26. Bühlmann, P. Bagging, boosting and ensemble methods. In *Handbook of Computational Statistics*; Springer: Berlin, Germany, 2012; pp. 985–1022.
27. Goodfellow, I.; Bengio, Y.; Courville, A. Deep Learning. MIT Press: Cambridge, UK, 2016.
28. Guan, N.; Lei, S.; Yang, C.; Xu, W.; Zhang, M. Delay Compensated Asynchronous Adam Algorithm for Deep Neural Networks. In Proceedings of the 2017 IEEE International Symposium on Parallel and Distributed Processing with Applications and 2017 IEEE International Conference on Ubiquitous Computing and Communications (ISPA/IUCC), Guangzhou, China, 12–15 December 2017; IEEE: Piscataway, NJ, USA, 2017.
29. Sriyanyong, P.; Song, Y.H. Unit commitment using particle swarm optimization combined with Lagrange relaxation. In Proceedings of the Power Engineering Society General Meeting, San Francisco, CA, USA, 16 June 2005.

processes

Article

Fault Diagnosis of Wind Turbine Main Bearing in the Condition of Noise Based on Generative Adversarial Network

Zhixin Fu, Zihao Zhou * and Yue Yuan

College of Energy and Electrical Engineering, Hohai University, Nanjing 211100, China
* Correspondence: 201306080047@hhu.edu.cn

Abstract: In order to solve the problem that the fault classification accuracy of the main bearing of the wind turbine is not high due to the unbalanced vibration signal data of the main bearing of the wind turbine under the background of noise, this article proposes a double-layer fault diagnosis model for the main bearing of the wind turbine that combines the auxiliary classifier generation adversarial network (ACGAN) and the deep residual shrinkage network (DRSN). First, the wind turbine main bearing data is sent into the ACGAN to learn the distribution features of fault data, and a particular type of fault data is generated to expand the original dataset to achieve balance conditions, and then the expanded dataset is sent to the DRSN to reduce noise to improve the fault classification accuracy. The simulation results show that, compared with the traditional deep learning model, the model proposed in this article can significantly improve the classification accuracy of the main bearing fault of wind turbines under noise conditions, and also has a strong diagnosis ability in a state of datasets with different loads.

Keywords: wind turbine; main bearing; fault diagnosis; noise; deep residual shrinkage network; auxiliary classifier generative adversarial network

Citation: Fu, Z.; Zhou, Z.; Yuan, Y. Fault Diagnosis of Wind Turbine Main Bearing in the Condition of Noise Based on Generative Adversarial Network. *Processes* **2022**, *10*, 2006. https://doi.org/10.3390/pr10102006

Academic Editor: Marco S. Reis

Received: 29 August 2022
Accepted: 1 October 2022
Published: 4 October 2022

Publisher's Note: MDPI stays neutral with regard to jurisdictional claims in published maps and institutional affiliations.

1. Introduction

As an important green renewable energy, wind energy has become a research hotspot in recent years, and the wind power industry has also developed rapidly. By the end of 2021, the cumulative installed capacity of wind turbines in the world has reached 328 million kilowatts, of which the installed capacity of offshore wind power has reached 26.39 million kilowatts [1]. If calculated according to the 5-year warranty period of wind turbines, about 3 GW capacity of offshore wind turbines are about to or have already gone out of the warranty period [2]. Therefore, the market potential for wind turbine operation and maintenance is huge. The operating environment of wind turbines is harsh, and the wind turbines are exposed to sand and snow for a long time [3]. In addition, there is also severe weather such as thunderstorms and fog at sea. According to statistics, the operation and maintenance cost of onshore wind farms is as high as 15% to 20% of the total wind farm revenue, while the operation and maintenance cost of offshore wind farms is much higher than that of onshore wind farms, accounting for about 20% to 25% of the total wind farm revenue [4]. The high fail rate of wind turbines brings great difficulties to the operation and maintenance of wind farms, and the failure to discover potential faults in time and repeated maintenance of components with a high fail rate will increase the operation and maintenance costs of wind farms.

During the operation stage of the wind farm, the faults of the generator, the gearbox, the transmission system, and the blades are the most common [5], of which the main bearing of the wind turbine plays a role in transmitting energy to the wind turbine [6]. As a rotating component, the main bearing is more prone to failure, and the entire unit will stop running after the failure, causing huge economic losses. The vibration signal contains all the useful information about the components and it is also one of the important indicators for analyzing the operating state. The fault diagnosis technology for analyzing the vibration signal of the main bearing is currently the most effective and widely used [7]. In recent years, some domestic and foreign scholars have introduced artificial intelligence methods into the field of fault diagnosis of rotating components of wind turbines, such as deep learning and so on. Compared with traditional fault diagnosis methods based on statistical analysis methods, the fault diagnosis method based on the neural network does not rely on a large number of signal processing related knowledge and rich expert experience, but the essential characteristics of faults are extracted from massive historical data, avoiding the randomness of manual selection of parameters, and the diagnosis process is more intelligent [8]. Cao et al. [9] use Long-Short Term Memory (LSTM) neural network to extract the fault characteristics of vibration signal of wind turbines and perform fault classification and compare this method with the support vector machine method to verify the superiority of the algorithm, the method proposed in this article achieves 97.2% of the classification accuracy of the gearbox. Wu et al. [10] adopt a convolutional neural network to study the one-dimensional vibration signal of the planetary gearbox. The conclusion shows that the accuracy of the one-dimensional convolutional neural network model for fault diagnosis of planetary gearbox is higher than that of traditional diagnosis methods. Yao et al. [11] propose a fault diagnosis method for rolling bearings based on a convolutional neural network and recurrent neural network.

The one-dimensional vibration signal is converted into a two-dimensional image signal by the Gram angle field method, and the image signal is input into the model for training, which has a higher fault classification accuracy, experiments show that the method proposed has an accuracy of more than 98.15% for the classification of rolling bearing faults. However, these references ignore that the fault data of offshore wind turbines is often difficult to be obtained, and there is a general problem of insufficient fault samples. In particular, the main bearing fault data of wind turbines accounts for a relatively low proportion of all fault data, and there is a serious unbalanced dataset problem. Therefore, it is difficult for deep learning methods to achieve high fault classification accuracy in this case. Zhou et al. [12] believe that when most classification algorithms classify unbalanced data, the obtained classification hyperplane will be biased toward a few types of data, which leads to the algorithm misjudging the minority type of data as the majority type of data. In order to obtain sufficient and balanced vibration signal samples, some scholars refer to generative adversarial networks in the field of rolling bearing fault diagnosis. Lu et al. [13] propose a data enhancement method for the vibration signal of the main bearing of wind turbines based on an auxiliary classification generation adversarial network, which can effectively extract the original data distribution characteristics and generate high-quality vibration signal samples, after using ACGAN to expand the original dataset, the fault classification accuracy of various models is improved by about 2%. Li et al. [14] improve the auxiliary classification generative adversarial network based on Bayesian optimization and Wasserstein distance, realize data enhancement, and obtained a higher fault classification accuracy of wind turbine planetary gearboxes. The classification accuracy of WAC-GAN could remain above 94% for various types of failures. In addition, due to the harsh operating environment of offshore wind turbines, the signal samples collected by sensors often contain noises. These noises will affect the feature extraction performance of neural networks during training. Traditional signal denoising methods often require a lot of statistical knowledge. Different noise interference is targeted for different noise reduction processing. Zhao et al. [15] propose a deep learning-based feature learning algorithm for noisy data, which integrates the attention mechanism and the idea of a soft threshold, effectively reducing the impact of noise interference on the model, experiments show that in the case of inserting various types of noise, the accuracy of DRSN with channel-wise thresholds (DRSN-CW) is about 3.32% higher than that of ResNet. In [16], a deep residual shrinkage network is added to the convolutional neural network to achieve signal noise reduction and solve the degradation problem of the multi-layer model. This method has a higher fault classification accuracy reaching 99.5% than the traditional neural network method. However, the above studies do not take into account the unbalanced data of the main bearing of the actual wind turbine, which limited the fault diagnosis capability of the model.

In order to improve the performance of the fault diagnosis model in practical applications, this article focuses on the research on the fault diagnosis method of wind turbine main bearing under noise conditions and proposes a fault diagnosis method of the main bearing of the wind turbine based on the auxiliary classification generate adversarial network and the deep residual shrinkage network. First, the auxiliary classification generative adversarial network is adopted to learn the data distribution of vibration signal samples with different signal-noise ratios, and the datasets of each fault are expanded. Then, the expanded dataset is sent to the deep residual shrinkage network for training. Finally, the test set is fed into the trained deep residual shrinkage network to test the fault classification accuracy of the model. Experimental results show that the proposed method has good fault diagnosis performance in the face of the vibration signal sample of the main bearing of the wind turbine when the actual operation contains noise interference, and the data is unbalanced.

The first chapter of this article is an introduction, the second chapter gives the structure of the model proposed in this article, the third chapter describes the basic principle of the auxiliary classification generative adversarial network, and the fourth chapter describes the basic principle of the deep residual shrinkage network, the fifth chapter uses two experiments to verify the effectiveness of the method proposed in this article, and the sixth chapter gives some conclusions and suggestions.

2. Fault Diagnosis Model of Wind Turbine Main Bearing

In this article, a fault diagnosis model of the main bearing of a wind turbine with a double-layer network structure is used, and the model is shown in Figure 1. The upper layer is a generative network based on the auxiliary classification generative adversarial network. The generator learns the data distribution characteristics of the original vibration signal dataset during training. Then, particular types of fault data are generated to expand the dataset to a balanced state. That is, the ratio of the sample of each type of fault data to health status data is 1:1; then the expanded dataset is fed into the classification network based on the deep residual shrinkage network, and the attention mechanism and soft threshold are used in the classification network to reduce the redundant noise in the signal adaptively, and the classifier can accurately identify the fault samples of the main bearing of the wind turbine through the training of the expanded dataset.

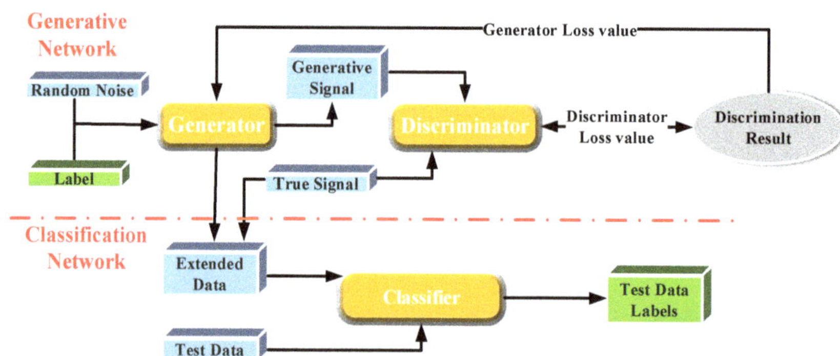

Figure 1. Fault Diagnosis Model of the Main Bearing of Wind Turbine.

3. Generative Adversarial Network

In 2014, Lan Goodfellow et al. proposed Generative Adversarial Networks (GAN) [17]. Since GAN can generate data with a specific distribution, it is an unsupervised deep learning model, so it is widely used in image inpainting, text generation, audio generation, and other fields, and is the research hotspot in the image field in recent years. Because the fault classification algorithm based on deep learning needs a large amount of fault data, and the actual wind turbine main bearing fault data samples are scarce, there is a serious sample imbalance problem between fault data and normal data. In order to improve the accuracy of fault classification, some scholars have introduced the generative adversarial network into the field of fault diagnosis. The generative adversarial network can learn the data distribution of real fault samples to generate new fault samples to supplement the original dataset. Studies have shown that [18], this method has great potential in the application of time series data generation.

3.1. Generative Adversarial Network Principle

GAN contains two networks, Generator and Discriminator. The application of the generator is to convert the random noise into data that is close to the one-dimensional vibration signal data distribution of the main bearing of the wind turbine as much as possible. The function of the discriminator is to judge the authenticity of the input samples. During the training, the generator and the discriminator game alternately. The game mechanism continuously improves the generation ability of the generator, so that the data generated by the generator is as real as possible and deceives the discriminator to achieve the purpose of being a genuine one.

The training of GAN mainly includes two stages: discriminator training and generator training. In each round of iteration, the generator and the discriminator compete with each other and finally reach the Nash equilibrium, that is, the discriminator classification accuracy rate reaches 50%. The real fault data of the main bearing of a wind turbine or the fake data generated by the generator has a 50% chance of being misjudged, and the generator completes the training. The structure of GAN is shown in Figure 2.

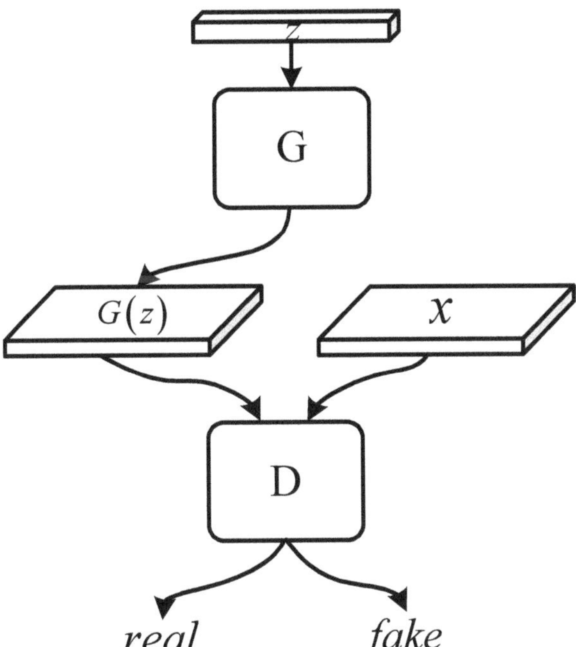

Figure 2. Generative Adversarial Network Structure Diagram.

In the initial stage of training, the capabilities of the generator and the discriminator are very weak. First, the random noise vector is sent to the generator to generate fake samples, fake samples, and the real signal samples are passed through the discriminator to generate scores. The loss function consists of samples and the labels corresponding to these samples, and the gradient is calculated to update the discriminator; then, the parameters of the discriminator are fixed, and fake samples will be sent to the discriminator to get the score since it is hoped that the samples generated by the generator are as real as possible and deceive the discriminator, The optimization goal at this time is to make the score reach 1 to update the generator by calculating the gradient. In the iteration, the generator and the discriminator have trained alternately, and finally, the two networks reach a Nash equilibrium state. At this time, the discriminator cannot distinguish between real samples and fake samples, and the accuracy of the discriminator is 50%. The network objective function is (1):

$$\min_{G}\max_{D} V(D,G) = E_{x \sim p_{data}(x)}[\log D(x)] + E_{z \sim p_z(z)}[\log(1 - D(G(z)))] \tag{1}$$

$$D(x) = \frac{p_{data}(x)}{p_{data}(x) + p_g(x)} \tag{2}$$

where E is the mathematical expectation, $p_{data}(x)$ and $p_g(x)$ is the probability that the sample is true or false, respectively, z is the random noise vector, and $G(z)$ is the fake sample generated by the generator. Our mission is to train the discriminator to maximize $\log D(x)$ and $\log(1 - D(G(z)))$, and train the generator to minimize $\log(1 - D(G(z)))$.

The generative model is essentially a maximum likelihood estimation. It is assumed that the initial distribution of the generator is $P_g(x|\theta)$, where θ is the parameter of the distribution. In order to make the generated data distribution close to the real data distribution, it is necessary to calculate the value $\hat{\theta}$ to maximize (3). Therefore, the calculation formula is (4):

$$L_g = \prod_{i}^{n} p_g\left(x^i \middle| \theta\right) \tag{3}$$

$$\hat{\theta} = \arg\max_{\theta} \prod_{i=1}^{n} p_g\left(x^i \middle| \theta\right) = \arg\min_{\theta} KL\left(P_{data}(x) \parallel P_g(x|\theta)\right) \tag{4}$$

where $p_g(x^i|\theta)$ is the likelihood function of the real data, $P_{data}(x)$ is the real data distribution, and $P_g(x|\theta)$ is the generated data distribution. GAN adopts KL divergence to measure the distance between two distributions. If the KL divergence reaches the minimum value 0, then the distribution $P_{data}(x)$ and $P_g(x|\theta)$ are equal everywhere.

3.2. Auxiliary Classifier Generative Adversarial Networks

Unlike traditional GAN, Auxiliary Classifier Generative Adversarial Networks (ACGAN) add labels to the random noises which are input to the generator and generate fake fault samples with a specific type of label. Then the true and false fault data samples are input into the discriminator to get the output results, and the output results include both true or false labels and classification labels. The network can be used to generate different types of wind turbine main-bearing fault data in a targeted manner, and the original fault dataset can be expanded into a balanced dataset.

It can be seen from Figure 3 that ACGAN not only outputs the probability that the fault sample is real data or not but also outputs the fault class probability of the sample. Since ACGAN has category labels when generating and judging samples, it makes the generated fault samples more controllable. The true or false judgment and classification loss functions are (5) and (6), respectively:

$$L_s = E_{x \sim P_{data}}[\log_2 D(x)] + E_{z \sim P_z}[\log_2(1 - D(G(z)))] \tag{5}$$

$$L_c = E_{c \sim P_{data}}[\log_2 D(c)] + E_{c \sim P_z}[\log_2(1 - D(G(c)))] \tag{6}$$

where, L_s is the probability that the sample is real, and L_c is the probability that the sample is correctly classified. Therefore, in training, the discriminator is trained to maximize $L_s + L_c$, and the generator is trained to maximize $L_c - L_s$.

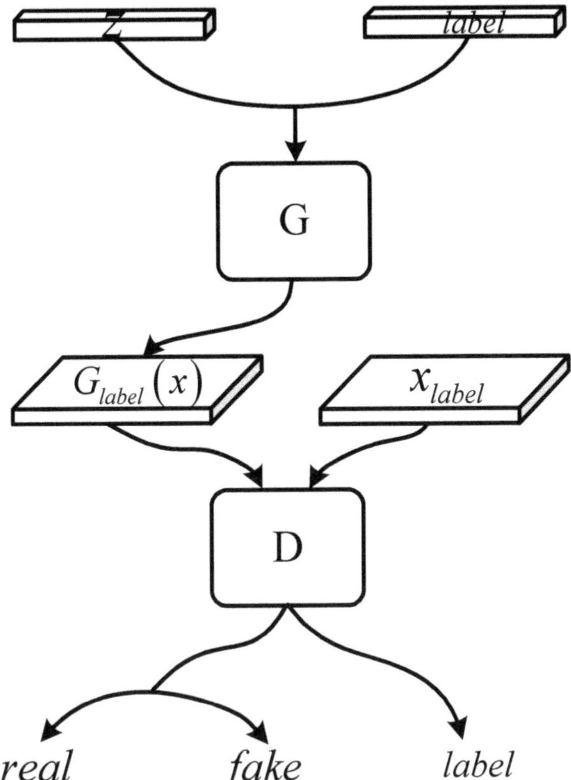

Figure 3. ACGAN Structure Diagram.

4. Deep Residual Shrinking Networks

The collected vibration signals of the main bearing of the wind turbine are often accompanied by noise actually, and the fault signal features are difficult to be extracted. The traditional signal noise reduction method is to transform the noisy signal (wavelet transform, empirical mode decomposition, etc.), and then use a soft threshold to reduce the noise, and finally, the signal is inverse transformed to obtain the signal after noise reduction. However, the noise signal of wind turbines may be different under different working conditions, and the selection of threshold is more complicated and requires a lot of relevant knowledge of signal processing. Therefore, this article selects the Deep Residual Shrinkage Network [15] as the classifier to diagnose the fault of the main bearing of the wind turbine. With the deepening of network layers, the ordinary convolutional neural network model training is difficult, the accuracy may be reduced. The residual network introduces the idea of an identity shortcut, the output of the previous layer of the network is directly transmitted to the next layer to achieve a smaller training error in the case of a larger number of network layers [19]. The DRSN is improved on the basis of the residual network to realize the function of noise reduction.

4.1. Attention

Attention Mechanism is widely used in the fields of natural language processing and pattern recognition. Its essence is similar to the human visual attention mechanism, that is, it selects the key information of the target task from many targets and suppresses useless information. Since the one-dimensional vibration data of the main bearing of the wind turbine is highly time-varying and the composition is complex, the introduction of an attention mechanism can adaptively weight different feature channels to highlight useful information. In this article, the channel threshold attention mechanism is used to apply weights to the signals on each channel to improve the fault classification ability of the network under various working conditions. For example, Squeeze-and-Excitation Networks (SENet) is a network that sets an attention mechanism for channels.

Figure 4 is a schematic diagram of the Squeeze-and-Excitation Networks module. The number of input channels is c_1, the number of output channels of the second layer is c_2, and then be compressed into a feature map of size $c_2 \times 1 \times 1$ by global average pooling. Finally, the *Softmax* activation function is used to obtain the weight of each channel and the second layer is weighted to obtain the output result. (7) is the *Softmax* function expression.

$$f(x_k) = \frac{\exp(x_k)}{\sum\limits_{k=1}^{c} \exp(x_k)} \tag{7}$$

where $f(x_k)$ represents the weight prediction value of the kth channel by the activation function *Softmax*, and c is the number of channels.

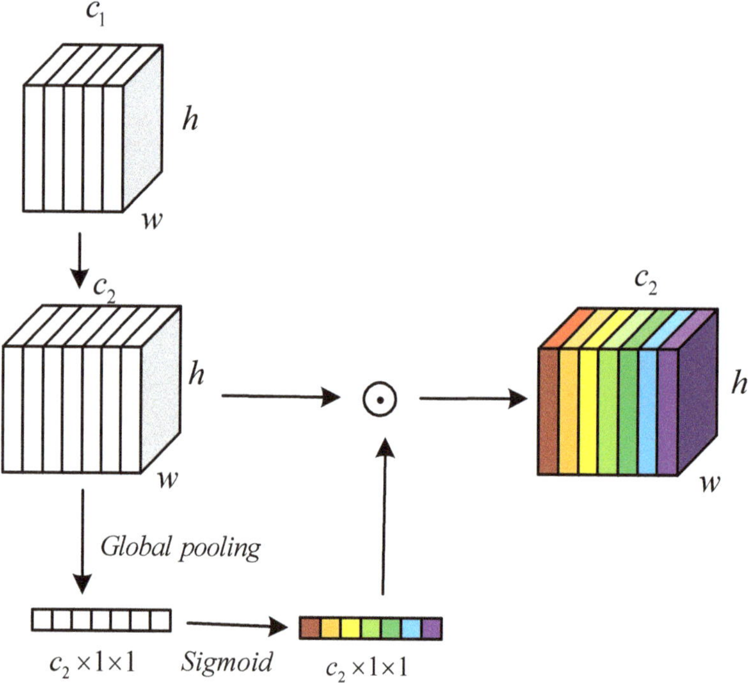

Figure 4. SENet Structure Diagram.

4.2. Soft Threshold Noise Reduction

The noise in the actual vibration signal greatly reduces the ability of model feature extraction. On the basis of the residual network, the DRSN adopts the method of soft threshold to denoise the signal. The soft threshold is to set a threshold, set the signal below the threshold to 0, and adjust the signal above the threshold to 0, that is, "shrink". The core of the DRSN is to notice the vibration of the main bearing of the wind turbine through the attention mechanism. The unimportant features in the signal are set to zero by the soft threshold, which enhances the neural network's ability to extract fault features from noisy signals. The selection of the threshold size has a direct impact on the noise reduction effect. (8) is the expression of the soft threshold.

$$S(x, \tau) = \begin{cases} x - \tau & x > \tau \\ 0 & -\tau < x < \tau \\ x + \tau & x < -\tau \end{cases} \tag{8}$$

where τ represents the size of the threshold, it can be seen from (8) that when the signal is within the threshold, the derivative is 0, otherwise, the derivative is 1.

4.3. Residual Shrinkage Module

In this article, a DRSN with a channel-wise thresholds module is used to build a classifier network. Different from ordinary DRSN, this module has independent thresholds in each channel [15]. The overall structure of the residual shrinkage module is shown in Figure 5.

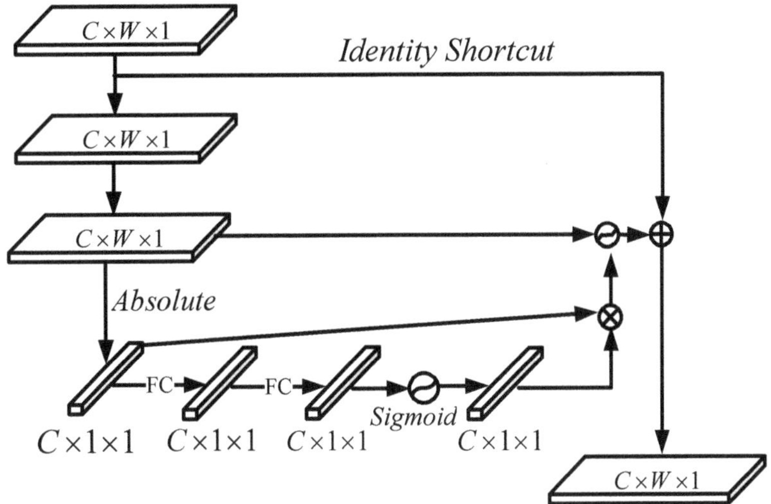

Figure 5. DRSN Module.

This module adds a denoising function to the traditional deep residual module. After the input data is passed through two layers of convolution, a one-dimensional vector is obtained after the global mean pooling layer. The attention weight of each channel is obtained by this vector through the two fully connected layers and the *Sigmoid* activation function, and the threshold is obtained by multiplying it with the corresponding average value of each channel. (9) is the expression of the threshold of the deep residual shrinkage module.

$$\tau_c = \omega_c \cdot average|x_c| \tag{9}$$

where τ_c is the threshold of channel c, ω_c is the weight of channel c, and $average|x_c|$ is the average value of the absolute value of each element of channel c.

5. Example Analysis

5.1. Model Framework

In order to test the effectiveness of the method mentioned above, the rolling bearing dataset [20] of Case Western Reserve University (CWRU) was selected as the simulation analysis object. The dataset comes from the sampled vibration data of the driving end of the wind turbine. The sampling frequency is 12 kHz, and each sample in the dataset contains 2048 sampling points. The final data is shown in Table 1. The simulation includes one normal state and nine different types of fault states, each state sample contains four different operating motor speeds. The dataset is randomly shuffled and divided into training and test sets.

Table 1. CWRU experimental data classification.

Fault Label	Bearing Status (Wear)	Fault Location	Fault Diameter/mm	Motor Load	Number of Training Samples	Number of Test Samples
0	Normal	/	0	0~3	700	120
1	Slight		0.18	0~3	160	70
2	Moderate	Inner Race	0.36	0~3	160	70
3	Heavy		0.54	0~3	160	70
4	Slight		0.18	0~3	160	70
5	Moderate	Ball	0.36	0~3	160	70
6	Heavy		0.54	0~3	160	70
7	Slight		0.18	0~3	160	70
8	Moderate	Outer Race	0.36	0~3	160	70
9	Heavy		0.54	0~3	160	70

Four different fault diagnosis models were implemented using TensorFlow 2.8.0, which is a machine learning toolkit released by Google. Experiments are conducted on the computer in which the CPU is AMD Ryzen 6 4800H, the GPU is Nvidia GeForce RTX 2060, and the memory is 32 GB.

In Table 1, load 0 represents the motor speed of 1797 rpm, load 1 represents the motor speed of 1772 rpm, load 2 represents the motor speed of 1750 rpm, and load 3 represents the motor speed of 1730 rpm.

The input of the generator in the generation network is a 100-dimensional normally distributed random vector and a label value. The label and the random vector are sent to the generator at the same time and become a fake sample of the same size as the input signal through a series of one-dimensional convolution operations. Each one-dimensional convolutional layer in the generator uses *LeakyReLU* as the activation function, adding a Dropout layer and a batch normalization layer to prevent overfitting and make the network easier to be trained. The structure of the discriminator is basically identical to the generator. The input data is a vibration signal sample, and the data after multi-layer one-dimensional convolution is passed through the fully connected layer and the *Softmax* activation function to obtain the true or false probability and the sample category probability respectively. The classification network structure is obtained by improvement on the basis of literature [15]. The improvement ideas are: (1) Appropriately reduce the depth of the network, which can significantly improve the training speed of the model and prevent the model from overfitting to a certain extent; (2) Increase the network width. It is found that increasing the number of channels of one-dimensional convolution can effectively improve the accuracy of the model. The RMSprop optimization algorithm is used in the model training, the hyperparameter is set to 0.9, and the adjustable learning rate with a lower limit of 0.00001 is used to speed up the convergence of the model.

5.2. Fault Diagnosis Ability under Noise Conditions

In order to test the denoising ability of the model in the condition of the unbalanced dataset, this experiment adds Gaussian white noise with different signal-noise ratios of -5 db, -2 db, 0 db, 2 db, and 5 db on the basis of the original dataset. CNN trained on the original dataset, DRSN trained on the original dataset, CNN trained on the dataset expanded by ACGAN, and DRSN trained on the dataset expanded by ACGAN are used to compare the classification accuracy for the test set. The CNN, DRSN, and ACGAN structures of different experimental groups in the simulation are the same, respectively.

The simulation results are shown in Figure 6a–e. The ordinate in the figure represents the model accuracy, and the abscissa represents the number of iterations.

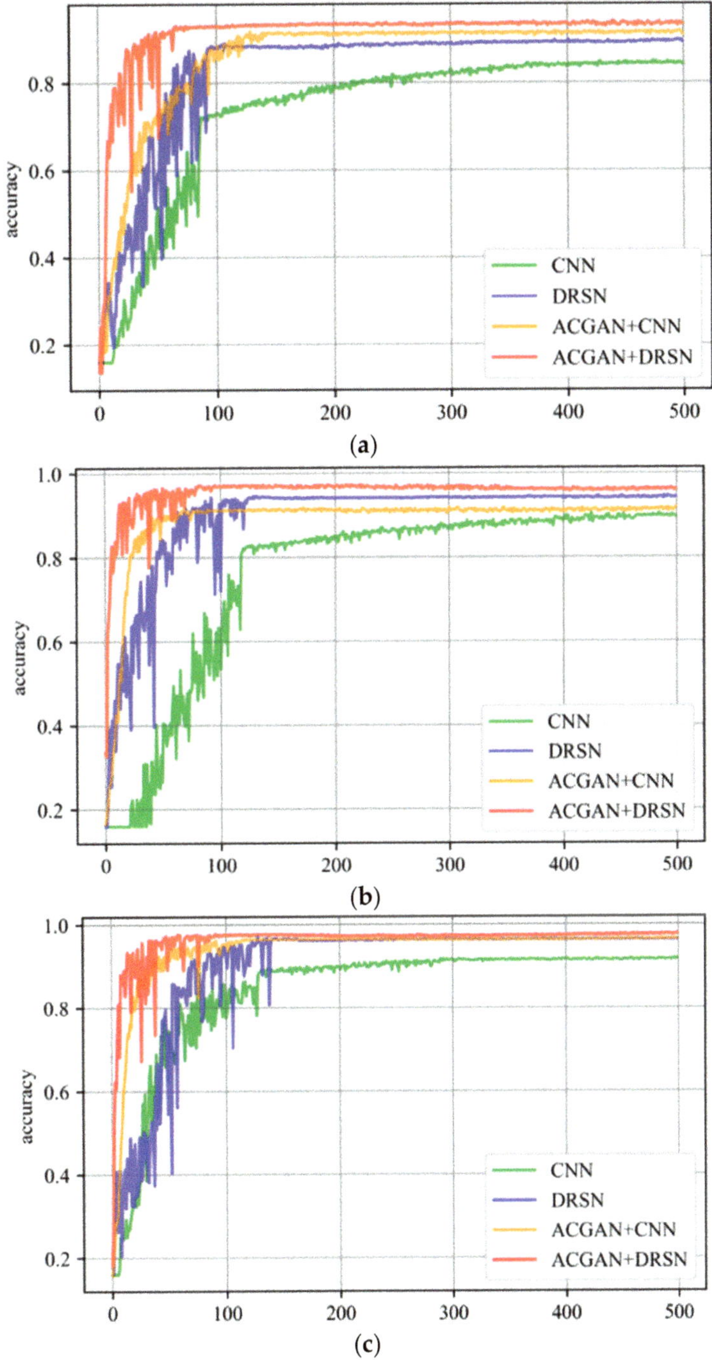

Figure 6. *Cont.*

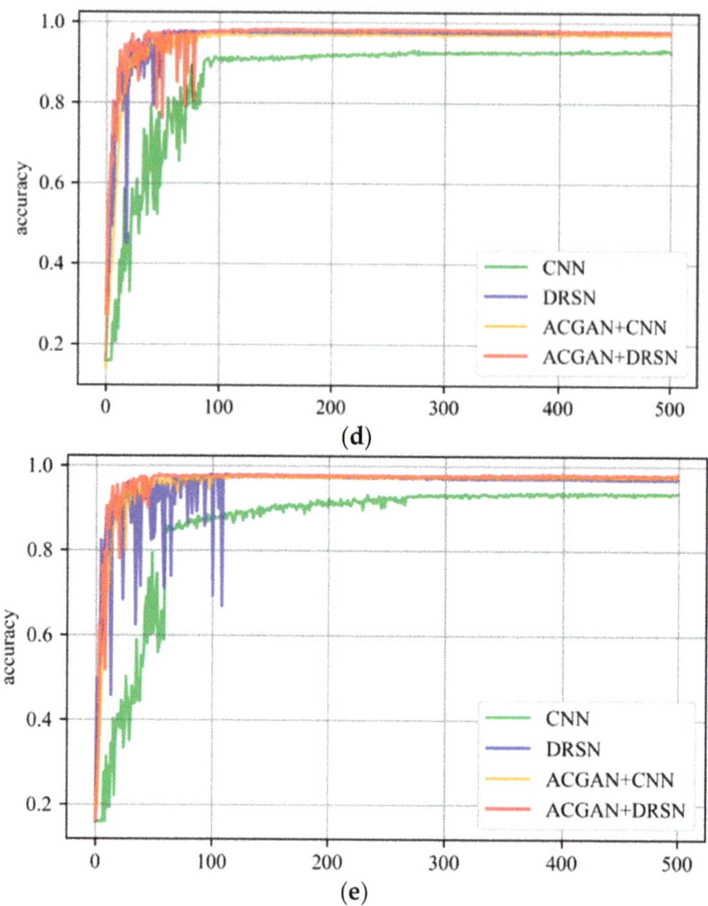

Figure 6. (**a**) Model fault classification accuracy in the condition of −5 db noise; (**b**) Model fault classification accuracy in the condition of −2 db noise; (**c**) Model fault classification accuracy in the condition of 0 db noise; (**d**) Model fault classification accuracy in the condition of 2 db noise; (**e**) Model fault classification accuracy in the condition of 5 db noise.

As can be seen from Figure 6a–e, in the initial stage of training, the training speed of the dataset improved by ACGAN is significantly faster than that of the ordinary dataset. When the noise in the original signal reaches −5 db, the fault classification accuracy of the model proposed in this article is improved by about 9.20% compared with the accuracy of the method only using CNN, and when the noise in the original signal is 5 db, the fault classification accuracy of the model proposed in this article is improved by about 4.53% compared with the accuracy of the method only use CNN. Therefore, the higher SNR of the noise contained in the original signal, the more obvious the improvement of the fault classification accuracy of the model proposed in this article compared with the model based on ordinary CNN only.

It can be seen from Figure 7 that when the vibration signal contains more noise, the classification accuracy of the three networks all has different degrees of degeneration in general. First, using DRSN to train a noisy fault dataset has a significant improvement in classification accuracy compared to the traditional CNN-based method. The classification accuracy of using traditional CNN to train the dataset expanded by ACGAN is significantly better than that of directly training the original dataset by CNN, and it implies the effectiveness of training the dataset expanded by ACGAN; In addition, using DRSN and CNN to train the dataset expanded by ACGAN are comparable in fault classification accuracy with different signal-noise ratios. Reference [21] added noise with a signal-to-noise ratio of −5~5 db to the vibration signal to simulate the complex working environment of rolling bearings in industrial production. It had concluded that when the signal-noise ratio is higher than 0 db, the model with DRSN and the ordinary model had a good performance of noise reduction, and when the signal-to-noise ratio reaches −5 db, the classification accuracy of CNN is only 79%, which is much lower than 86% of DRSN. Reference [22] proved by experiments that the expansion of the original unbalanced dataset by ACGAN can reduce the influence of unbalanced data on the classification accuracy and the misjudgment rate of fault diagnosis. Therefore, previous studies are consistent with the experimental results in this article. In the case of training the dataset expanded by AC-GAN, the classification accuracy is obviously better compared with using CNN when selecting the DRSN as a classifier. For vibration signals with different signal-noise ratio noises, the classification accuracy of the model proposed in this article changes relatively gently. The accuracy can be maintained above 90%. Generally, the classification accuracies of the other three methods are inferior to the method proposed in this article. Table 2 shows the fault classification accuracy of each model in conditions of different noises. The fault classification accuracy of the model proposed in this article is the highest under different noises, and the fault classification accuracy changes smoothly in the condition of −5~5 db noises. In conclusion, the ACGAN + DRSN model has good classification accuracy and stability.

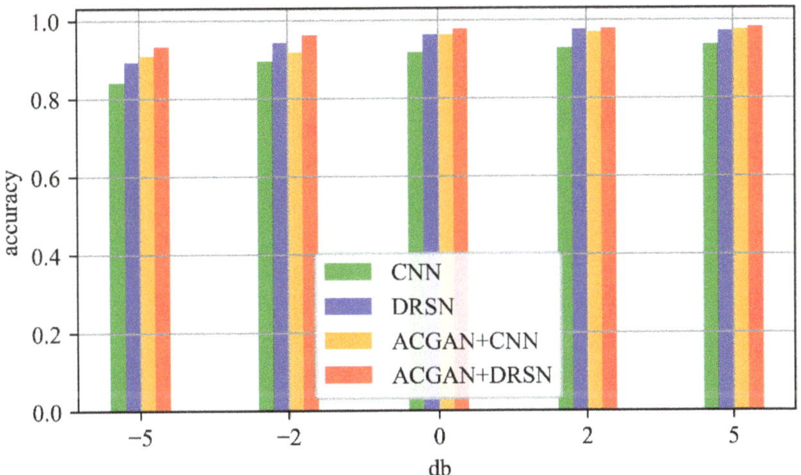

Figure 7. Fault classification ability under the circumstance of different noise.

Table 2. Fault classification accuracy in the condition of different noises.

SNR	Mean Failure Recognition Rate		
	CNN	ACGAN + CNN	ACGAN + DRSN
−5 db	84.133%	91.067%	93.333%
−2 db	89.600%	91.867%	96.133%
0 db	91.733%	96.400%	97.733%
2 db	92.800%	96.933%	97.733%
5 db	93.600%	97.467%	98.133%

5.3. Fault Diagnosis Capability under Variable Working Conditions

In this example, a high amount of noise by $-10\sim-5$ db is randomly added to the original vibration signal to simulate the actual working conditions of the main bearing of the offshore wind turbine [6]. The vibration data with the motor speed of 1797 rpm, 1772 rpm, and 1750 rpm are used as the training set, and the vibration data with the motor speed of 1730 rpm are used as the test set to compare the generalization ability of the three fault classification models in the noise background. Figure 8a–d is the accuracy confusion matrices of the four models which use traditional CNN to train the original dataset, using DRSN to train the original dataset, using traditional CNN to train ACGAN-expanded data, and using DRSN to train ACGAN-expanded data.

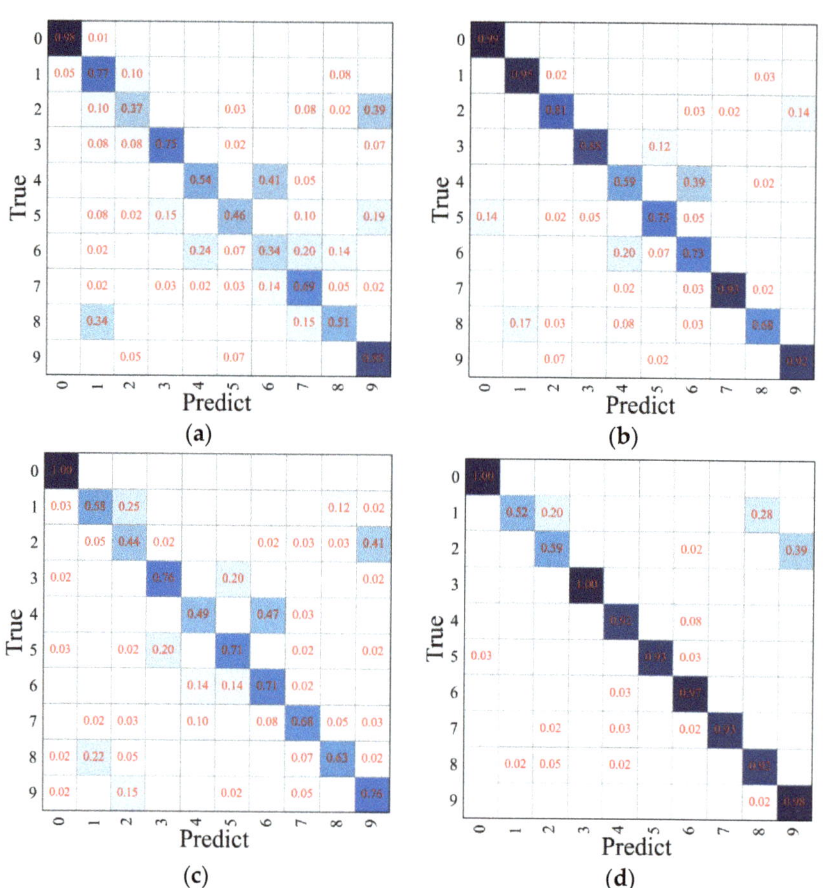

Figure 8. (**a**) CNN model fault classification ability; (**b**) DRSN model fault classification ability; (**c**) ACGAN + CNN model fault classification ability; (**d**) ACGAN + DRSN model fault classification ability. The darker the blue squares in these figures, the higher the probability that the fault will be correctly classified.

It can be seen that when training the unbalanced dataset of the vibration signal of the main bearing of the wind turbine under the noise conditions, the classification accuracy of the health state samples of the main bearing of the wind turbine is the highest, and the classification accuracy of the remaining samples are low. When using CNN alone to diagnose a test set, the accuracy of each fault sample is difficult to maintain above 50%, and the accuracy of the test set is about 71%; Compared with the method only using traditional CNN, the classification accuracies of the other two methods have been significantly improved; While the accuracy of each fault classification of the model proposed in this article is higher than 52%, and the accuracy of the test set is higher than 89%. Therefore, the ACGAN + DRSN model still has stronger fault classification ability when tested under different working conditions, and the classification accuracy is significantly higher than that of the ordinary CNN model. In order to visually demonstrate the fault classification ability of ACGAN+DRSN, the dimensionality reduction visualization of the model feature extraction effect is carried out.

Figure 9a–d is the effect diagrams of dimensionality reduction visualization using t-SNE for the original dataset, the output dataset of the last layer of the traditional CNN model, the traditional DRSN model, the ACGAN + CNN model, and the ACGAN + DRSN model, respectively. In Figure 9a, all the fault states of the main bearings of wind turbines are crossed and difficult to be classified; Figure 9b shows that after using the traditional CNN model, the health data of the main bearing of the wind turbine has been effectively classified, and the fault data initially shows the boundary, but it is still difficult to be classified; Figure 9c,d show that various types of fault samples have obvious boundaries, but each cluster obviously contains more than two different fault states, and it's hard to see which approach is better; In Figure 9e, with the model proposed in this article, various states clustering is enhanced. The red area is the healthy state of the main bearing of the wind turbine, this area has the best clustering effect. The slight wear and heavy wear of the rolling elements are not very separable in this example, but the rest of the fault states can be well classified effectively.

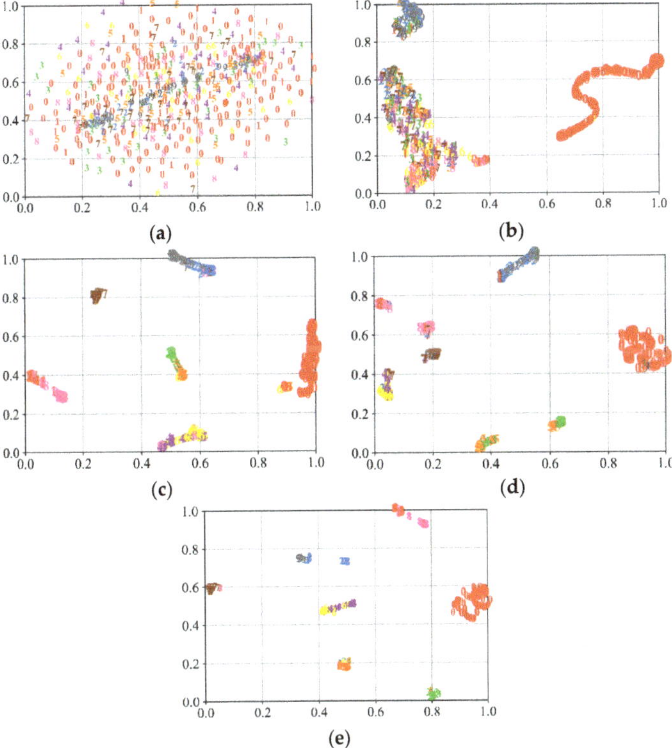

Figure 9. (**a**) Visualization of Raw dataset; (**b**) Visualization of CNN model dataset; (**c**) Visualization of DRSN model dataset; (**d**) Visualization of the dataset of ACGAN + CNN; (**e**) Visualization of the dataset of ACGAN + DRSN.

Table 3 shows the fault classification accuracy under different motor speeds. The motor speed of the test set of Experiment 1, Experiment 2, Experiment 3, and Experiment 4 are 1730 rpm, 1750 rpm, 1772 rpm, and 1798 rpm, respectively. The other three motor speeds are used as the training set. The accuracy of the model proposed in this article is higher than the other models obviously. The accuracy is always maintained above 85%.

Table 3. Fault classification accuracy under variable working conditions.

Experiment	CNN	DRSN	ACGAN + CNN	ACGAN + DRSN
1	71.13%	86.22%	75.16%	89.34%
2	74.71%	87.35%	75.23%	91.13%
3	72.49%	86.57%	79.53%	88.66%
4	78.62%	83.85%	81.85%	87.38%

6. Discussion

In order to solve the problem that the vibration fault data of wind turbine main bearing is difficult to be obtained in the condition of noise, resulting in the low fault diagnosis accuracy, this article proposes a fault classification method for wind turbine main bearing based on ACGAN and DRSN. This method has the following advantages:

(1) In view of the problem that the model training accuracy is not high due to insufficient fault data of the main bearing of the wind turbine, this article uses ACGAN to learn the distribution characteristics of fault data from the limited vibration signal samples, and generate high-quality fault samples to achieve data enhancement and improve the classification accuracy of the model;

(2) The use of attention mechanism and soft threshold of deep residual shrinkage network as a classification network can effectively reduce the different degrees of noise interference contained in the dataset, and fully explore the data fault characteristics of the main bearing of the wind turbines;

(3) Compared with the traditional CNN model, the ACGAN + DRSN model still has a stronger fault classification ability in the face of an unbalanced dataset containing noise under the variable working motor speed of the main bearing of the wind turbine.

The limitations of the methodological approach adopted are as follows:

(1) The research in this article is based on ACGAN. Many scholars have found that the original GAN has the problem of unstable training and poor ability to generate data.

(2) This model needs to train two models successively, which takes a longer time, so the training efficiency needs to be improved.

The simulation data of this experiment is still based on laboratory data and artificially added random noise. Considering the slight error of the wind turbine itself and changes in the operating environment, it is recommended that the wind turbine operation and maintenance manufacturers fully collect the different types of actual vibration data of the main bearing of the wind turbine when using the model proposed by this article. In addition, since the fault classification accuracy of the model decreases when the wind turbine runs at a new motor speed, the different working conditions of the wind turbine should be fully considered based on the model to improve its robustness of the model.

Author Contributions: Conceptualization, Z.F. and Z.Z.; methodology, Z.Z.; software, Z.Z.; validation, Z.F., Z.Z. and Y.Y.; formal analysis, Z.F. and Z.Z.; investigation, Z.F. and Z.Z.; resources, Z.F. and Z.Z.; data curation, Z.F. and Z.Z.; writing—original draft preparation, Z.F. and Z.Z.; writing—review and editing, Z.F., Z.Z. and Y.Y.; visualization, Z.Z.; supervision, Z.F.; project administration, Z.F.; funding acquisition, Y.Y. All authors have read and agreed to the published version of the manuscript.

Funding: This research was funded by [Jiangsu Coastal Development and Collaborative Innovation Center Project] grant number [B2106118].

Institutional Review Board Statement: The study did not require ethical approval.

Informed Consent Statement: Not applicable.

Data Availability Statement: The study did not report any data.

Conflicts of Interest: The authors declare no conflict of interest.

References

1. Polaris Wind Power Network. In-Depth Report on China's Wind Power Industry in 2021. Available online: https://news.bjx.com.cn/html/20220512/1224507.shtml (accessed on 31 May 2022).
2. Polaris Wind Power Network. A Weekly Review of Wind Power. Available online: https://news.bjx.com.cn/html/20220114/1199608.shtml (accessed on 25 January 2022).
3. Long, X.; Yang, P.; Guo, H.; Wu, X. Review of Fault Diagnosis Methods for Large Wind Turbines. *Power Syst. Technol.* **2017**, *41*, 3480–3491.
4. Chen, X.F.; Guo, Y.J.; Xu, C.; Shang, H. Review of Fault Diagnosis and Health Monitoring for Wind Power Equipment. *China Mech. Eng.* **2020**, *31*, 175–188.
5. Zheng, X.X.; Ye, C.J.; Yang, F. Research and Development of Operation and Maintenance for Offshore Wind Farms. *Power Syst. Energy* **2012**, *28*, 90–94.
6. Zeng, Z. Research on Fault Diagnosis Method of Wind Turbine Transmission System Based on Deep Learning. Master's Thesis, Northeast Electric Power University, Jilin, China, 2021.
7. Zhang, L.; Lang, Z.Q. Wavelet Energy Transmissibility Function and its Application to Wind Turbine bearing Condition Monitoring. *IEEE Trans. Sustain. Energy* **2018**, *9*, 1833–1843. [CrossRef]
8. Wang, J. Research on Fault Diagnosis Technology of Planetary Gearboxes Based on Deep Learning. Master's Thesis, Nanjing University of Aeronautics and Astronautics, Nanjing, China, 2018.
9. Cao, L.; Zhang, J.; Wang, J.; Qian, Z. Intelligent fault diagnosis of wind turbine gearbox based on Long short-term memory networks. In Proceedings of the IEEE International Symposium on Industrial Electronics, Vancouver, BC, Canada, 12–14 June 2019; pp. 890–895.
10. Wu, C.Z.; Jiang, P.C.; Ding, C.; Feng, F.; Chen, T. Intelligent fault diagnosis of rotating machinery based on one-dimensional convolutional neural network. *Comput. Ind.* **2019**, *108*, 53–61. [CrossRef]
11. Yao, L.; Sun, J.J. Fault Diagnosis Method for Rolling Bearing Based on Gramian Angular Fields and CNN-RNN. *Bearing* **2022**, *2*, 61–67.
12. Zhou, Y.; Sun, H.Y.; Fang, Q.; Xia, H. Review of imbalanced data classification methods. *Appl. Res. Comput.* **2022**, *39*, 1615–1621.
13. Lu, J.L.; Zhang, X.G. Fault Diagnosis of Main Bearing based of Wind Turbine Based on Improved Auxiliary Classifier Generative Adversarial Network. *Autom. Electr. Power Syst.* **2021**, *45*, 148–154.
14. Li, D.D.; Liu, Y.H. Fault Diagnosis Method of Wind Turbine Planetary Gearbox Based on Improved Generative Adversarial Network. *Proc. CSEE* **2021**, *41*, 7496–7506.
15. Zhao, M.; Zhong, S. Deep Residual Shrinkage Networks for Fault Diagnosis. *IEEE Trans. Ind. Inform.* **2020**, *16*, 4681–4689. [CrossRef]
16. Che, C.C.; Wang, H.W. Fault diagnosis of rolling bearing based on deep residual shrinkage network. *J. Beijing Univ. Aeronaut. Astronaut.* **2021**, *47*, 1399–2021.
17. Goodfellow, I.; Pouget-Abadie, J.; Mirza, M.; Xu, B.; Warde-Farley, D.; Ozair, S.; Courville, A.; Bengio, Y. Generative adversarial nets. *Adv. Neural Inf. Process. Syst.* **2014**, *3*, 2672–2680.
18. Wang, P.; Hou, B.R. ECG arrhythmias detection using auxiliary classifier generative adversarial network and residual network. *IEEE Access* **2019**, *7*, 100910–100922. [CrossRef]
19. He, K.M.; Zhang, X.Y.; Ren, S.Q.; Sun, J. Deep Residual Learning for Image Recognition. In Proceedings of the 2016 IEEE Conference on Computer Vision and Pattern Recognition, Las Vegas, NV, USA, 27 June 2016.
20. Xu, M.; Wang, P. Research on Rolling Bearing Fault Diagnosis Based on FB-LSTM Resnet. Available online: http://kns.cnki.net/kcms/detail/41.1148.th.20220507.1708.006.html (accessed on 22 July 2022).
21. Yang, S.L. Bearing Fault Diagnosis Based on Hybrid Domain Attention Mechanism Convolutional Network and Residual Contraction Network. Master's Thesis, School of Information and Communication Engineering, Beijing, China, 2020.
22. Lu, J.L.; Zhu, C.F. Power Transformer Fault Diagnosis Technology Based on Improved ACGAN Data Augmentation. *Electr. Power Sci. Eng.* **2021**, *37*, 42–51.

processes

Article

Time-of-Use Pricing Strategy of Integrated Energy System Based on Game Theory

Xiaoling Yuan *, Yi Guo, Can Cui and Hao Cao

College of Energy and Electrical Engineering, Hohai University, Nanjing 211100, China
* Correspondence: lingx@hhu.edu.cn

Abstract: The integrated energy system is the mainstream energy utilization form of integrating a power system, natural gas system and thermal system, which provides a new way to solve the problem of renewable energy accommodation. The integrated energy system includes a variety of energy generation and conversion equipment, and its internal electricity, gas, cooling and thermal systems must balance the multiple energy supplies required by users. The integrated energy supplier (IES) and integrated energy user (IEU), as different stakeholders, pursue the maximization of their own profit. However, integrated energy suppliers should consider their market share and the sustainability of participating in market competition. Based on the constraints of energy access, conversion and accommodation, and the equipment for energy generation, conversion and consumption, we established an energy flow model. Constrained by the dynamic equilibrium of the supply of integrated energy suppliers and the demand of integrated energy users, a Stackelberg game model of integrated energy suppliers and users was established, and the existence of a Nash equilibrium solution of the game was proved. A genetic algorithm was used to solve the Nash equilibrium solution under two conditions aiming at the integrated energy supplier's maximum profit and target profit. Considering the demand of integrated energy users in different time periods, we analyzed the time-of-use pricing strategy of the integrated energy based on the balance of the energy supply and demand. The results of a case study show that if integrated energy suppliers adopt the time-of-use pricing strategy of maximum profit, the energy load distribution of integrated energy users can be smoothed, and energy utilization and economic benefits of the system can be improved. If integrated energy suppliers adopt the time-of-use pricing strategy of target profit, enlarge the market by limiting their own profit and obtain the purchase willingness of integrated energy users by reducing the energy price, they can have a larger market share, a more reliable profit and a guarantee of long-term participation in market transactions.

Keywords: integrated energy system; Stackelberg game; time-of-use pricing; demand response; Nash equilibrium

Citation: Yuan, X.; Guo, Y.; Cui, C.; Cao, H. Time-of-Use Pricing Strategy of Integrated Energy System Based on Game Theory. *Processes* **2022**, *10*, 2033. https://doi.org/10.3390/pr10102033

Academic Editor: Enrique Rosales-Asensio

Received: 27 September 2022
Accepted: 4 October 2022
Published: 8 October 2022

1. Introduction

With the expansion of the scale of renewable energy development, power systems have new requirements on the intermittent and fluctuating nature of renewable energy. Traditional power systems usually increase investment on both the supply and demand sides, such as the use of deep peak shaving technology, and distributed energy storage technologies and the allocation of renewable energy based on demand-side response alleviate the intermittent and fluctuation problems of renewable energy, and high investment costs and insufficient energy consumption should be faced at the same time. Therefore, improving energy utilization and realizing the large-scale use of renewable energy have become the inevitable choices in the process of integrated energy system development [1].

The integrated energy supplier (IES) is based on an integrated energy system and uses renewable energy as the main energy source. The energy supply methods tend to diversify, which provides great power supply flexibility to the traditional power system.

The grid power supply, renewable energy system power supply and gas-driven combined cooling and heating and power (CCHP) system are conducive to the use of energy in the cascade transfer process, while improving the integrated energy utilization rate and reducing pollutant emissions [2]. After considering the needs of the integrated energy users (IEUs), the supply and demand relationship can be balanced through the price incentive mechanism. While stimulating the IEU to adjust the allocation of energy demand to achieve a balanced optimization of the supply and demand side, the utility function is maximized to ensure IEU satisfaction. IEU uses different energy sources in different periods of time. According to IEU's demand for integrated energy, IES obtains the current maximum profit through the establishment of time-of-use prices.

With the continuous improvement of the Energy Internet [3], IES can supply energy to multiple groups of IEUs at the same time, achieving energy interconnection [4–6]. In traditional research, energy consumption is optimized separately from the perspective of the participants, without considering the interaction between them. The game strategy can solve the rational optimal strategy set of market participants, which is of great significance to the formulation of energy prices and the planning and operation of IES. Reference [7] pointed out the effectiveness and validity of price linkage in the game. Reference [8] proposed a regional integrated energy game optimization strategy considering load demand response and aiming at optimal comprehensive profits. Reference [9] proposed a double-layer Stackelberg game model aiming at maximizing the profits of IES and minimizing the costs of IEU. Reference [10] proposed to take potential function as the solution method and obtained the game strategy of a multi-energy market including electricity, thermal and gas. Reference [11] took the maximum profit as the comparison index, and analyzed different operation strategies under the non-cooperative game, semi-cooperative game and cooperative game of the integrated energy system. Reference [12] established a game model by using two-level optimization, and compared and analyzed the profit difference of IES in the retail and wholesale markets. Reference [13] proposed a two-level collaborative control strategy model of "electricity–thermal–gas" integrated energy system based on multi-agent deep reinforcement learning to improve the energy efficiency of the integrated energy system and reduce costs. Reference [14], taking energy efficiency and cost as optimization objectives, proposed a two-stage energy management method of a thermal–electricity integrated energy system considering dynamic pricing of the Stackelberg game and operation strategy optimization. However, the above dynamic game models related to energy trading are all aimed at maximizing profits or minimizing costs. They only consider economics without considering the energy trading volume of IES after the game or analyzing the willingness of IEU to purchase energy in the face of energy prices after the game. These two are related to whether IES can participate in the market operation and maintain reliable profit for a long time. In order to analyze this problem, we consider the demand response of IEU and the time-of-use price of IES, and compare the profit, energy trading volume and price dynamic curve after equilibrium pricing from the perspective of maximum profit and target profit.

When considering IEU demand response, IES first initializes the energy price. As a follower, IEU can determine the energy consumption according to the price strategy and its own demand constraints [15]. The game relationship between IES and IEU is in order. According to this characteristic, this paper focuses on the equilibrium pricing and quantitative energy interaction between IES and IEU. Firstly, a multi-energy flow Stackelberg game model with a master–slave relationship is established, and then the objective functions with the maximum profit and target profit of IES are constructed. Secondly, it is proved that the equilibrium solution of the established game model exists in the process of multi-agent participation. Finally, the time-of-use pricing strategy is solved for different time periods under the two different objective profits. Integrated energy trading volume and pricing curves are obtained and the integrated energy interaction between IES and IEU is analyzed. If the IES adopts the time-of-use pricing strategy of maximum profit, the load distributions of IEUs are smoothed, and the energy utilization rate and economic efficiency of the system are improved. If IES adopts the time-of-use pricing strategy of target profit, integrated energy trading volume is increased and energy unit price is decreased. For IES, increased integrated energy trading volume means more market share. For IEU, decreased energy unit price means more energy trading volume at the same price. This strategy is helpful for integrated energy suppliers to participate in market operations for a long time and maintain reliable profit.

The remainder of this paper is organized as follows. In Section 2, game model subjects of IESs and IEUs are established. In Section 3, the Stackelberg game model is presented and the existence of a Nash equilibrium solution is proved. In Section 4, a case study is detailed to demonstrate the proposed game strategy. In Section 5, several important conclusions are summarized.

2. The IES and IEU Model

2.1. IES Model

The profit of IES is determined by operating costs, spot market price and integrated energy trading volume as IES enters the integrated energy market. In this study, energy trading between IES and IEU was treated as a completely competitive market behavior, and the bidding behavior within IESs was not considered. The IES formulates a price strategy based on the IEU's demand for integrated energy. The interaction between IES (leader) and IEU (follower) was based on Stackelberg game theory [16–18]. An integrated energy interaction Stackelberg game model considering IEU's energy consumption behavior was established. Figure 1 is the decision model of IES.

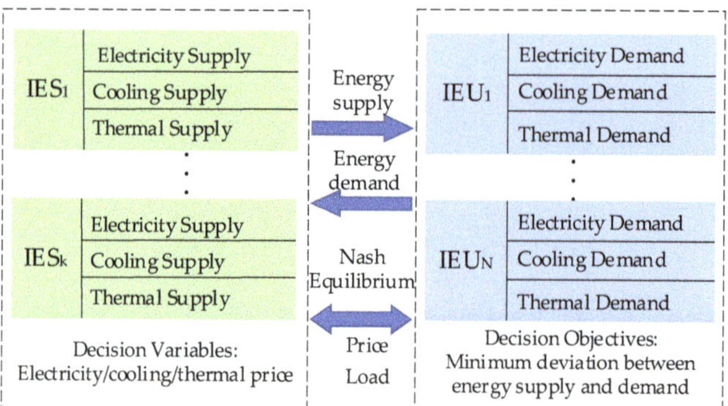

Figure 1. Framework of decision model of IES.

2.1.1. Multi-Energy Flow Modeling of Integrated Energy Systems

The integrated energy system studied in this paper is mainly composed of an energy supply network, energy exchange link and terminal integrated energy consumption unit. It is a multi-energy flow system consisting of cooling, thermal energy and electricity. The system has low cost, high flexibility and reliability in the actual application process. When the integrated energy system is operating, the waste thermal generated during electricity generation can be reused. The gas turbine consumes the natural gas input from the gas supply system to generate electricity and high-temperature steam. The steam is recovered by the waste heat boiler to generate high-pressure steam and low-pressure steam. The former enters the unit steam turbine to generate electricity, and the latter is used by the deaerator of the waste heat boiler. There is a heating surface at the end of the waste heat boiler, which can generate heat medium water to meet the heating energy requirements. In addition, steam turbines provide extraction services for thermal power needs. The heating medium water is first supplied with hot water, and the remaining absorption refrigeration unit is used for cooling. The extracted steam can be used for cooling the hot water heat exchanger. The electricity generated by gas and steam turbines is used to meet electrical loads. In addition, electric refrigeration equipment and absorption refrigeration machines supplement the insufficient cooling. The combined cooling, heating and electricity system based on natural gas can effectively improve the integrated energy utilization rate by combining electric generation and heating. The input power of the grid power supply system and the renewable energy power supply system is used for the user's electricity demand. At the same time, the insufficient cooling/thermal demand of the IEU can be supplemented by electric refrigeration equipment and electric heating equipment through energy allocation.

In order to accurately reflect the characteristics of the system energy flow conversion, the energy flow function is used to characterize the balance between electricity/cooling/thermal energy [19]. A regional integrated energy system based on grid power supply, renewable energy supply and natural gas input is shown in Figure 2.

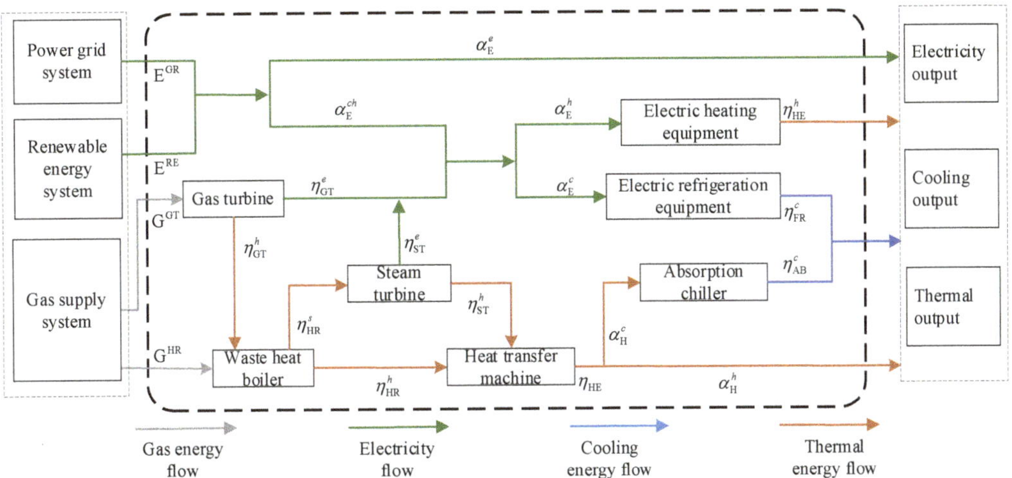

Figure 2. Schematic diagram of regional integrated energy system.

Equation (1) is the energy conversion equation of the regional integrated energy system.

$$
\begin{cases}
O^e &= \alpha_E^e \left(E^{GR} + E^{RE} \right) \\
O^c &= \alpha_H^c \eta_{AB}^c \eta_{HE} \left(\eta_{HR}^h + \eta_{ST}^h \eta_{HR}^s \right) \left(\eta_{GT}^h G^{GT} + G^{HR} \right) + \alpha_E^c \eta_{FR}^c \\
& \quad \left[\eta_{GT}^e G^{GT} + \eta_{ST}^e \eta_{HR}^h \left(\eta_{GT}^h G^{GT} + G^{HR} \right) + \alpha_E^{ch} \left(E^{GR} + E^{RE} \right) \right] \\
O^h &= \alpha_H^h \eta_{HE} \left(\eta_{ST}^h \eta_{HR}^s + \eta_{HR}^h \right) \left(\eta_{GT}^h G^{GT} + G^{HR} \right) + \alpha_E^h \eta_{HE}^h \\
& \quad \left[\eta_{GT}^e G^{GT} + \eta_{ST}^e \eta_{HR}^h \left(\eta_{GT}^h G^{GT} + G^{HR} \right) + \alpha_E^{ch} \left(E^{GR} + E^{RE} \right) \right]
\end{cases}
\tag{1}
$$

where O^e, O^c and O^h are electricity output, cooling output and thermal output of the integrated energy system, respectively. α_E^e is the energy distribution coefficient of direct electricity output, and α_E^{ch} is the energy distribution coefficient supplementing cooling and thermal supply. α_E^h is the energy distribution coefficient of electric heating equipment, and α_E^c is the energy distribution coefficient of electric refrigeration equipment. α_H^c is the energy distribution coefficient of the thermal exchanger supplying the absorption chiller, and α_H^h is the energy distribution coefficient of the thermal exchanger directly outputting thermal energy. η_{GT}^e and η_{GT}^h are the electrical efficiency and thermal efficiency of the gas turbine, respectively. η_{HR}^s and η_{HR}^h are the steam efficiency and thermal efficiency of the waste heat boiler, respectively. η_{ST}^e and η_{ST}^h are the electrical efficiency and thermal efficiency of the steam turbine, respectively. η_{HE} is the thermal energy efficiency of the heat transfer machine. η_{HE}^h is the thermal energy conversion efficiency of electric heating equipment, η_{FR}^c

is the cooling energy conversion efficiency of electric refrigeration equipment and η_{AB}^c is the conversion efficiency of thermal energy to cooling energy of the absorption refrigeration machine. G^{GT} is the volume of natural gas input to the gas turbine, and G^{HR} is the volume of natural gas input to the waste heat boiler. E^{GR} is the power supply of the grid, and E^{RE} is the power supply of the renewable energy.

2.1.2. IES Objective Function

As the energy producer and converter in the integrated energy system, the IES needs to price the unit-integrated energy. We consider the pricing strategy of IES under the conditions of maximum profit and target profit.

(1) Maximum profit. IES adjusts the input of power supply and natural gas according to users' demand, prices unit-integrated energy and obtains maximum profit. Equation (2) represents the objective function of IES with maximum profit.

$$
\begin{aligned}
\max U^{IES} = {}& p^e O^e + p^c O^c + p^h O^h - (c^{re} E^{RE} + \\
& f^{re} + c^{gr} E^{GR} + f^{gr} + c^g (G^{GT} + G^{HR}) + f^g)
\end{aligned}
\tag{2}
$$

where p^e is the unit electricity price, p^c is the unit cooling price, p^h is the unit thermal price, c^{re} is the variable cost coefficient of the renewable energy system and f^{re} is the fixed cost of the renewable energy system. c^{gr} is the variable cost coefficient of the grid system, f^{gr} is the fixed cost of the grid system, c^g is the variable cost coefficient of the gas supply system and f^g is the fixed cost of the gas supply system.

(2) Target profit. IES adjusts the power input and volume of natural gas required for production according to users' demand, appropriately reduces the overall energy price, enlarges the market and obtains the target profit. Equation (3) represents the objective function of IES with the target profit.

$$
\begin{aligned}
\min |U^{IES} - U^{IES*}| = {}& |p^e O^e + p^c O^c + p^h O^h - (c^{re} E^{RE} + f^{re} + \\
& c^{gr} E^{GR} + f^{gr} + c^g (G^{GT} + G^{HR}) + f^g) - U^{IES*}|
\end{aligned}
\tag{3}
$$

where U^{IES*} is the target profit. IES can maintain its own operation under the target profit. IES supplies integrated energy to satisfy the energy demand of IEUs after the energy loss in the transmission and conversion.

$$
\begin{cases}
O^e = (1 + \alpha^e) d^e \\
O^c = (1 + \alpha^c) d^c \\
O^h = (1 + \alpha^h) d^h
\end{cases}
\tag{4}
$$

where α^e, α^c and α^h are network loss parameters of electrical, cooling and thermal energy, respectively. d^e is electricity demand, d^c is cooling demand and d^h is thermal demand.

2.2. IEU Model

The IEU is different from traditional electrical energy users. The energy consumption pattern is more complicated, and it requires the supply of multiple types of energy (cooling/thermal/electricity). From the perspective of the IEU's own interests, on the premise of ensuring satisfaction with energy consumption, energy prices will affect the IEUs' initiative to adjust their energy consumption strategy to minimize energy costs [20].

The implementation of price incentive strategy enables IEUs to maximize the satisfaction of energy consumption. A quadratic function U_n^{IEU} is used to represent the consumption satisfaction of IEUs. The quadratic utility function U_n^{IEU} is described as Equation (5) [21].

$$
U_n^{IEU} = \left[v_n^e d_n^e - \frac{u_n^e}{2} (d_n^e)^2 \right] + \left[v_n^c d_n^c - \frac{u_n^c}{2} (d_n^c)^2 \right] + \left[v_n^h d_n^h - \frac{u_n^h}{2} (d_n^h)^2 \right]
\tag{5}
$$

where v_n^e and u_n^e are the constant coefficients of preference of IEU n concerning electricity, v_n^c and u_n^c are the constant coefficients of preference of IEU n regarding cooling and v_n^c and u_n^c are the constant coefficients of preference of IEU n in terms of thermal energy, which are used to describe the relationship between utility function and demand. d_n^e is the electricity demand of IEU n, d_n^c is the cooling demand of IEU n and d_n^h is the thermal demand of IEU n.

Each IEU determines the demand for electricity supply, cooling supply and thermal supply based on unit electricity price p^e, unit cooling price p^c and unit thermal price p^h. For IEUs, the consumption function C_n^{IEU} of purchasing energy is described as Equation (6).

$$C_n^{\text{IEU}} = p^e d_n^e + p^c d_n^c + p^h d_n^h \tag{6}$$

The objective function of IEUs, W_n, is defined as the difference between the consumption function and the quadratic utility function.

$$\min W_n = C_n^{\text{IEU}} - U_n^{\text{IEU}} \tag{7}$$

Calculating the first-order partial derivative of W_n, the optimal demand of IEU is obtained as Equation (8).

$$\begin{cases} d_n^e = \dfrac{v_n^e}{u_n^e} - \dfrac{1}{u_n^e} p^e \\[6pt] d_n^c = \dfrac{v_n^c}{u_n^c} - \dfrac{1}{u_n^c} p^c \\[6pt] d_n^h = \dfrac{v_n^h}{u_n^h} - \dfrac{1}{u_n^h} p^h \end{cases} \tag{8}$$

$$\text{s.t.} \begin{cases} d_n^e \geqslant d_{n,f}^e \\[4pt] d_n^c \geqslant d_{n,f}^c \\[4pt] d_n^h \geqslant d_{n,f}^h \end{cases} \tag{9}$$

where $d_{n,f}^e$, $d_{n,f}^c$ and $d_{n,f}^h$ are the user's base load on electricity, cooling and thermal energy, respectively.

3. Stackelberg Game Model

In the process of solving the Stackelberg equilibrium solution, the game subjects restrict each other to coordinate the equilibrium state of the Stackelberg game model. When IES evaluates the original integrated energy price, the base energy demand of the IEU in each time period is fully considered. Energy prices are updated continuously until the supply and the demand are balanced. The i-th integrated energy price is updated as Equation (10).

$$\begin{cases} p_{i+1}^e = \max\left\{ p_i^e + \tau_i^e \left[(1 + \alpha^e) \sum_{n=1}^{N} d_{n,i}^e - O_i^e \right], p_{min}^e \right\} \\[10pt] p_{i+1}^c = p_i^c + \tau_i^c \left[(1 + \alpha^c) \sum_{n=1}^{N} d_{n,i}^c - O_i^c \right] \\[10pt] p_{i+1}^h = p_i^h + \tau_i^h \left[(1 + \alpha^h) \sum_{n=1}^{N} d_{n,i}^h - O_i^h \right] \end{cases} \tag{10}$$

where p_{min}^e is the IEU's willingness price of electricity, and τ_i^e, τ_i^c and τ_i^h are the dynamic speed adjustment parameters of electricity price, cooling price and thermal price, respectively. N is the number of IEUs. $d_{n,i}^c$, $d_{n,i}^e$ and $d_{n,i}^h$ are the cooling demand, electricity demand and thermal demand of IEU n in the i-th update, respectively. O_i^c, O_i^e and O_i^h are the cooling output, electricity output and thermal output of the integrated energy system in the i-th update, respectively. p_i^c, p_i^e and p_i^h are the unit cooling price, unit electricity price and unit thermal price of the integrated energy system in the i-th update, respectively. p_{i+1}^c, p_{i+1}^e and

p_{i+1}^{h} are the unit cooling price, unit electricity price and unit thermal price of the integrated energy system in the $i+1$-th update, respectively.

Dynamic speed adjustment parameters related to the current number of iterations are used as Equation (11) [21] to improve the convergence speed of the algorithm.

$$\begin{cases} \tau_i^e = \frac{1}{\lambda^e + \mu^e i} \\ \tau_i^c = \frac{1}{\lambda^c + \mu^c i} \\ \tau_i^h = \frac{1}{\lambda^h + \mu^h i} \end{cases} \tag{11}$$

where λ^e, μ^e, λ^c, μ^c, λ^h and μ^h are constants.

3.1. Equilibrium Existence in Stackelberg Game

In a multi-subject strategic game, if the utility function of each participant is continuous, and the utility function of its own strategy is quasi-concave, then the Nash equilibrium of pure strategy must exist [22].

The utility function of the IEU is continuous. There is an optimal response strategy for the utility function of IEU $U_n^{\mathrm{IEU}}(p_n^e, p_n^c, p_n^h)$, if the unit electricity price p_n^e, the unit cooling price p_n^c and the unit thermal price p_n^h are satisfied (Equation (12)).

$$U_n^{\mathrm{IEU}}(p^{e*}, p^{c*}, p^{h*}) \geq U_n^{\mathrm{IEU}}(p^e, p^c, p^h), \forall p^e, p^c, p^h \in R^+ \tag{12}$$

where U_n^{IEU} is the quadratic utility function value of IEU n. p_n^c, p_n^e and p_n^h are the unit cooling price, unit electricity price and unit thermal price of IEU n, respectively. p^{e*}, p^{c*} and p^{h*} are the target unit electricity price, cooling price and thermal price, respectively.

The Hessian matrix of $U_n^{\mathrm{IEU}}(p^e, p^c, p^h)$ is as in Equation (13).

$$\nabla^2_{p^e, p^c, p^h}(U^{\mathrm{IEU}}) = diag \begin{bmatrix} \frac{\partial^2 U_n^{\mathrm{IEU}}}{\partial p_n^{e2}} & \frac{\partial^2 U_n^{\mathrm{IEU}}}{\partial p_n^e \partial p_n^c} & \frac{\partial^2 U_n^{\mathrm{IEU}}}{\partial p_n^e \partial p_n^h} \\ \frac{\partial^2 U_n^{\mathrm{IEU}}}{\partial p_n^c \partial p_n^e} & \frac{\partial^2 U_n^{\mathrm{IEU}}}{\partial p_n^{c2}} & \frac{\partial^2 U_n^{\mathrm{IEU}}}{\partial p_n^c \partial p_n^h} \\ \frac{\partial^2 U_n^{\mathrm{IEU}}}{\partial p_n^h \partial p_n^e} & \frac{\partial^2 U_n^{\mathrm{IEU}}}{\partial p_n^h \partial p_n^c} & \frac{\partial^2 U_n^{\mathrm{IEU}}}{\partial p_n^{h2}} \end{bmatrix}^N_{n=1} \tag{13}$$

Then,

$$\nabla^2_{p^e, p^c, p^h}(U^{\mathrm{IEU}}) = diag \begin{bmatrix} -u_n^e & 0 & 0 \\ 0 & -u_n^c & 0 \\ 0 & 0 & -u_n^h \end{bmatrix}^N_{n=1} \tag{14}$$

According to Equation (14), the Hessian matrix of utility function is a negative definite matrix; that is, the utility function of IEU is strictly a concave function of p_n^e, p_n^c and p_n^h. Therefore, the equilibrium solution of the Stackelberg game model exists.

3.2. Algorithm Flow

Under the constraints of price and renewable energy output, IES maximizes the profit by adjusting the power supply of the grid, the renewable energy output, the gas volume of the gas turbine and the waste heat boiler. The energy demand of IEU adjusts the overall energy price. A genetic algorithm (GA) was used to solve the game model. The number of populations is set to 10, the maximum number of iterations is 5000, the crossover rate between chromosomes is 60% and the random mutation rate is 1% to prevent the model from falling into a local optimum solution. The game model updates the next integrated energy unit price based on the difference in the integrated energy price for each iteration, and accelerates convergence through dynamic speed adjustment parameters related to the number of iterations. According to the existence of the Nash equilibrium solution, there is a solution balance the integrated energy demand and supply, and the corresponding unit price of the integrated energy is the transaction price at that moment.

Based on the above process, the algorithm flowchart of the Stackelberg game model is shown in Figure 3.

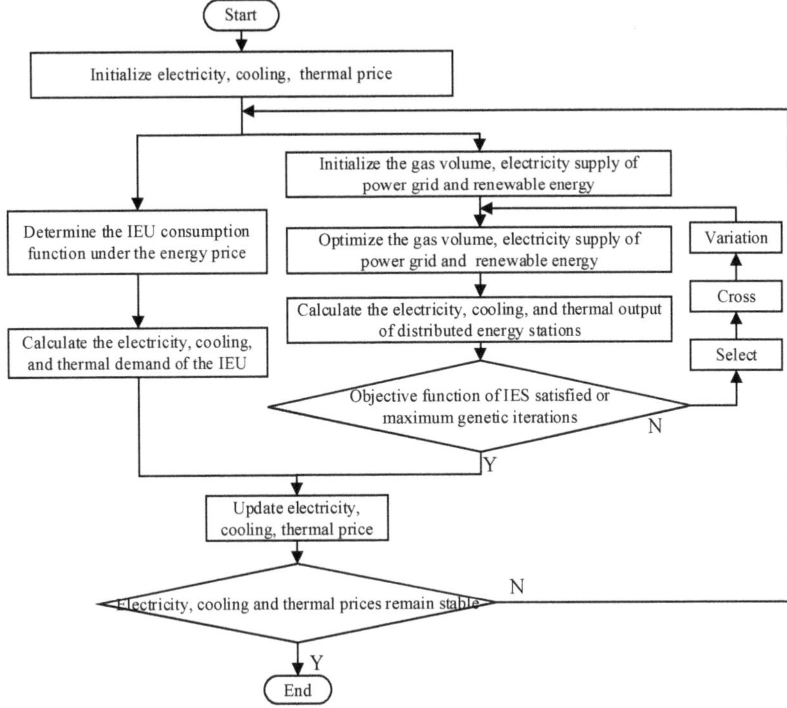

Figure 3. Flowchart of decision model of IES.

The unit prices are initialized randomly before the game begins. IEU balances consumption behavior based on the base load in various time periods and different preferences of integrated energy. For IES, the output of renewable energy is uncertain and fluctuating, so it is assumed that the power supply of power grid and the gas supply of the gas supply system are sufficient to satisfy the energy demand of the IEU at any time. IES iterates on the target of profit after each price adjustment.

In order to verify the correctness and effectiveness of the model and algorithm proposed in this paper, 200 IEUs were set for simulation analysis on a typical day with a unit operation period of 1 h. We analyzed the changes in energy trading volume and price after the game model reaches the Nash equilibrium considering the IEU's energy consumption behavior.

4. Case Study

4.1. Case Data

For the renewable energy system, the IES has a 150 kW wind farm and a 50 kW photovoltaic farm as electricity supply. Assuming the system has sufficient capacity, the renewable energy output is shown in Figure 4.

The main types and efficiency parameters of the production equipment of the integrated energy system are shown in Table 1.

The energy distribution coefficient, cost coefficient of equipment, loss coefficient in the process of energy transmission and user preference constant for integrated energy are shown in Table 2 [21].

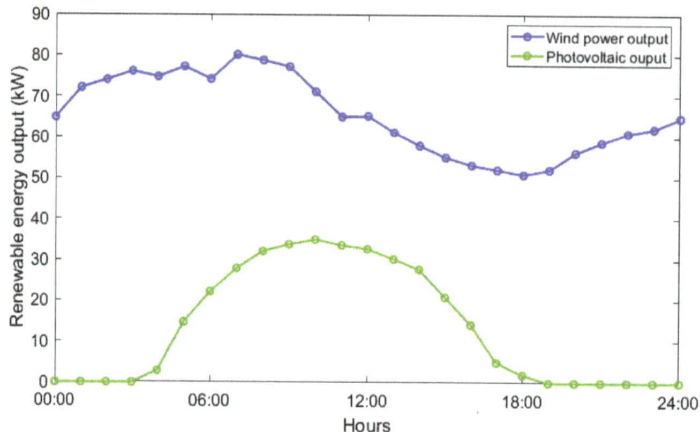

Figure 4. Hourly power output curve of the wind and photovoltaic farms on a typical day.

Table 1. Experimental data classification of cooling, thermal and electricity-generating units.

Equipment	Parameter	Value
Gas turbine	Capacity /kW	400
	Electricity generation efficiency η_{GT}^{e}	0.35
	Heat production efficiency η_{GT}^{h}	0.5
Waste heat boiler	Capacity /kW	200
	Boiler steam efficiency η_{HR}^{s}	0.1
	Heat production efficiency η_{HR}^{s}	0.7
Steam turbine	Capacity /kW	160
	Electricity generation efficiency η_{ST}^{e}	0.42
	Heat production efficiency η_{ST}^{h}	0.38
Heat transfer machine	Capacity /kW	300
	Heat production efficiency η_{HE}	0.8
Absorption refrigerator	Capacity /kW	100
	Cooling efficiency η_{AB}^{c}	1.3
Electric heating equipment	Capacity /kW	250
	Heat production efficiency η_{HE}^{h}	0.8
Electric refrigeration equipment	Capacity /kW	200
	Cooling efficiency η_{FR}^{c}	4

Table 2. Loss, cost and preference coefficients.

α_{H}^{c}	0.15	c^{g}	0.18	α^{e}	0.04	λ^{e}	0.05	v_{e}	0.05
α_{H}^{h}	0.85	c^{re}	0.05	α^{c}	0.08	λ^{c}	0.04	v_{c}	0.03
α_{E}^{c}	0.5	c^{gr}	0.25	α^{h}	0.06	λ^{h}	0.03	v_{h}	0.04
α_{E}^{h}	0.5	f^{g}	10	-	-	μ^{e}	4	u_{e}	4
α_{E}^{e}	0.9	f^{re}	2	-	-	μ^{c}	4	u_{c}	4
α_{E}^{ch}	0.1	f^{gr}	5	-	-	μ^{h}	4	u_{h}	4

Figure 5 is the base load of the IEU for electrical/cooling/thermal energy on a typical day. During the day, the IEU has a high demand for electricity between 9:00 and 18:00, and the demand for electricity peaks at 12:00. Cooling/thermal demands of IEU rise from 6:00 to 12:00 and stabilize after 12:00. The demand for the thermal load decreases after 19:00 and the demand for the cooling load declines after 22:00.

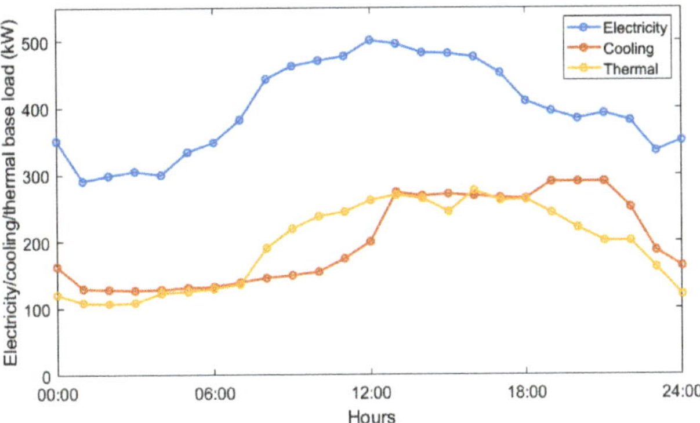

Figure 5. Hourly base load curves of IEU on a typical day.

4.2. Result Analysis in the Case of IES Maximizing Profit

From the perspective of the power grid, during the peak period of the electrical demand of the IEU, the power output of renewable energy is constrained by the equipment itself, and cannot fully satisfy the demand of the IEU. The shortage of power can be supplemented by the power grid.

The cooling/thermal demand of the IEU is mainly a transferable load in the valley period, and the electric refrigeration equipment and electric heating equipment with electric energy as input can dynamically adjust the input gas turbine and waste heat for part of the cooling/thermal demand of the IEU. The natural gas volume in the boiler is adjusted. Figure 6 shows the natural gas volume input into the gas turbine and waste heat boiler.

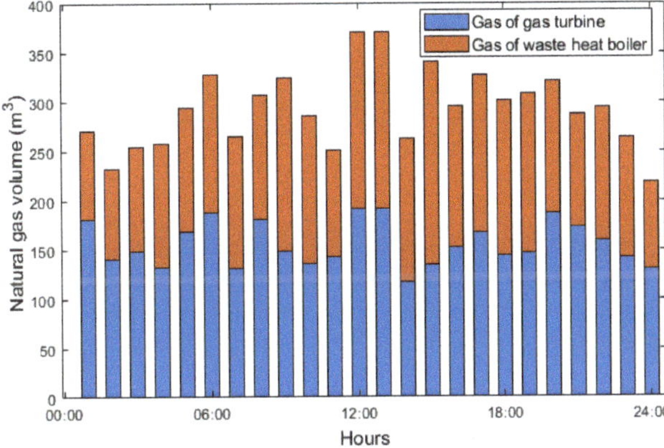

Figure 6. Hourly gas input volume of IES.

After considering the balance of the game model of the demand-side response, the energy use of the integrated energy system appears in the form of complementary mutually beneficial. The cost of renewable energy generation is extremely low, and the supply of energy is preferentially based on renewable energy. For the insufficient demand, when the electricity price is high, natural gas is used first, and the power grid is used for power supply assistance. When the electricity price is low, the power grid is used first, and natural gas is used for assistance.

In the process of energy supply, IES's energy input and output are always in a dynamic balance. Figure 7 shows the amount of interactive energy in the integrated energy system in a game.

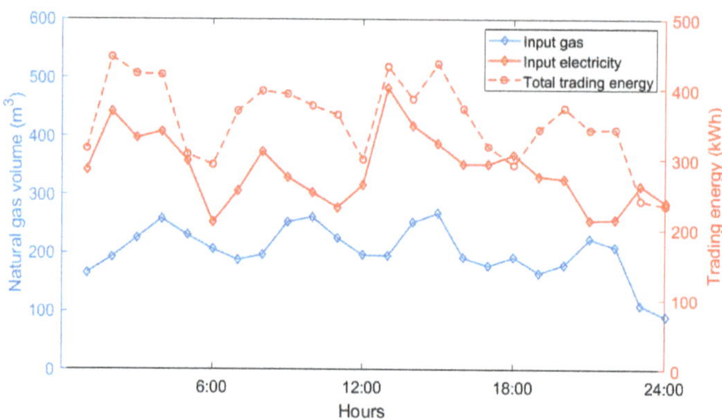

Figure 7. Input and output volume of the integrated energy system in a game.

As energy prices continue to adjust, the IEU responds optimally with the goal of maximizing the utility function, thereby determining the amount of energy demand. IES formulates time-of-use pricing of integrated energy for maximum profit. The unit-integrated energy price changes before and after the game equilibrium are shown in Figure 8.

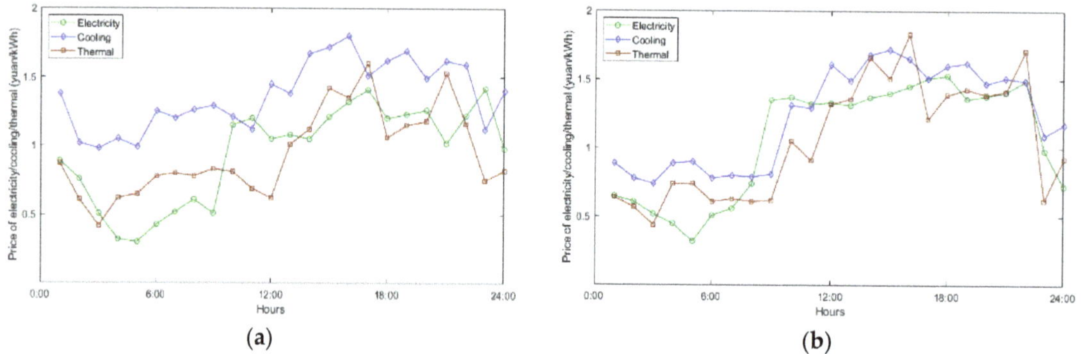

Figure 8. Comparison of optimal unit electricity/cooling/thermal price before and after Nash equilibrium. (**a**) Optimal unit electricity/cooling/thermal price before Nash equilibrium. (**b**) Optimal unit electricity/cooling/thermal price after Nash equilibrium.

The IEU has a small demand for energy from 0:00 to 8:00, and the overall unit energy price is lower. The IEU's base load grows from 6:00 to 12:00. The IES maximizes profit by satisfying IEU's consumer demand satisfaction and increasing unit energy prices. During the period from 12:00 to 22:00, the IEU's energy demand is high, and IES maintains the balance between supply and demand by maintaining a high overall unit energy price during that period to maximize profit. During the period from 22:00 to 24:00, the IEU's base load gradually decreases, while the unit price of integrated energy gradually decreases and stabilizes. After the game reaches equilibrium, the unit-integrated energy price during the peak period of IEU demand rises, and the unit-integrated energy price during the low period of IEU demand declines.

Considering the demand for integrated energy and the base load of IEU in different time periods, IES adopts the strategy of time-of-use pricing to maximize profit. Figure 9 shows the integrated energy volume on a typical day before and after the game equilibrium.

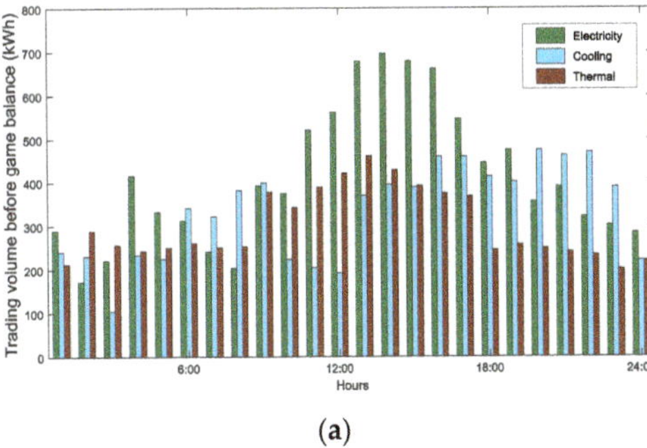 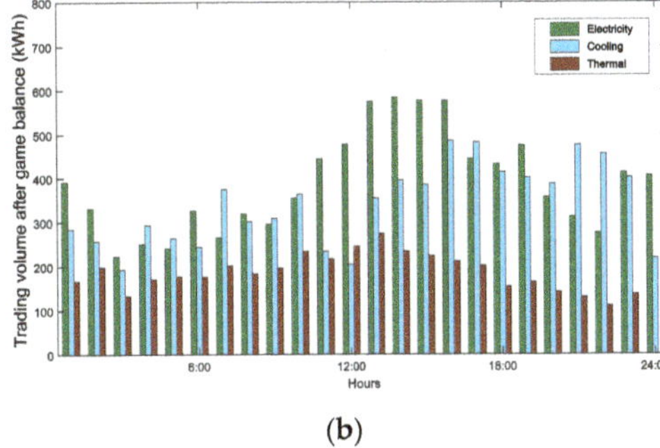

(a)   **(b)**

Figure 9. Comparison of integrated energy trading volume before and after Nash equilibrium. (**a**) Integrated energy trading volume before Nash equilibrium. (**b**) Integrated energy trading volume after Nash equilibrium.

In this case, the IEU's demand for electrical load is mainly concentrated in the period from 12:00 to 22:00. Due to the formulation of the time-of-use electricity price, the IEU changes the electricity demand of the low valley period, and the electricity volume decreases during the peak demand period.

The IEU's demand for thermal energy and cooling is mainly distributed between 11:00 and 24:00. During this period, the trading price of cooling/thermal energy has little fluctuation, and both sides of the game need to trade as much cooling/thermal energy as possible to balance the supply and demand. The IEU's cooling/thermal demand declines during the 0:00–11:00 time period, and the total trading volume of cooling/thermal energy also declines.

After the game balance, due to the low-price strategy incentives during the low demand period and the high price strategy during the peak demand period, on a typical day, the energy volume curves of IES and IEU are relatively smoother than before the game balance. After the game reached equilibrium, the peak-to-valley difference between electricity volume decreased by 44.3%, the peak-to-valley difference between cooling volume decreased by 36.9% and the peak-to-valley difference between thermal volume decreased by 49.5%, effectively achieving peak shaving and valley filling so that the energy supply in each period tends to be balanced.

In order to verify the impact of IEU's demand on economic benefits, the profits are compared before and after game equilibrium. Figure 10 is the hourly profit of IES in a typical day. Figure 11 is the hourly consumption of IEU in a typical day.

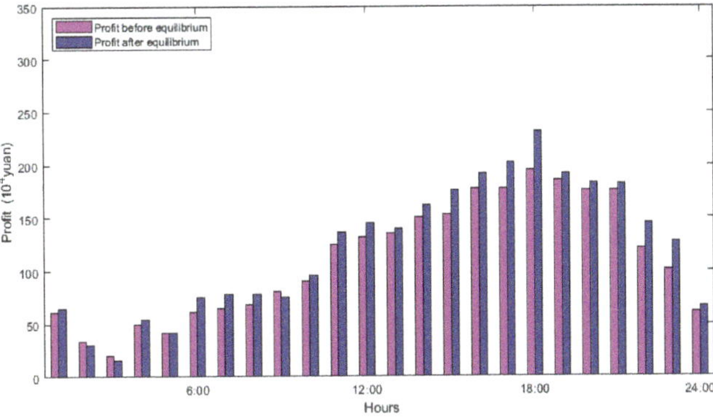

Figure 10. Comparison of IES profit before and after equilibrium in a typical day.

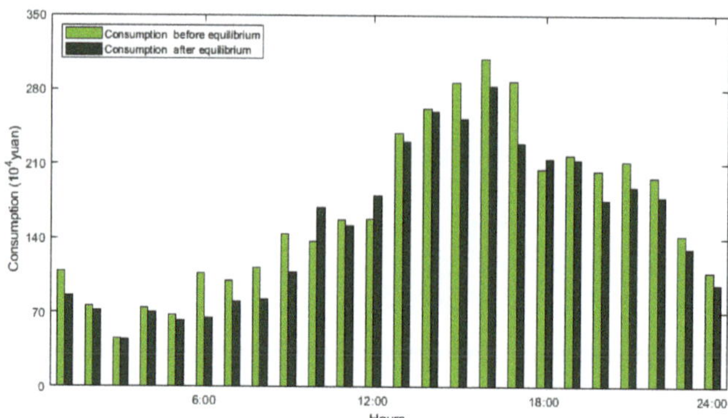

Figure 11. Comparison of IEU consumption before and after equilibrium in a typical day.

Before and after the game equilibrium considering IEU's behavior, and on the premise of meeting the IEU's energy demand, by implementing the time-of-use pricing strategy, IES improves the profit by 8.9% while smoothing the load distribution. IEUs reduce consumption by 11.4% while guaranteeing their own satisfaction.

4.3. Result Analysis in the Case of IES Target Profit

When the objective function of the integrated energy system is different, the equilibrium state of the game model will change. At the beginning of the game, the IES in the multi-subject game can also enlarge the market by reducing the unit energy price to gain a greater share in the market. At this time, the IES strategic objectives have changed, from maximizing profit to acquiring target profit to maintain daily operations. If the maximum daily target profit is set to 10 million yuan, Figure 12a,b show the changes in unit-integrated energy prices before and after the Nash equilibrium.

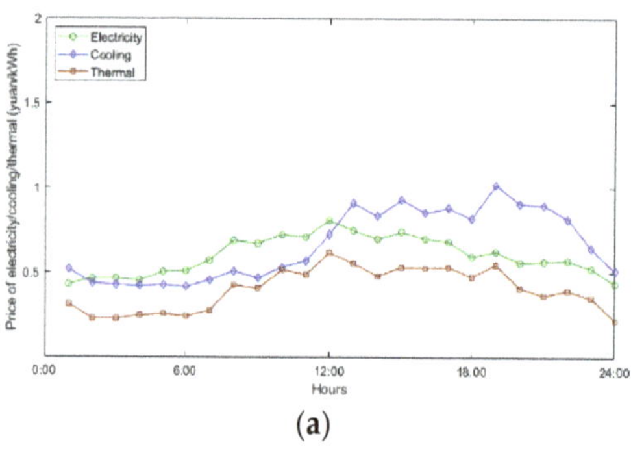
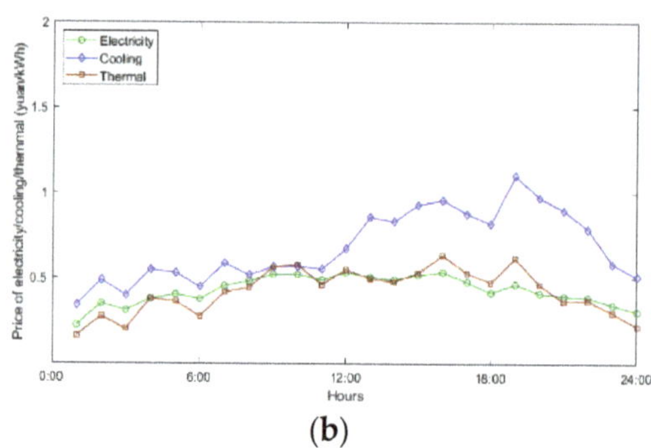

(a)   **(b)**

Figure 12. Comparison of optimal unit electricity/cold/thermal prices in the case of IES target profit. (**a**) Optimal unit electricity/cold/thermal prices before Nash equilibrium. (**b**) Optimal unit electricity/cold/thermal prices after Nash equilibrium.

After the game balance, the unit price of electric energy decreases, and the unit prices of cooling energy and thermal energy basically remain unchanged. There are two ways for IES to control profit by adjusting the integrated energy price, raising the integrated energy price or reducing it. For IEU, the decline in unit-integrated energy price means that more integrated energy can be traded at the same price. The increase in integrated energy trading volume helps IES to enlarge the market. Changes in integrated energy trading volume before and after the game balance are shown in Figure 13a,b.

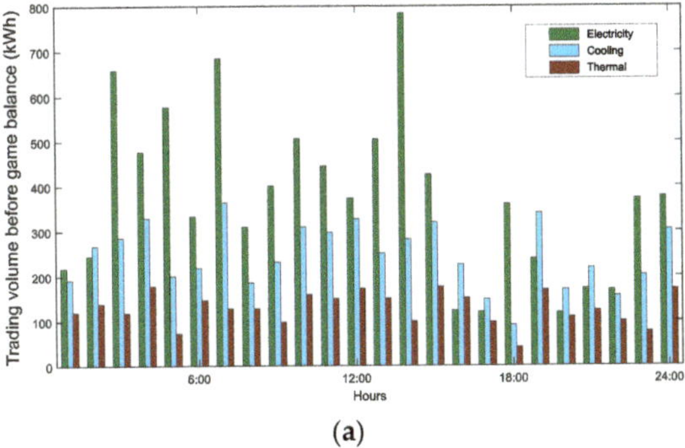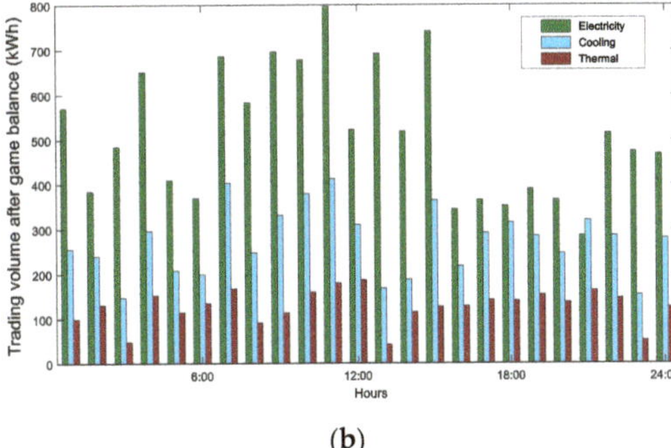

(a)

(b)

Figure 13. Comparison of integrated energy trading volume before and after Nash equilibrium. (**a**) Integrated energy trading volume before Nash equilibrium. (**b**) Integrated energy trading volume after Nash equilibrium.

IES reduces prices appropriately to increase integrated energy trading volume. After the game balance, the trading volume of electricity increased by 37.04%, the trading volume of cooling increased by 10.65% and the trading volume of thermal energy decreased by 0.6%. The price of electricity has a large reduction, the IEU's demand for electricity increased and more electricity was purchased in the interaction with IES. The prices of cooling and thermal have little changes, and the IEU only makes appropriate adjustments according to the corresponding prices and demand. After the game reached balance, the peak-to-valley difference between electricity volume decreased by 23.45%, the peak-to-valley difference between cooling volume decreased by 3.9% and the peak-to-valley difference between thermal volume decreased by 7.1%. The peak sharing and valley filling are not as effective as the strategy under the objective of maximizing profit, but they also have a certain effect.

The profit function of IES and the consumption function of IEU before and after the game balance are shown in Figures 14 and 15.

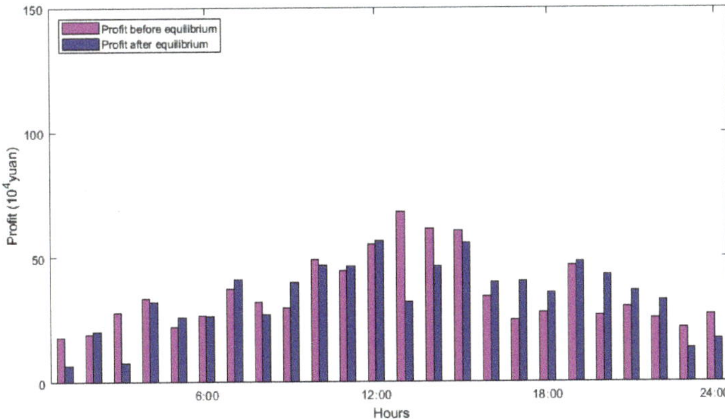

Figure 14. Comparison of IES profit before and after equilibrium.

While IES expanded the energy trading volume, the profit decreased by 3.5%. In the case of IES fixing the profit and reducing the price, IEUs reduce consumption by 17.5%.

The main purpose of the IES controlling profit is to expand the trading volume of integrated energy and become the target of energy purchase priority for users in the game. As the final trading volume is difficult to determine directly, IES participates in the game by controlling the target profit. Finally, the unit price of integrated energy decreased in each time period. Due to the target profit, the unit price of integrated energy and the trading volume mainly showed a negative correlation. Part of the trading volume of the integrated energy in the peak time period shifts to the valley time period.

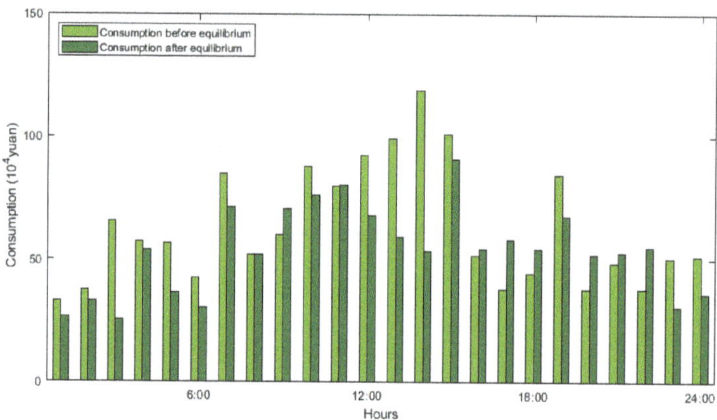

Figure 15. Comparison of IEU consumption before and after equilibrium.

Comparing Figures 8 and 9 with Figures 12 and 13, the unit electricity price, unit cooling price and unit thermal price after game balance under the objective of target profit are reduced by 58.54%, 46.37% and 60.09%, respectively, and the total energy trading volume is increased by 32.06% compared with the objective of maximum profit. It is obvious that the energy price of the game strategy aimed at maximizing profit is significantly higher than that of the game strategy aimed at target profit, and the energy trading volume is also significantly lower than the latter. It can be concluded that the IES with target profit as objective function of the game strategy can enlarge the market by limiting its own profit, and can obtain the purchase willingness of IEUs by reducing the energy price, so that it can have a larger market share, ensure its participation in market transactions for a long time and ensure reliable profit.

5. Conclusions

We analyzed and modeled the game linkage between IES and IEU, which are the participants in the integrated energy market, and constructed the objective functions with maximum profit and target profit of IES. It is proved that the equilibrium solution of the established game model exists in the process of multi-agent participation. Finally, the time-of-use pricing strategy is solved for different time periods under two different objective profits. Integrated energy trading volume and pricing curves are obtained under the two different profit objectives. According to the comparison and analysis of the game strategies under the two different objectives, several conclusions can be drawn.

(1) IES and IEU are linked by participating in the game. Typical energy equipment is used as the basic unit, and the energy input/output model of the IES is established. The time-of-use pricing strategy is the IES's strategy, and the objective function is to maximize its own profit. When the game reaches the Nash equilibrium, IES's operating profit significantly increases.

(2) As the main body of energy consumption, consumption function and utility function of IEU are considered to establish the IEU model. While satisfying the utility function, the IEU influences the time-of-use pricing strategy of IES by changing its own energy consumption behavior. When the game reaches the Nash equilibrium, the utility function of the IEU is satisfied and the consumption is reduced.

(3) Under the two objectives, the peak and valley difference in the electricity/cooling/thermal trading volume after the game balance decreased by 30–50%. It can be concluded

that time-of-use pricing will reduce peak load and fill valley load, smooth the load distribution and improve the stability of IES's energy supply.

(4) The game strategy aiming at target profit has a larger market share and user audience, and it has the potential to participate in the market operation for a long time and obtain reliable profit. The case results show that the energy trading volume of the game strategy with target profit is about 32.06% higher than that of the strategy with the maximum profit, and the pricing is also 54% lower than the latter. The IES can fix its own profit to improve market competitiveness.

Author Contributions: Conceptualization, X.Y. and Y.G.; methodology, X.Y. and Y.G.; software, Y.G. and C.C.; validation, H.C. and C.C.; formal analysis, X.Y.; investigation, H.C.; resources, Y.G.; data curation, C.C.; writing—original draft preparation, Y.G.; writing—review and editing, X.Y.; visualization, C.C.; supervision, X.Y.; project administration, X.Y.; funding acquisition, X.Y. All authors have read and agreed to the published version of the manuscript.

Funding: This research received no external funding.

Conflicts of Interest: The authors declare no conflict of interest.

References

1. Gao, X.; Knueven, B.; Siirola John, D.; Miller David, C.; Dowling Alexander, W. Multiscale simulation of integrated energy system and electricity market interactions. *Appl. Energy* **2022**, *316*, 119017:1–119017:16. [CrossRef]
2. Zhao, J.; Chen, L.; Wang, Y.; Liu, Q. A review of system modeling, assessment and operational optimization for integrated energy systems. *Sci. China Inf. Sci.* **2021**, *64*, 19201:1–19201:23. [CrossRef]
3. Joseph, A.; Balachandra, P. Energy internet, the future electricity system: Overview, concept, model structure, and mechanism. *Energies* **2020**, *13*, 4242. [CrossRef]
4. Kai, Y.; Yi, S.; Zhang, H.; Sun, C.; Wang, S. Development of multi-energy cooperative optimization configuration for park energy internet. In Proceedings of the 2018 2nd IEEE Conference on Energy Internet and Energy System Integration (EI2), Beijing, China, 20–22 October 2018; pp. 1–9.
5. Wang, H.; Wang, C.; Khan, M.Q.; Zhang, G.; Xie, N. Risk-averse market clearing for coupled electricity, natural gas and district heating system. *CSEE J. Power Energy Syst.* **2019**, *5*, 240–248. [CrossRef]
6. Jiang, T.; Zhang, R.; Li, X.; Chen, H.; Li, G. Integrated energy system security region: Concepts, methods, and implementations. *Appl. Energy* **2021**, *283*. [CrossRef]
7. Zhao, X.; Zhu, L.; Ding, L. Evolutionary game analysis of government, new energy industry and traditional energy industry in energy structure adjustment. *Wuhan University J. (Philos. Soc. Sci.)* **2018**, *1*, 13.
8. Wu, H.; Liu, Y.; Yang, Q.; Xu, L.; Zhong, L. Optimal RIES operation strategy based on sistributionally robust game considering demand response. *Electric Power Constr.* **2022**, *43*, 108–118.
9. Lu, P.; Feng, C.; Wu, W.; Wang, T.; Xie, L.; Ma, R. Two-Layer Stackelberg game model for energy supply and demand in the park. *Electr. Power* **2022**, *55*, 74–79.
10. Wu, B.; Zhang, S.; Wang, X.; Liu, S. Equilibrium Strategy Analysis of Demand Response for Integrated Energy Service Provider Participating in Multi-energy Market Transaction. *Power Syst. Technol.* **2022**, *46*, 1800–1811.
11. Heidari, A.; Bansal, R.C. Probabilistic correlation of renewable energies within energy hubs for cooperative games in integrated energy markets. *Electr. Power Syst. Res.* **2021**, *199*, 107397:1–107397:13. [CrossRef]
12. Aras, S.; Sepideh, K.; Mohamad, R.; Alireza, A. A Stackelberg game approach for multiple energies trading in integrated energy systems. *Appl. Energy* **2017**, *200*, 315–329.
13. Yan, J.; Duan, Z.; Gao, J.; Chen, S.; Zhou, B.; Wang, Y. Coordinated Control Strategy of Electricity-Heat-Gas Integrated Energy System Considering Renewable Energy Uncertainty and Multi-Agent Mixed Game. *Front. Energy Res.* **2022**, *10*, 943213:1–943213:11.
14. Huang, Y.; Wang, Y.; Liu, N. A two-stage energy management for heat-electricity integrated energy system considering dynamic pricing of Stackelberg game and operation strategy optimization. *Energy* **2022**, *244*, 122576:1–122576:20. [CrossRef]
15. Ma, T.; Wu, J.; Hao, L.; Yan, H.; Li, D. A real-time pricing scheme for energy management in integrated energy systems: A Stackelberg game approach. *Energies* **2018**, *11*, 2858. [CrossRef]
16. Ma, T.; Du, Y.; Gou, Q.; Peng, L.; Wang, C.; He, P. Non-cooperative competition game model of multiple subjects in electricity market based on Berge-NS equilibrium. *Electr. Power Autom. Equip.* **2019**, *39*, 192–204.
17. Lin, K.; Wu, J.; Liu, D.; Li, D.; Gong, T. Energy management optimization of micro energy grid based on hierarchical Stackelberg game theory. *Power Syst. Technol.* **2019**, *43*, 974–981.
18. Zhang, Y.; Zhao, H.; Li, B.; Wang, X. Research on dynamic pricing and operation optimization strategy of integrated energy system based on Stackelberg game. *Electr. Power Energy Syst.* **2022**, *143*, 108446:1–108446:18. [CrossRef]
19. Fan, H.; Wang, H.; Zhang, S. Hierarchical modeling and energy-information flow hybrid calculation of integrated energy cyber physical system. *Power Syst. Technol.* **2022**, *46*, 3028–3039.

20. Chen, L.; Tang, H.; Wu, J.; Li, C.; Wang, Y. A robust optimization framework for energy management of CCHP users with integrated demand response in electricity market. *Electr. Power Energy Syst.* **2022**, *141*, 108181:1–108181:20. [CrossRef]
21. Wu, L.; Jing, Z.; Wu, Q.; Deng, S. Equilibrium strategies for integrated energy systems based on Stackelberg game model. *Autom. Electr. Power Syst.* **2018**, *42*, 142–150.
22. Carbonell-Nicolau, O.; McLean, R.P. Nash and Bayes–Nash equilibria in strategic-form games with intransitivities. *Econ. Theory* **2019**, *68*, 935–965. [CrossRef]

processes

Article

Predictive Commutation Failure Suppression Strategy for High Voltage Direct Current System Considering Harmonic Components of Commutation Voltage

Xiaolin Liu [1], Zeyu Cao [1], Bingtuan Gao [1,*], Zhuan Zhou [2], Xingang Wang [2] and Feng Zhang [2]

[1] School of Electrical Engineering, Southeast University, Nanjing 210096, China
[2] State Grid Xinjiang Electric Power Co., Ltd., Urumqi 830011, China
* Correspondence: gaobingtuan@seu.edu.cn; Tel.: +86-025-83792260

Abstract: The commutation failure of high voltage direct current (HVDC) systems could lead to unstable operation of the alternating current/direct current (AC/DC) hybrid power grid. The commutation voltage distortion caused by harmonics is a considerable but unclear factor of commutation failure. According to the control switching process of HVDC systems, the commutation voltage-time area method is employed to analyze and reveal the influence mechanism of harmonic components of commutation voltage on first and subsequent commutation failures. Considering the distortion characteristics of AC voltage, a predictive commutation failure suppression strategy considering multiple harmonics of commutation voltage is proposed. In this strategy, the new extinction angle and the zero-crossing offset angle after voltage distortion are calculated considering the harmonic components so as to obtain the compensation margin of the lag trigger angle by combining the correction margin with the voltage change rate. Moreover, the tuning method of parameters of extinction angle and voltage prediction variables are provided. Finally, a case study based on CIGRE standard HVDC system is performed and analyzed by using power systems computer-aided design (PSCAD) software. Compared with the International Council on Large Electric systems (CIGRE) standard test model and traditional commutation failure prevention (CFPREV) control model, the results verify that the proposed strategy can effectively reduce the risk of first and subsequent commutation failures and improve the sensitivity of CFPREV control.

Keywords: high voltage direct current; harmonic components; voltage prediction; commutation failure suppression; commutation failure prevention

Citation: Liu, X.; Cao, Z.; Gao, B.; Zhou, Z.; Wang, X.; Zhang, F. Predictive Commutation Failure Suppression Strategy for High Voltage Direct Current System Considering Harmonic Components of Commutation Voltage. *Processes* **2022**, *10*, 2073. https://doi.org/10.3390/pr10102073

Academic Editors: Haoming Liu, Jingrui Zhang and Jian Wang

Received: 22 August 2022
Accepted: 11 October 2022
Published: 13 October 2022

Publisher's Note: MDPI stays neutral with regard to jurisdictional claims in published maps and institutional affiliations.

1. Introduction

High voltage direct current (HVDC) transmission with grid converters are widely used in long-distance power transmission, underground and submarine cable power transmission and regional power grid interconnection due to their advantages, such as long transmission distance, small transmission loss and good economy. They are an important means to solve the uneven distribution of energy load in China [1–3]. Semi-controlled thyristors are mostly used as the converter elements. When the AC system at the sending and receiving ends fails, it is easy to cause commutation failure [4]. The commutation failure will lead to a sharp increase in DC current and a sharp drop in DC voltage. In serious cases, it will cause a DC blocking fault, which will interrupt the DC transmission power and eventually crash the system [5,6]. If the fault is not cleared in time after the first commutation failure, it is easy to cause subsequent commutation failure. After the subsequent commutation failure, the DC voltage, current, power and other electrical quantities change dramatically, which will have multiple impacts on the AC system. When the AC system is weak, it will cause a DC blocking fault and even cascading failure, bringing great risks and challenges to the safe and stable operation of AC/DC hybrid system [7].

There are many reasons for commutation failure. Considering the fundamental voltage, the increase of DC current, the decrease of AC voltage, the increase of commutation reactance and the decrease of lead trigger angle β will lead to the decrease of extinction angle γ. When γ is less than the inherent critical trigger angle γ_{min}, the commutation failure will occur [8,9]. After an AC system fault, the harmonic component and phase angle offset of commutation voltage will also cause commutation failure. Therefore, voltage distortion is also an important reason for commutation failure [10]. At present, the optimal control of HVDC transmission is mostly based on the protection control modules of converter stations, including voltage dependent current order limiter (VDCOL), commutation failure prevention (CFPREV), constant extinction angle control and power coordination control. The control system has a poor response ability to commutation voltage distortion at present [11,12].

Most of the existing methods to suppress commutation failure are based on improving the converter topology or optimizing the system control strategy [13]. Connecting capacitors and inductors in series and parallel on both sides of the AC transformer and adding reactive power compensation devices to change the structure of the converter can improve the ability to suppress commutation failure to a certain extent. However, adding power electronic devices has the disadvantages of high investment cost and high operation risk [14]. VSC-HVDC uses fully controlled devices to construct converters that are not dependent on AC system commutation, which can effectively inhibit the subsequent commutation failures, but it is not economical and operable [15,16]. Therefore, the optimization control strategy comes more from improving VDCOL and CFPREV modules, with less investment and greater operability. The basic principle of CFPREV is that when an AC fault is detected, a trigger angle margin is generated as the decrease of the lag trigger angle to achieve early triggering and reduce the risk of commutation failure. The harmonic component of the converter bus in the inverter side can cause frequent fluctuation of the CFPREV output, which leads to abnormal fluctuation of DC power. By distinguishing the zero-sequence component of inrush current and AC fault, the output of CFPREV can be controlled to be constant [17]. The literature [18] points out that CFPREV can exacerbate the negative impact of the initial fault voltage angle on commutation failure. When the fault phase voltage crosses zero, the single-phase short-circuit fault will significantly delay the start of CFPREV. Based on this, a calculation method of commutation failure probability is proposed. The literature [19] designs a coordinated controller, which can change the output of CFPREV according to the change of the extinction angle of the remote inverter, so that the trigger angle can be adaptively adjusted. But the accuracy is affected by signal transmission. The existing CFPREV control module has the problems of low output margin accuracy and poor flexibility of parameter adjustment, which may aggravate the subsequent commutation failure in the actual system.

In the process of commutation failure, the control behavior of the controller is affected by many factors. Therefore, the comprehensive effect of various electrical quantities after fault should be considered in the formulation of control and protection strategies. By suppressing the drop of DC current after faults, a control strategy for optimizing VDCOL is proposed in the literature [20], which can suppress AC overvoltage to a certain extent, but the effect of commutation failure suppression needs to be verified. The literature [21] points out that the waveform distortion of commutation voltage is the main reason for commutation failure under slight fault. By considering the time-domain effects of harmonics and DC current, the literature [22] proposes a prediction method of commutation voltage and current to realize the prediction of commutation failure, but the fault suppression strategy needs further study. A trigger angle prediction algorithm is proposed in the literature [23] by using the harmonic component to calculate the extinction angle to detect whether the commutation failure occurred. The commutation-time area method can be used to analyze the influence of harmonics on the commutation process. In the literature [24], the control parameters of each harmonic are calculated by using the commutation-time area method, and an additional harmonic control module is added to VDCOL module. But it cannot suppress the first commutation failure. Based on the commutation-time area, the dynamic adjustment of the extinction angle command value is realized in the literature [25] and the virtual resistor is used to trigger VDCOL in advance, but the suppression effect for some fault scenarios needs to be verified.

In view of the above research deficiencies, this paper analyzes the influence mechanism of the harmonic component of the commutation voltage on the commutation failure and proposes a predictive commutation failure suppression strategy considering the multiple harmonics of the commutation voltage based on the variation characteristics of the AC voltage after the fault. When it is applied to different AC fault scenarios, it can reduce the risk of the first and subsequent commutation failures more sensitively by considering the harmonic component of the voltage distortion and the dynamic change of the AC voltage. Finally, the simulation comparison between the proposed strategy and the traditional CFPREV strategy shows that it can reduce the number of commutation failures.

The main contributions of this paper are summarized as follows:

1. Based on the commutation-time area method, the critical commutation area of harmonic components in fault recovery process is provided. And the mechanism of first and subsequent commutation failures is analyzed in combination with the change of electrical quantity.
2. A novel commutation failure suppression strategy considering multiple harmonics of commutation voltage and voltage prediction is proposed. The new extinction angle and zero-crossing offset angle after voltage distortion are given based on the harmonic components, so as to obtain the compensation margin of the lag trigger angle by combining the correction margin with the voltage change rate.
3. Tuning method of parameters of extinction angle and voltage prediction variables are provided. Extensive case studies based on a CIGRE standard HVDC system are performed and analyzed and compared with a CIGRE standard test model and commutation failure prevention (CFPREV) control model. Simulation results verify that the proposed strategy can suppress the first and subsequent commutation failures and reduce the number of commutation failures effectively when different degrees of faults occur.

The rest of this paper is organized as follows: Section 2 analyzes the principle of the inverter under the influence of harmonics. In Section 3, a commutation failure suppression strategy considering harmonics and voltage prediction is proposed. Based on the strategy, the tuning method of parameters are provided in Section 4. Simulation results based on PSCAD software and discussion are presented in Section 5. Finally, we conclude this paper in Section 6.

2. Commutation Failure and Harmonic Influence

The topology of the line-commutated converter based HVDC (LCC-HVDC) system is depicted in Figure 1, including sending-end grid, rectifier station, DC transmission lines, inverter station and receiving-end grid. Semi-controlled thyristors are widely used as the converter elements in HVDC, which can easily cause commutation failure at the receiving-end grid. Commutation failure is a common fault for HVDC systems, and it is easily affected by harmonics of commutation voltage during the AC fault occurrence. Therefore, it is meaningful to investigate the principle of commutation failure affected by the harmonics.

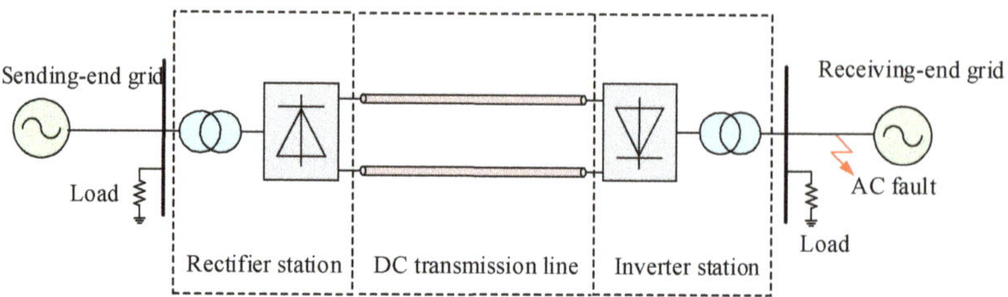

Figure 1. Topology of LCC-HVDC.

2.1. The Principle of Commutation under the Influence of Harmonics

When the AC fault occurs in the sending and receiving end power grid, the converter station near the fault point will fail at commutation. The essence of commutation failure is that the extinction angle is smaller than the minimum extinction angle γ_{min} (about 7°) corresponding to the recovery blocking time of the converter valve at the reverse voltage (about 0.4 s for high-power thyristor). In engineering, the converter unit is generally a six-pulse converter. Figure 2a analyzes the commutation process of thyristor VT3 to VT5, and Figure 2b is the simplified diagram of the inverter commutation process.

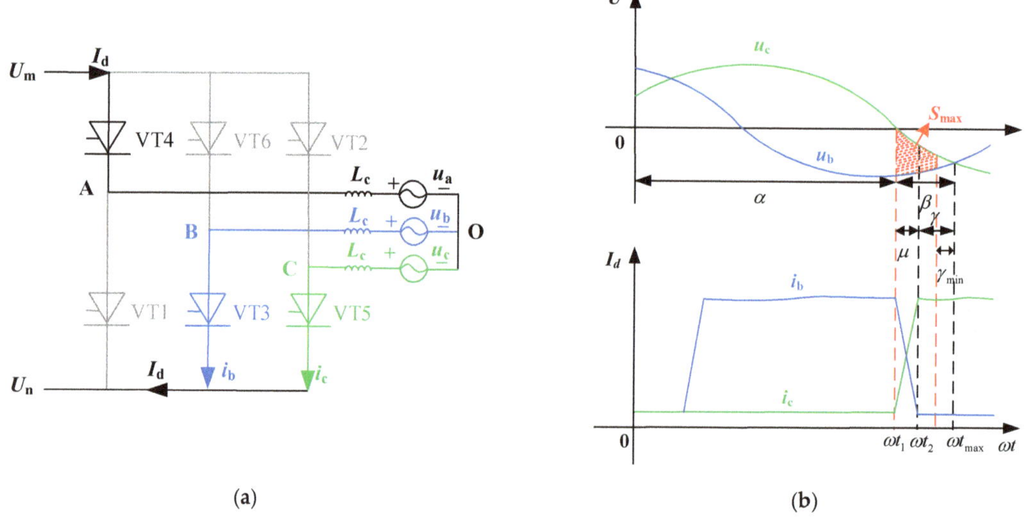

(a)　　　　　　　　　　　　　　　　　(b)

Figure 2. Mechanism analysis of commutation process. (**a**) Equivalent model of inverter side 6-pulse converter. (**b**) Simplified diagram of inverter commutation process.

Before commutation, the thyristor VT3 and VT4 are turned on to form a DC current circuit. When VT5 receives the trigger pulse, VT3 begins to convert to VT5 and the three thyristors are simultaneously turned on. Ignoring the resistance of the large capacity transformer, the default three-phase commutation inductance is equal. The positive directions of current i_a, i_b and i_c are the conduction directions of thyristor. The positive directions of voltage u_a, u_b and u_c are shown in Figure 2a. The voltage equation during the commutation from valve VT3 to valve VT5 can be written as [26]:

$$L_c \frac{di_b}{dt} - L_c \frac{di_c}{dt} = -u_b + u_c \tag{1}$$

where L_c is the commutation inductance. The commutation voltage u_{bc} can be represented by the line voltage amplitude U_L on the side of the commutation valve:

$$u_{bc} = U_L \sin(\omega t) \tag{2}$$

Considering the effect of flat wave reactor, the DC current I_d before and after commutation is basically unchanged. By substituting $i_d = i_b + i_c$ into Formula (1) and integrating both sides from t_1 to t_2, the expression of commutation voltage time area (S) can be obtained:

$$S = 2L_c I_d = \int_{t_1}^{t_2} u_{bc} dt = \int_{(\pi-\beta)/\omega}^{(\pi-\gamma)/\omega} U_L \sin(\omega t) dt \tag{3}$$

where t_1 is the trigger time and can be expressed by $(\pi - \beta)/\omega$; t_2 is the end time of commutation, which can be expressed by $(\pi - \gamma)/\omega$.

By further simplification, the following formula can be obtained:

$$\gamma = \arccos\left(\cos\beta + \frac{2\omega L_c I_d}{U_L}\right) \tag{4}$$

The critical extinction angle γ_{min} is the minimum extinction angle required for normal turn-off of the thyristor. If the extinction angle corresponding to the commutation end time t_2 is the critical extinction angle γ_{min}, namely $(\pi - \gamma_{min})/\omega$, the maximum commutation area that the system can provide is defined as $S_{max} = \int_{(\pi-\gamma_{min})/\omega}^{(\pi-\beta)/\omega} U_L \sin(\omega t) dt$. The required commutation area for normal commutation of the system is defined as $S_{need} = 2L_c I_d$. When the required commutation area $S_{need} > S_{max}$, the extinction angle γ will be less than the critical extinction angle γ_{min}, the system cannot provide enough of a commutation margin, and commutation failure will occur. When the commutation voltage is distorted, the voltage distortion waveform is shown in Figure 3. During normal operation of the system, the commutation voltage is U. After an AC fault, the voltage waveform is distorted, the commutation voltage changes from U to U', the amplitude decreases and the phase angle shifts φ. According to the parameter changes in Figure 3, the commutation area before the fault is $S_x + S_y$. After the fault, the commutation area provided by the system is reduced to S_y due to the distortion of voltage waveform. In order to provide sufficient a commutation margin, it is necessary to increase the commutation area S_z. At this time, γ decreases to γ'. When the extinction angle is smaller than the critical extinction angle γ_{min}, commutation failure occurs.

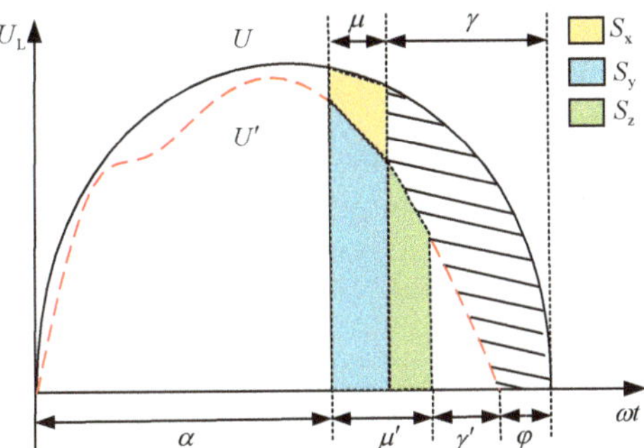

Figure 3. Diagram of voltage waveform distortion.

After the voltage distortion occurs, the commutation voltage u_{bc} will contain the harmonic components:

$$u_{bc} = U_{1L} \sin(\omega t) + \sum_{n=2}^{N} U_{nL} \sin(n\omega t + \varphi_n) \tag{5}$$

Substitute Formula (5) into Formula (3) can get the following formula:

$$2L_c I_d = \int_{(\pi-\beta)/\omega}^{(\pi-\gamma)/\omega} [U_{1L}\sin(\omega t) + \sum_{n=2}^{N} U_{nL}\sin(n\omega t + \varphi_n)]dt = S_1 + \sum_{n=2}^{N} S_n \tag{6}$$

where S_1 and S_n are defined as the commutation area provided by the fundamental voltage and the harmonic voltage, respectively, and further simplification of Formula (6) can be obtained:

$$S_1 = U_{1L}[\cos(\pi - \beta) - \cos(\pi - \gamma)] \tag{7}$$

$$S_n = \frac{U_{nL}}{n}[\cos(n\pi - n\beta + \varphi_n) - \cos(n\pi - n\gamma + \varphi_n)] \tag{8}$$

The condition for the normal switching off of thyristors is as follows: the commutation area provided by the system is larger than the required commutation area. Therefore, under the influence of harmonics, the condition for the normal commutation of the system is as follows:

$$S_{need} \leq S_1 + \sum_{n=2}^{N} S_n \tag{9}$$

2.2. Influence of Harmonics on Subsequent Commutation Failure

When faults occur in the AC system of the HVDC transmission system, the control system at the rectifier and inverter side switches the control blocks to increase the trigger angle and ensure a certain commutation margin. As shown in Figure 4, the basic control blocks of the inverter side of the HVDC transmission system include the voltage dependent current order limiter (VDCOL), constant current (CC), constant extinction angle (CEA) and current error control (CEC) [27].

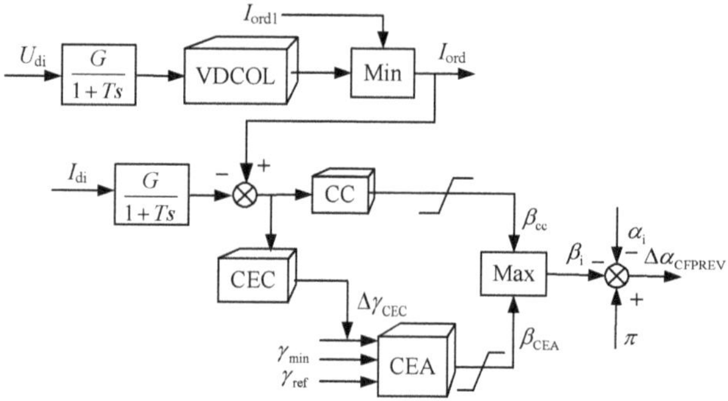

Figure 4. Control block diagram of HVDC inverter side.

In Figure 4, U_{di} and I_{di} are the measured values of DC voltage and DC current on the inverter side after passing through the first-order low-pass filter; I_{ord1} is the DC current instruction of the upper level; when the DC voltage is lower than the low-voltage current limiting control threshold value, VDCOL starts and outputs the DC current command value I_{ord}; γ_{min} and γ_{ref} are the minimum measured extinction angle and the reference value of extinction angle (15°), respectively; when the DC current reaches the current deviation control starting threshold, CEC outputs $\Delta\gamma_{CEC}$, which is added to the input of CEA as an increase. In the actual control of the system, CEA and CC control are started at the same time, and output β_{CEA} and β_{CC}, respectively. The maximum values of the two angles are taken as the output of the inverter side advance trigger angle β_i, and α_i is obtained through transformation.

When a short-circuit fault occurs in the AC system on the inverter side, the second commutation failure will occur if the fault degree is deeper. Take the CIGRE standard test system as an example, the three-phase short-circuit grounding fault on the inverter side is set up. The starting time of the fault is 2 s and lasts for 0.2 s. The changes of electrical quantities during the fault process are shown in Figure 5.

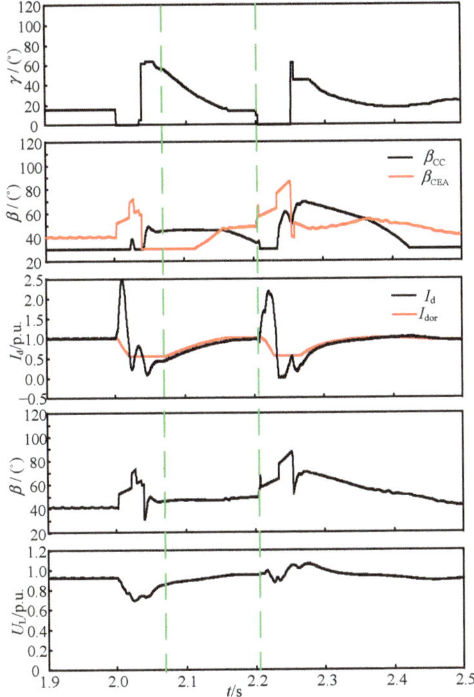

Figure 5. Response waveform of inverter side after three-phase short-circuit grounding fault.

On the one hand, it can be seen from Figure 5 that in the fault recovery process of the first commutation failure (process between green dotted lines), the advance trigger angle $\beta_{cc} > \beta_{CEA}$, and the system switches from CEA control to CC control. After the first commutation failure, the DC voltage decreases and the DC current rises sharply, which makes the inverter side enter the VDCOL control and the DC current command value decreases. When the DC current decreases to a certain extent, the system is normally commutated, and the DC current command value I_{ord} rises slowly. Therefore, in the process of fault recovery, the DC current I_d increases slowly with I_{ord}, resulting in the increase of commutation demand area S_{need}. On the other hand, in the process of fault recovery, AC voltage and DC current are basically restored to the standard value, but a large number of harmonic components appear at the receiving end. Therefore, it can be preliminarily judged that the subsequent commutation failure is not due to the commutation voltage drop caused by insufficient reactive power, but the commutation voltage distortion caused by harmonics. The critical area of commutation voltage time (γ is 7°) calculated by Formulas (7) and (8) in the fault process is shown in Figure 6. S_1 is the commutation area of fundamental voltage, S_{1-15} and S_{2-15} are the commutation areas of 1st–15th and 2nd–15th harmonic voltages, respectively. It can be seen that after the fault occurs, although the 2nd–15th harmonic commutation area increases, the fundamental voltage commutation area S_1 drops greatly, which causes the S_{1-15} to drop sharply, so that the Formula (9) cannot be satisfied and the subsequent commutation failure of the system occurs. The waveforms of critical commutation area S_1 and S_{1-15} before the first and subsequent commutation failure are amplified, as shown in Figure 6b,c. One can see that before the first commutation failure occurs, the harmonic deteriorates the commutation condition and the insufficient commutation margin leads to commutation failures. In the process of fault recovery after the first commutation failure, the harmonics reduce the commutation area, and, finally, the subsequent commutation failure occurs.

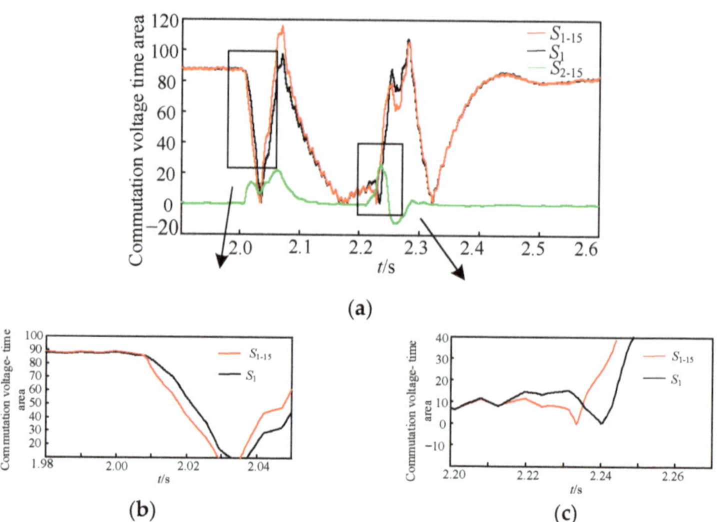

(a)

(b)

(c)

Figure 6. Commutation voltage time area after commutation bus fault. (**a**) Commutation voltage time area of the whole simulation. (**b**) Commutation voltage time area before the first commutation failure. (**c**) Commutation voltage time area before the subsequent commutation failure.

3. Suppression Strategy Considering Harmonics and Voltage Prediction

The key idea of CFPREV is that after the AC system fault is detected, the reduction of the output angle margin is triggered in advance to increase the extinction angle γ, so as to avoid the abnormal extinction of the thyristor and the commutation failure of the system. The basic control block diagram of CFPREV is shown in Figure 7, including zero sequence detector and abc-$\alpha\beta$ three-phase fault detector. The traditional CFPREV has the disadvantages of single detection, insufficient prevention accuracy and lack of flexibility in parameter adjustment. Under some fault conditions, it may even aggravate the fault and cause subsequent commutation failure [28,29].

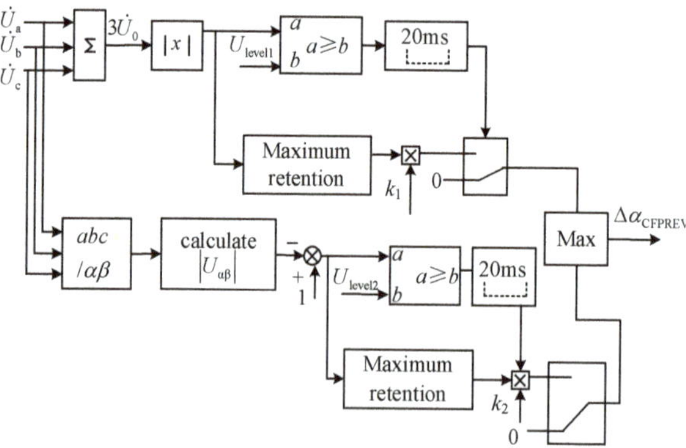

Figure 7. Schematic diagram of CFPREV control.

In order to alleviate the problem of poor speed and sensitivity caused by traditional CFPREV, a commutation failure suppression strategy considering harmonic and voltage prediction is proposed.

3.1. Suppression Strategy Considering Multiple Harmonics of Commutation Voltage

The commutation voltage distortion caused by harmonics will lead to voltage amplitude decrease, phase angle change and even zero-crossing offset. The three-phase short-circuit grounding fault at the inverter side is set in the CIGRE standard test system.

The fault time is 2 s and the duration is 0.2 s. The voltage distortion rate of 2nd–15th and 2nd–31st harmonics (the fundamental voltage distortion rate is 100%) can be obtained by the fast Fourier transform (FFT) decomposition of the commutation voltage, as shown in Figure 8. It can be seen that the voltage has a large harmonic distortion after the fault occurs, and the distortion rate reaches 42.4%. After the fault, the second large harmonic distortion rate appeared, reaching 26.1% at 2.24 s. The two harmonic distortions are both caused by the saturation of converter transformer caused by the substantial increase of DC current. By amplifying the waveform of Figure 8, it can be seen that there is little difference between 2nd–15th harmonic distortion rate and 2nd–31st harmonic distortion rate, the maximum is only 2.07%. Therefore, the subsequent calculation can consider the harmonic number to 15 times.

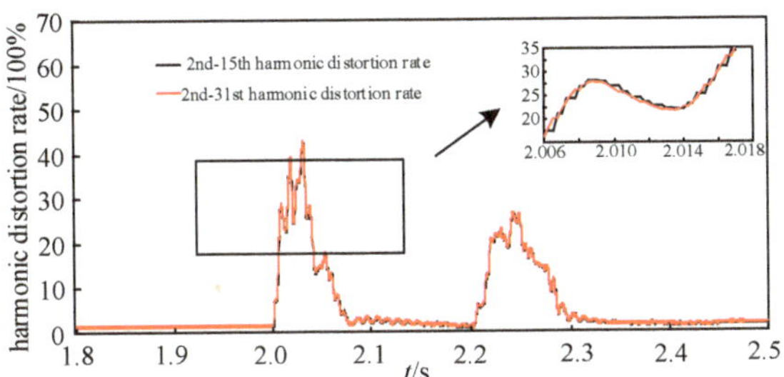

Figure 8. Harmonic voltage distortion rate under fault.

According to Formulas (7) and (8), the commutation voltage-time area corresponding to fundamental wave and each harmonic can be calculated and the total commutation area can be obtained by the summation:

$$\sum_{n=2}^{N} S_n = \sum_{n=2}^{N} \frac{U_{nL}}{n} [\cos(n\pi - n\beta + \varphi_n) - \cos(n\pi - n\gamma + \varphi_n)] \tag{10}$$

The extinction angle γ can be obtained by Formula (4). Harmonics not only bring changes in the voltage amplitude and phase, but also cause zero crossing offset. If the offset angle is $\Delta\varphi$, the extinction angle should meet the Formula (11) to turn off normally:

$$\arccos\left(\cos\beta + \frac{2\omega L_c I_d}{U_L}\right) - \Delta\varphi \geq \gamma_{\min} \tag{11}$$

It is assumed that the voltage variation vector diagram of single-phase short-circuit fault occurs on the inverter side of the AC system, as is shown in Figure 9 [18], and the harmonics distort the voltage of each phase. U_A, U_B and U_C are the three-phase voltage amplitudes before the fault; $U_A{'}$ and $U_B{'}$ are the voltage amplitudes of phase A and phase B, respectively, after fault; U_{AB} and $U_{AB}{'}$ are line voltage amplitude before and after fault; $\Delta\varphi_A$ and $\Delta\varphi_B$ are the voltage phase deviation of phase A and phase B before and after the fault; $\Delta\varphi_{AB}$ is phase deviation of the line voltage. Expression of $\Delta\varphi_{AB}$ can be obtained by cosine theorem and phasor relation:

$$\Delta\varphi_{AB} = \left| \arccos\frac{U_{AB}^2 + U_B^2 - U_A^2}{2U_{AB}U_B} - (30° - \Delta\varphi_B) \right| \tag{12}$$

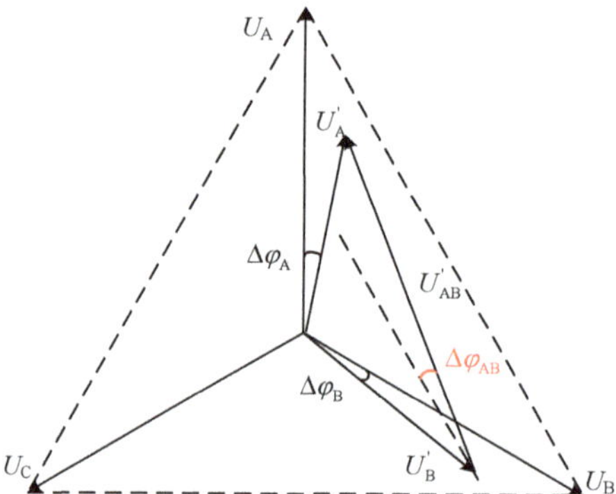

Figure 9. Three-phase voltage phasor diagram of short circuit fault.

In the same way, the formulas for calculating the phase deviation of other line voltages can be obtained:

$$\Delta\varphi_{BC} = \left| \arccos\frac{U_{BC}^2 + U_C^2 - U_B^2}{2U_{BC}U_C} - (30° - \Delta\varphi_C) \right| \tag{13}$$

$$\Delta\varphi_{CA} = \left| \arccos\frac{U_{CA}^2 + U_A^2 - U_C^2}{2U_{CA}U_A} - (30° - \Delta\varphi_A) \right| \tag{14}$$

It can be seen from Formulas (12)–(14) that the voltage distortion will cause the phase shift of the commutation voltage, which will be unfavorable to the commutation process of the converter valve. In order to ensure the normal switching off of the thyristor, it should be triggered in advance to ensure a certain commutation margin.

3.2. Compensation Correction Based on Voltage Change Rate

If the output trigger angle margin of CFPREV is too large, it will lead to an increase in the reactive power consumption of the converter at the inverter side, thus causing the voltage drop of the converter bus, which is not conducive to the recovery of the DC system [30]. If the change trend of voltage can be considered in the fault process, the CFPREV output trigger angle can be compensated and corrected, which can promote the recovery of voltage. The design of dynamic voltage prediction variable K_i is shown in Formula (15):

$$K_i = 1 - \frac{dU_{aci}}{dt}\cdot B, \quad B = \begin{cases} B_1 & U_{aci} \geq U_{level} \\ B_2 & U_{aci} < U_{level} \end{cases} \tag{15}$$

where U_{aci} is the effective value of the line voltage of the AC system at the inverter side; B_1 and B_2 are correction coefficients; U_{level} represents the reference value of AC voltage.

When the voltage drops, the voltage change rate is negative. At this time, K_i is greater than 1, and the output trigger angle is increased, that is, the commutation margin is increased; when the voltage rises, the voltage change rate term is positive, and K_i is less than 1, which reduces the output trigger angle and promotes the voltage recovery. In the subsequent commutation failure process, the coefficient can also dynamically adjust the commutation margin according to the voltage variation trend.

3.3. Suppression Strategy Considering Harmonic and Voltage Prediction

The principle block diagram of commutation failure suppression strategy considering harmonic and voltage prediction is shown in Figure 10, which mainly includes three links: start-up block, voltage prediction block and trigger block.

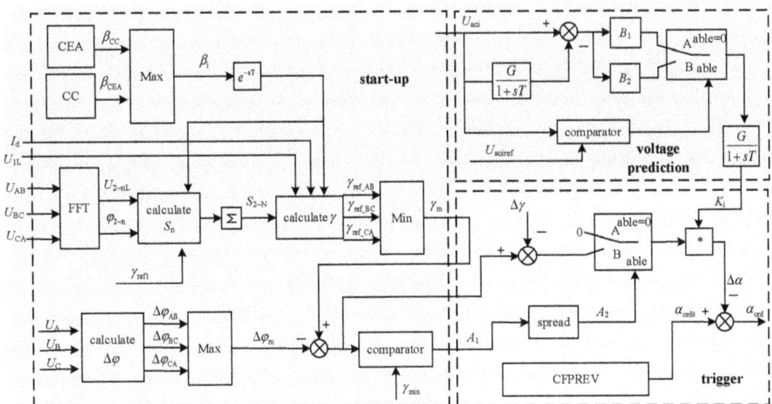

Figure 10. Principle block diagram of commutation failure suppression strategy considering harmonic and voltage prediction.

Considering the start-up link of the multiple harmonics of the commutation voltage, the commutation voltage-time area corresponding to the 2nd–15th harmonics is calculated by Formula (10). Then, it is substituted into Formula (16) to obtain the calculated values γ_{ref_AB}, γ_{ref_BC} and γ_{ref_CA} corresponding to the extinction angles of each commutation voltage in the fault process, and take the minimum value of the extinction angle corresponding to each commutation voltage to obtain γ_m:

$$\gamma = \arccos\left(\cos\beta + \frac{2\omega L_c I_d - \sum\limits_{n=2}^{N} S_n}{U_L}\right) \tag{16}$$

$$\gamma_m = \min\{\gamma_{ref\_AB}, \gamma_{ref\_BC}, \gamma_{ref\_CA}\} \tag{17}$$

Combined with the zero-crossing offset after the distortion of the commutation voltage, the zero-crossing offset of each commutation voltage is calculated by Formulas (12)–(14), and the maximum value is taken to obtain $\Delta\varphi_m$. The difference between γ_m and $\Delta\varphi_m$ is defined as the risk judgment value of commutation failure considering the distortion of commutation voltage after fault, as shown in Formula (18):

$$\gamma_m - \Delta\varphi_m < \gamma_{min} \tag{18}$$

where γ_{min} is the critical extinction angle of thyristor (7°). After meeting the starting criterion, the comparator output A_1 is equal to 1, and A_2 is obtained as the starting criterion of the trigger link after broadening.

In the voltage prediction link, the effective value U_{aci} of the line voltage of the AC system on the inverter side is collected in real time, compares the voltage at the current time with the voltage value passing through the first-order inertia section to obtain the voltage change rate, which is substituted into Formula (15) to calculate the compensation coefficient K_i. In the actual simulation process, U_{aci} will have a small range of fluctuations, a low-pass filter can be used to reduce noise. When the startup criterion is established, the output of A_2 is 1, which determines that the system has the risk of commutation failure and starts the trigger link to achieve early trigger. The trigger margin $\Delta\alpha$ is shown in Formula (19), where $\Delta\gamma$ is the compensation margin.

$$\Delta\alpha = K_i \cdot (\gamma_m - \Delta\varphi_m - \Delta\gamma) \tag{19}$$

After detection of commutation failure risk, start the trigger link and output the lag trigger angle margin $\Delta\alpha$, as CFPREV output trigger angle reduction. The index of the proposed control strategy is more sensitive to the harmonic component generated by the commutation voltage distortion, and the response speed is faster.

4. Tuning Method of Parameters

4.1. Calculation of Reference Value of Extinction Angle

In order to determine the pre-trigger time and pre-trigger margin of this strategy, the fast Fourier transform FFT is used to decompose each commutation voltage and the U_{nL} and phase angle φ_n of each harmonic voltage is obtained. The commutation-time areas corresponding to 2nd–15th harmonic voltage are calculated by substituting Formula (10). The reference value γ_0 of the extinction angle here is taken as the critical value $\gamma_{\min}$ (7°). Due to the controller switching requiring a certain time, in order to achieve early trigger, the pre-trigger angle takes the actual value of the trigger angle of the previous cycle β_0.

After summing the commutation area corresponding to each harmonic, the calculated value of the corresponding extinction angle is substituted into Formula (16), where U_L takes the amplitude of the fundamental commutation voltage U_{1L}. The relevant formulas are shown in (20) and (21):

$$\sum_{n=2}^{N} S_n = \sum_{n=2}^{N} \frac{U_{nL}}{n}\left[\cos(n\pi - n\beta_0 + \varphi_n) - \cos(n\pi - n\gamma_0 + \varphi_n)\right] \tag{20}$$

$$\gamma = \arccos\left(\cos\beta_0 + \frac{2\omega L_c I_d - \sum\limits_{n=2}^{N} S_n}{U_{1L}}\right) \tag{21}$$

4.2. Voltage Change Rate Module Parameter Tuning

Collect the effective value U_{aci} of the converter bus voltage in real time and make a difference between the voltage value at the current moment and the voltage value after the first-order low-pass filter [31], so as to obtain the calculated value of the change rate of the converter bus voltage, as shown in Figure 10.

The reason for the first commutation failure is that the commutation margin is insufficient due to the commutation bus voltage drop. In order to suppress the occurrence of the first commutation failure, the value of B_2 should be increased to obtain a larger trigger margin, and trigger in advance to suppress the failure. When the voltage exceeds the converter bus voltage threshold U_{level}, the value of B_1 is reduced to avoid too small a margin to suppress subsequent commutation failure. After the simulation test, the reference values of B_1 and B_2 are 3–10 and 11.5–18.5, respectively. The actual parameters will be adjusted according to the fault scenario. The voltage threshold U_{level} is 0.928 p.u.

When the fault occurs, the measurement time of FFT is 2–3 ms [32], and the calculation time of voltage change rate of the converter bus is about 1 ms. The calculation and discrimination of the advanced firing angle can be realized by a multi-core digital signal processor (DSP) with a nanoseconds processing speed [33,34]. The time scale from the fault of the converter bus to the local commutation failure is generally 6–8 ms, so the proposed control strategy can meet the requirements of the rapidity of commutation failure suppression.

5. Case Study

Based on the PSCAD/EMTDC simulation platform, the proposed suppression strategy is built on the CIGRE standard test model. The CFPREV parameters are consistent with the literature [35]. The CIGRE model parameters are shown in Table 1.

Table 1. Test system parameters.

Parameter	Rectifier	Inverter
AC voltage/kV	345	230
DC voltage/kV	507	497
DC current/kA	2	2
trigger angle/(°)	20	18
DC power/MW	1014	994

5.1. Validation of Commutation Failure Suppression

A three phase short-circuit fault is set on the AC bus at the inverter side, the fault occurs in 2 s and lasts for 0.2 s. The aim is to compare and analyze the fault response of the following three control strategies: control strategy I: CIGRE standard test system; control strategy II: CFPREV control; control strategy III: the strategy proposed in this paper. The responses of extinction angle, AC voltage, trigger angles of inverter side, predictive variable of dynamic voltage K_i and currents at valve side of converter transformer under three strategies are shown in Figure 11.

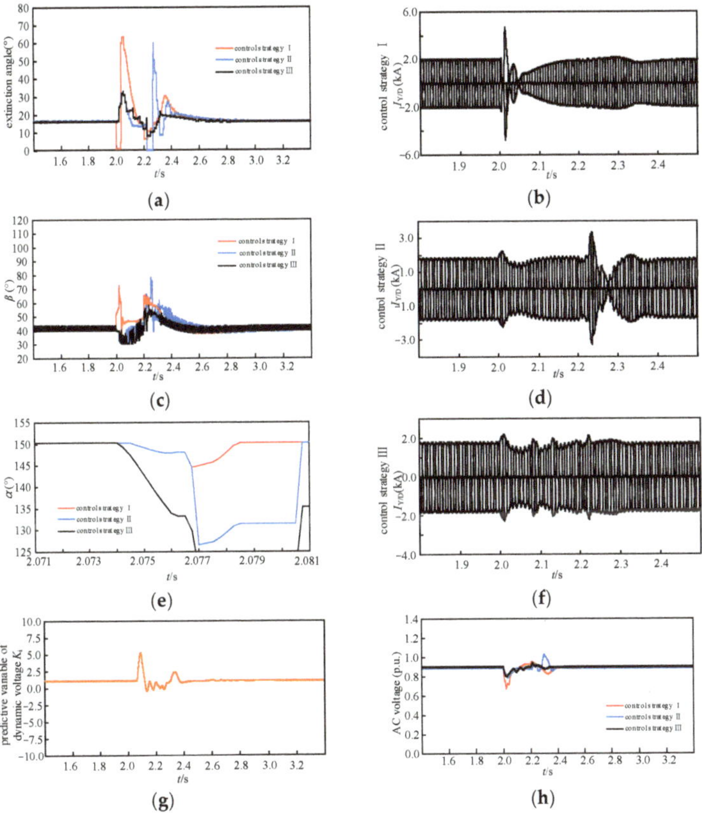

Figure 11. Waveforms of three-phase short-circuit fault under different strategies. (**a**) Extinction angle γ under three strategies. (**b**) Current at valve side of converter transformer under strategy I. (**c**) Trigger advance angle β under three strategies. (**d**) Current at valve side of converter transformer under strategy II. (**e**) Trigger delay angle α under three strategies. (**f**) Current at valve side of converter transformer under strategy III. (**g**) Predictive variable of dynamic voltage K_i. (**h**) AC voltage at inverter side.

According to the valve side currents of converter transformer in Figure 11, after a three-phase short-circuit fault in inverter side of AC system, both control strategies I and II failed to commutation failure once and the extinction angle decreased to 0. The reason for the commutation failure of the CFPREV strategy at 2.21 s is that the CFPREV output leads to the small trigger angle α at the inverter side, the increase of reactive power consumption of the converter and the voltage drop of the AC system, which deteriorate the commutation conditions. The strategy in this paper suppresses the first commutation failure and the system has no subsequent commutation failure. When the fault occurs, the voltage change rate is less than 0, the predictive variable of dynamic voltage K_i is greater than 1 and the trigger angle margin is increased to suppress the first commutation failure. By comparing the lag trigger angle of Figure 11, it can be seen that the strategy in this paper realizes the early trigger and the trigger margin is greater than the other two strategies, which can suppress commutation failure to a certain extent.

In order to further verify the adaptability of the strategy in this paper, single-phase short-circuit and three-phase short-circuit grounding faults are set on the inverter side and the electrical responses are shown in Figures 12 and 13. It can be seen from Figure 12 that the single-phase short-circuit fault leads to two consecutive commutation failures in the original CIGRE system, and both CFPREV and the proposed strategy inhibit the first and subsequent commutation failures. However, the trigger angle margin of this strategy is obtained in advance compared with CFPREV strategy, and the advance trigger angle margin increases and the voltage recovery speed is faster. At the same time, in the steady state of the fault, the value of the extinction angle of the strategy in this paper does not increase, and the converter consumes less reactive power, which is conducive to the rapid recovery of the DC system. It can be seen from Figure 13 that under the three-phase short-circuit grounding fault, strategy I has two commutation failures and the CFPREV strategy has deteriorated the commutation conditions, resulting in one commutation failure. The proposed strategy realizes the early triggering and obtains a large trigger angle margin, which inhibits the first and subsequent commutation failures.

5.2. Subsequent Commutation Failure Suppression Verification

In order to fully verify the ability of the proposed strategy to suppress the first and subsequent commutation failures at different fault levels, further simulations are performed in different fault scenarios. First, it is necessary to define the fault level F_L as shown in (22). The larger the F_L value is, the more serious the fault is.

$$F_L = \frac{U_L^2}{\omega L_f} \frac{1}{P_{dN}} \times 100\% \tag{22}$$

where P_{dN} is the rated DC transmitted power; U_L is the AC bus voltage on the inverter side; L_f is the grounding inductance; and ω is the rated angular frequency.

Two fault scenarios of three-phase short-circuit fault and single-phase short-circuit fault are set up with different fault severity levels to compare and analyze the fault suppression capability of the strategy in this paper, the CFPREV strategy and the original CIGRE system. The results are shown in Figure 14.

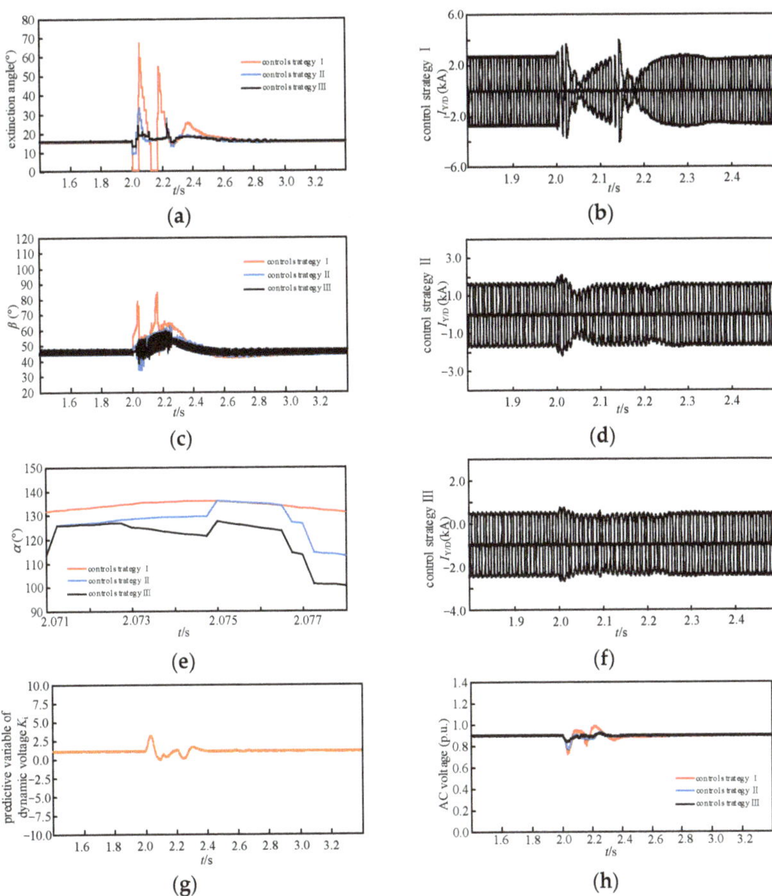

Figure 12. Waveforms of single-phase short-circuit fault under different strategies. (**a**) Extinction angle γ under three strategies. (**b**) Current at valve side of converter transformer under strategy I. (**c**) Trigger advance angle β under three strategies. (**d**) Current at valve side of converter transformer under strategy II. (**e**) Trigger delay angle α under three strategies. (**f**) Current at valve side of converter transformer under strategy III. (**g**) Predictive variable of dynamic voltage K_i. (**h**) AC voltage at inverter side.

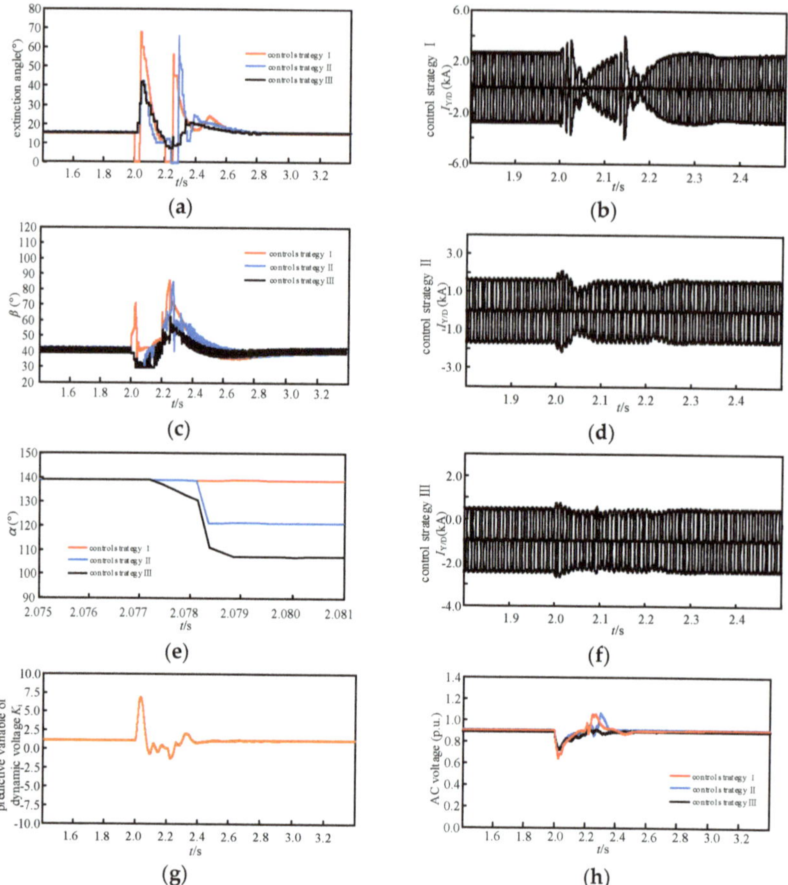

Figure 13. Waveforms of three-phase short-circuit grounding fault under different strategies. (**a**) Extinction angle γ under three strategies. (**b**) Current at valve side of converter transformer under strategy I. (**c**) Trigger advance angle β under three strategies. (**d**) Current at valve side of converter transformer under strategy II. (**e**) Trigger delay angle α under three strategies. (**f**) Current at valve side of converter transformer under strategy III. (**g**) Predictive variable of dynamic voltage K_i. (**h**) AC voltage at inverter side.

The simulation results show that the CFPREV and the proposed strategy can suppress the subsequent commutation failure compared with the original CIGRE system when the three-phase short-circuit fault occurs in the inverter side AC system, but the suppression effect of the proposed strategy is better. CFPREV control and the strategy in this paper begin to fail in the first commutation when F_L is 17.9% and 23.8%, respectively. Therefore, the strategy in this paper can inhibit the first commutation failure to a certain degree of failure. When a single-phase short-circuit fault occurs, the original CIGRE system and CFPREV strategy have different degrees of single and subsequent commutation failures, while the strategy in this paper has no commutation failure below 67.28% of the fault level. When F_L exceeds 67.28%, the single-phase short-circuit fault is more serious, and the restraining ability of commutation failure of the strategy in this paper is the same as the other two strategies.

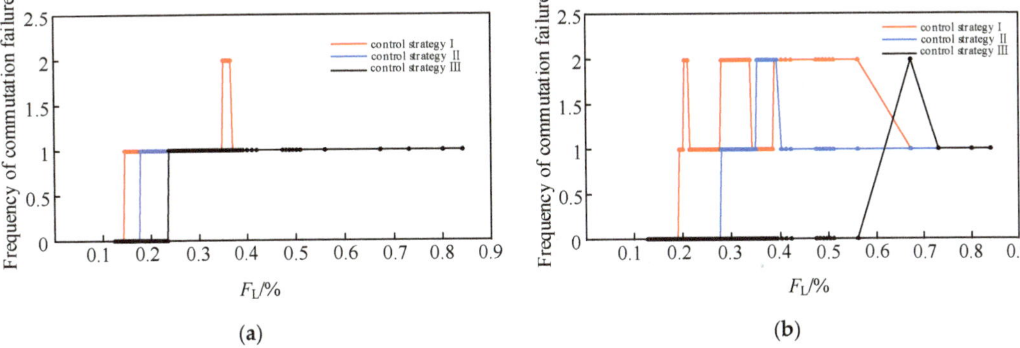

Figure 14. The suppression ability under different strategies to commutation failure. (**a**) Frequency of commutation failure under three-phase short-circuit fault. (**b**) Frequency of commutation failure under single-phase short-circuit fault.

Compared with CFPREV control, the proposed strategy can track the dynamic changes of voltage in real time. When the commutation voltage drops, the compensation margin of the trigger angle is increased, so as to realize the early trigger and suppress the first commutation failure. In the stage of fault recovery, the compensation margin of the trigger angle is dynamically adjusted to reduce, which is conducive to the rapid recovery of the DC system and inhibits the occurrence of subsequent commutation failures.

6. Conclusions

In this paper, a predictive commutation failure suppression strategy considering multiple harmonics of commutation voltage is proposed combined with the first and subsequent commutation failure mechanisms in HVDC transmission. The proposed strategy comprehensively considers the harmonic component of commutation voltage distortion and the dynamic change of AC bus voltage, and triggers in advance to increase the commutation supply area. Through theoretical and simulation analyses, the conclusions are as follows:

1. During the fault recovery process after the first commutation failure, the voltage distortion of the commutation bus caused by harmonics leads to a significant reduction in the commutation supply area, which is an important factor leading to the subsequent commutation failure.
2. The proposed strategy calculates the extinction angle and offset angle based on the harmonic components and constructs a comparator to judge whether there is a risk of commutation failure. It is more accurate to obtain the early trigger amount in advance and improves the sensitivity of commutation failure suppression.
3. The prediction module of proposed strategy considers the dynamic change rate of AC bus voltage during the process of faults and suppresses the first and subsequent commutation failures by calculating different indicators. Compared with the traditional control methods, the strategy in this paper can suppress the first commutation failure under the three-phase short-circuit fault degree of 23.8% and the subsequent commutation failure, and the first and subsequent commutation failures with single-phase short-circuit fault degree below 67.28%.

The future works will focus on interaction principle and suppression strategy of commutation failure in multi-infeed HVDC system.

Author Contributions: Conceptualization: X.L., B.G. and Z.C.; validation: X.L. and Z.C.; investigation: X.L. and B.G.; methodology: X.L.; original draft: X.L.; writing—review and editing: X.L. and B.G.; data provision: Z.Z., X.W. and F.Z. All authors have read and agreed to the published version of the manuscript.

Funding: This research was funded by State Grid Corporation of China Science and Technology Project: "Research on DC control optimization for improving dynamic and transient performance of AC / DC hybrid power grid", grant number 5230HQ21000S.

Institutional Review Board Statement: Not applicable.

Informed Consent Statement: Not applicable.

Data Availability Statement: The data presented in this study are available in article.

Conflicts of Interest: The authors declare no conflict of interest.

Nomenclature

α	Trigger delay angle of LCC.
β	Trigger advance angle of LCC.
γ	Extinction angle.
μ	Overlap angle.
u	Alternating voltage.
L_c	Equivalent inductance.
U	Commutation voltage.
U_L	Amplitude of commutation voltage.
I	Direct current.
S	Commutation area.
i	Alternating current.
φ	Current error control.
Δ	Perturbational component of variables.
B	Correction coefficients of voltage change rate.
$I_{Y/D}$	Current at valve side of converter transformer.
K	Dynamic voltage prediction variable.
F_L	Fault level.
	Superscripts and Subscripts
N	Maximum harmonic numbers.
ref	Reference value of variables.
d	Variables at direct current side.
level, 0	Steady-state value of variables.
ac	Variables at alternating current side.
n	Harmonic numbers.
ord	Command value of variables.
i	Variables at inverter side.
ab,bc,ca/AB,BC,CA	Phase to phase of variables.
a,b,c/A,B,C	Phase to neutral of variables.

References

1. Wang, M.; An, T.; Ergun, H.; Lan, Y.; Andersen, B.; Szechtman, M.; Leterme, W.; Beerten, J.; Hertem, D.V. Review and Outlook of HVDC Grids as Backbone of Transmission System. *CSEE J. Power Energy Syst.* **2021**, *7*, 797–810.
2. Yang, Z.; Gao, B.; Cao, Z. Optimal Current Allocation Strategy for Hybrid Hierarchical HVDC System with Parallel Operation of High-Voltage and Low-Voltage DC Lines. *Processes* **2022**, *10*, 579. [CrossRef]
3. Xue, Y.; Zhang, X.; Yang, C. AC Filterless Flexible LCC HVDC With Reduced Voltage Rating of Controllable Capacitors. *IEEE Trans. Power Syst.* **2018**, *33*, 5507–5518. [CrossRef]
4. Li, C.; Li, K.; Jiang, F.; Tai, W.; Wu, X.; Tang, Y. Improvement Method of Commutation Failure Predictive Control Based on Voltage Waveform Fitting. In Proceedings of the 2021 IEEE Sustainable Power and Energy Conference (iSPEC), Nanjing, China, 23–25 December 2021.
5. Wei, J.; Cao, Y.; Wu, Q.; Li, C.; Huang, S.; Zhou, B.; Xu, D. Coordinated Droop Control and Adaptive Model Predictive Control for Enhancing HVRT and Post-Event Recovery of Large-Scale Wind Farm. *IEEE Trans. Sustain. Energy* **2021**, *12*, 1549–1560. [CrossRef]
6. Yin, C.; Li, F. Analytical Expression on Transient Overvoltage Peak Value of Converter Bus Caused by DC Faults. *IEEE Trans. Power Syst.* **2021**, *36*, 2741–2744. [CrossRef]
7. Hong, L.; Zhou, X.; Xia, H.; Liu, Y.; Luo, A. Mechanism and Prevention of Commutation Failure in LCC-HVDC Caused by Sending End AC Faults. *IEEE Trans. Power Deliv.* **2021**, *36*, 473–476. [CrossRef]

8.   Jing, W.; Jun, L.; Yang, Z.; Hao, Z.; Cong, X.; Hu, Y. A Novel Control Strategy for Suppression of HVDC Subsequent Commutation Failure. In Proceedings of the 2020 IEEE 3rd Student Conference on Electrical Machines and Systems (SCEMS), Jinan, China, 4–6 December 2020.

9.   Liu, L.; Lin, S.; Sun, P.; Liao, K.; Li, X.; Deng, Y.; He, Z. A Calculation Method of Pseudo Extinction Angle for Commutation Failure Mitigation in HVDC. *IEEE Transactions on Power Delivery.* **2019**, *34*, 777–779. [CrossRef]

10.   Yang, W.; Miao, S.; Zhang, S.; Li, Y.; Han, J.; Xu, H.; Zhang, D. A Commutation Failure Risk Analysis Method Considering the Interaction of Inverter Stations. *Int. J. Electr. Power Energy Syst.* **2020**, *120*, 106009. [CrossRef]

11.   Liu, B.; Chen, Z.; Yang, S.; Lu, C.; Deng, X.; Wang, R. Research on methods of measuring extinction angle and measures to suppress repetitive commutation failures through equivalent DC input resistance. *Int. J. Electr. Power Energy Syst.* **2021**, *133*, 107326. [CrossRef]

12.   Chen, M.; Kang, Z.; Gan, D.; Wu, H. Impact of Commutation Failure Preventions on HVDC System Based on a Multi-Index Value Set Approach. *IEEE Access.* **2021**, *9*, 48080–48090. [CrossRef]

13.   Xue, Y.; Zhang, X.; Yang, C. Commutation Failure Elimination of LCC HVDC Systems Using Thyristor-Based Controllable Capacitors. *IEEE Trans. Power Deliv.* **2018**, *33*, 1448–1458. [CrossRef]

14.   Wang, H.; Dong, X. Commutation Failure Mitigation Method Based on Improved Commutation Area in HVDC system. In Proceedings of the 2019 IEEE 8th International Conference on Advanced Power System Automation and Protection (APAP), Xi'an, China, 21–24 October 2019.

15.   Xu, C.; Yu, Z.; Chen, Z.; Zhao, B.; Wang, Z.; Ren, C.; Zeng, R. Comprehensive Analysis and Experiments of RB-IGCT, IGCT with Fast Recovery Diode and Standard Recovery Diode in Hybrid Line Commutated Converter for Commutation Failure Mitigation. *IEEE Trans. Ind. Electron.* **2022**, *70*, 1126–1139. [CrossRef]

16.   González-Cabrera, N.; Castro, L.M.; Gutiérrez-Alcaraz, G.; Tovar-Hernandez, J.H. Alternative approach for efficient OPF calculations in hybrid AC/DC power grids with VSC-HVDC systems based on shift factors. *Int. J. Electr. Power Energy Syst.* **2021**, *124*, 106395. [CrossRef]

17.   Gao, K.; Li, C.; Li, X.; Wang, H.; Rao, Y.; Liu, W. Transformer Inrush Current Impact on Commutation Failure Prevention of UHVDC Transmission system. In Proceedings of the 2019 IEEE 3rd Conference on Energy Internet and Energy System Integration (EI2), Changsha, China, 8–10 November 2019; pp. 2730–2734.

18.   Yao, W.; Liu, C.; Fang, J.; Ai, X.; Wen, J.; Cheng, S. Probabilistic Analysis of Commutation Failure in LCC-HVDC System Considering the CFPREV and the Initial Fault Voltage Angle. *IEEE Trans. Power Deliv.* **2020**, *35*, 715–724. [CrossRef]

19.   Yao, W.; Wang, L.; Xiong, Y.; Tan, Y.; Li, D.; Li, C.; When, J. Interaction Mechanism and Coordinated Control of Commutation Failure Prevention in Multi-Infeed Ultra-HVDC System. *Int. Trans. Electr. Energy Syst.* **2022**, *2022*, 7088114. [CrossRef]

20.   Wang, T.; Pei, L.; Wang, J.; Wang, Z. Overvoltage Suppression Under Commutation Failure Based on Improved Voltage-Dependent Current Order Limiter Control Strategy. *IEEE Trans. Ind. Appl.* **2022**, *58*, 4914–4922. [CrossRef]

21.   Zhang, G.; Jing, L.; Wang, B. Study on HVDC Consecutive Commutation Failures Mitigation Method Caused by Harmonics. *Proc. CSEE* **2020**, *48*, 33–41.

22.   Wang, Q.; Zhang, C.; Wu, X.; Tang, Y. Commutation Failure Prediction Method Considering Commutation Voltage Distortion and DC Current Variation. *IEEE Access* **2019**, *7*, 96531–96539. [CrossRef]

23.   Xiao, H.; Li, Y.; Duan, X. Enhanced Commutation Failure Predictive Detection Method and Control Strategy in Multi-Infeed LCC-HVDC Systems Considering Voltage Harmonics. *IEEE Trans. Power Syst.* **2021**, *36*, 81–96. [CrossRef]

24.   Chen, J.; Wang, Q.; Zhu, Z. Evaluation of Commutation failure Risk for HVDC Caused by Harmonic Voltage. In Proceedings of the 2019 IEEE 3rd Conference on Energy Internet and Energy System Integration (EI2), Changsha, China, 8–10 November 2019.

25.   Wei, Z.; Fang, W.; Liu, J. Variable Extinction Angle Control Strategy Based on Virtual Resistance to Mitigate Commutation Failures in HVDC System. *IEEE Access* **2020**, *8*, 93692–93704. [CrossRef]

26.   Mirsaeidi, S.; Tzelepis, D.; He, J.; Dong, X.; Said, D.M.; Booth, C. A Controllable Thyristor-Based Commutation Failure Inhibitor for LCC-HVDC Transmission Systems. *IEEE Trans. Power Electron.* **2021**, *36*, 3781–3792. [CrossRef]

27.   Xiao, H.; Li, Y.; Lan, T. Sending End AC Faults can Cause Commutation Failure in LCC-HVDC Inverters. *IEEE Trans. Power Deliv.* **2020**, *35*, 2554–2557. [CrossRef]

28.   Sun, W.; Wang, Q.; Gu, R.; Zheng, C.; Wang, Y. Coordinated Control Method of Multi-infeed DC System for Commutation Failure Prevention. In Proceedings of the 2021 IEEE Sustainable Power and Energy Conference (iSPEC), Nanjing, China, 23–25 December 2021.

29.   Wei, Z.; Yuan, Y.; Lei, X.; Wang, H.; Sun, G.; Sun, Y. Direct-Current Predictive Control Strategy for Inhibiting Commutation Failure in HVDC Converter. *IEEE Trans. Power Syst.* **2014**, *29*, 2409–2417. [CrossRef]

30.   Tian, D.; Xiong, X. Corrections of Original CFPREV Control in LCC-HVDC Links and Analysis of its Inherent Plateau Effect. *CSEE J. Power Energy Syst.* **2022**, *8*, 10–16.

31.   Zhang, J.; Lin, W.; Wen, J. DC Fault Protection of DC Grid Based on DC Voltage Change Rate. *South. Power Syst. Technol.* **2017**, *11*, 14–22.

32.   Tiwari, R.; Gupta, O.; Ansari, S. Frequency Spectral Analysis of 6-Pulse LCC-HVDC in Single Conductor Ground Return Configuration Using FFT. In Proceedings of the 2022 2nd International Conference on Power Electronics & IoT Applications in Renewable Energy and its Control (PARC), Mathura, India, 1–5 January 2022.

33. Psaras, V.; Vozikis, D.; Adam, G.; Burt, G. Real Time Evaluation of Wavelet Transform for Fast and Efficient HVDC Grid Non-Unit Protection. In Proceedings of the 2020 IEEE Power & Energy Society General Meeting (PESGM), Montreal, QC, Canada, 1–5 August 2020.
34. Zhu, Y.; Chang, D.; Guo, Q.; Guo, H.; Hu, B.; Wu, M.; Zhu, Y.; Yu, D. Research and Implementation of Real-Time Accurate Simulation Platform for Large-Scale Stability Control System. In Proceedings of the 2021 IEEE 5th Conference on Energy Internet and Energy System Integration (EI2), Taiyuan, China, 22–24 October 2021.
35. Meah, K.; Sadrul Ula, A.H.M. Simulation Study of the CIGRE HVDC Benchmark Model with the WSCC Nine-bus Power System Network. In Proceedings of the 2009 IEEE/PES Power Systems Conference and Exposition, Seattle, WA, USA, 15–18 March 2009.

MDPI AG
Grosspeteranlage 5
4052 Basel
Switzerland
Tel.: +41 61 683 77 34

Processes Editorial Office
E-mail: processes@mdpi.com
www.mdpi.com/journal/processes